Plant Disease

AN ADVANCED TREATISE

VOLUME II
How Disease Develops in Populations

Advisory Board

Plant Disease

AN ADVANCED TREATISE

VOLUME II

How Disease Develops in Populations

Edited by

JAMES G. HORSFALL

The Connecticut Agricultural
Experiment Station
New Haven, Connecticut

ELLIS B. COWLING

Department of Plant Pathology
and School of Forest Resources
North Carolina State University
Raleigh, North Carolina

ACADEMIC PRESS   New York   San Francisco   London   1978

A Subsidiary of Harcourt Brace Jovanovich, Publishers

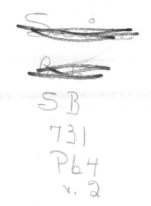

ACADEMIC PRESS, INC.
111 Fifth Avenue, New York, New York 10003

United Kingdom Edition published by
ACADEMIC PRESS, INC. (LONDON) LTD.
24/28 Oval Road, London NW1   7DX

Library of Congress Cataloging in Publication Data

Main entry under title:

Plant disease.

    Includes bibliographies and index.
    CONTENTS: v. 1. How disease is managed.--v. 2. How
disease develops in populations.
    1. Plant diseases. I. Horsfall, James Gordon,
Date   II. Cowling, Ellis Brevier, Date
SB731.P64   632   76-42973
ISBN 0-12-356401-8 (v. 2)

To the theoreticians of our science—
especially to that pioneering theoretician
whose book *Pflanzliche Infectionslehre*
for the first time treated
Plant Pathology as a sophisticated science,

ERNST GÄUMANN

who signed his treasured letters

Contents

CHAPTER 1 PROLOGUE: HOW DISEASE DEVELOPS IN POPULATIONS

ELLIS B. COWLING AND JAMES G. HORSFALL

CHAPTER 2 SOME EPIDEMICS MAN HAS KNOWN

JAMES G. HORSFALL AND ELLIS B. COWLING

CHAPTER 3 COMPARATIVE ANATOMY OF EPIDEMICS

J. KRANZ

CHAPTER 4   METHODOLOGY OF EPIDEMIOLOGICAL RESEARCH
J. C. ZADOKS

CHAPTER 5   INSTRUMENTATION FOR EPIDEMIOLOGY
S. P. PENNYPACKER

CHAPTER 6   PATHOMETRY: THE MEASUREMENT OF PLANT DISEASE
JAMES G. HORSFALL AND ELLIS B. COWLING

CHAPTER 7   INOCULUM POTENTIAL
RALPH BAKER

CHAPTER 8   DISPERSAL IN TIME AND SPACE: AERIAL PATHOGENS
DONALD E. AYLOR

CHAPTER 9   DISPERSAL IN TIME AND SPACE: SOIL PATHOGENS
H. R. WALLACE

CHAPTER 14 DISEASES IN FOREST ECOSYSTEMS: THE IMPORTANCE OF FUNCTIONAL DIVERSITY

ROBERT A. SCHMIDT

CHAPTER 15 CLIMATIC AND WEATHER INFLUENCES ON EPIDEMICS

JOSEPH ROTEM

CHAPTER 16 GEOPHYTOPATHOLOGY

HEINRICH C. WELTZIEN

CHAPTER 17 AGRICULTURAL AND FOREST PRACTICES THAT FAVOR EPIDEMICS

ELLIS B. COWLING

CHAPTER 18   PEOPLE-PLACED PATHOGENS: THE
EMIGRANT PESTS

RUSSELL C. McGREGOR

List of Contributors [1]

Numbers in parentheses indicate the pages on which the authors' contributions begin.

DONALD E. AYLOR (159), The Connecticut Agricultural Experiment Station, New Haven, Connecticut 06504

RALPH BAKER (137), Department of Botany and Plant Pathology, Colorado State University, Fort Collins, Colorado 80521

ELLIS B. COWLING (1, 17, 119, 361), Department of Plant Pathology and School of Forest Resources, North Carolina State University, Raleigh, North Carolina 27607

P. R. DAY (263), Genetics Department of The Connecticut Agricultural Experiment Station, New Haven, Connecticut 06504

JAMES G. HORSFALL (1, 17, 119), The Connecticut Agricultural Experiment Station, New Haven, Connecticut 06504

J. KRANZ (33), Tropishe Phytopathologie, Jŭstŭs Liebig-Universität, 6300 Giessen, Federal Republic of Germany

RUSSELL C. McGREGOR (383), National Association of State Universities and Land-Grant Colleges, Washington, D.C. 20036

S. P. PENNYPACKER (97), Department of Plant Pathology, Pennsylvania State University, University Park, Pennsylvania 16802

C. POPULER (239), Station de Phytopathologie, Centre de Recherches Agronomiques de l'Etat, Gembloux, Belgium

JOSEPH ROTEM (317), Division of Plant Pathology, Agricultural Research Organization, The Volcani Center, Bet Dagan, Israel and Department of Life Sciences, Bar-Ilan University, Ramat Gan, Israel

ROBERT A. SCHMIDT (287), School of Forest Resources and Conservation, Institute of Food and Agricultural Sciences, University of Florida, Gainesville, Florida 32611

ROBERT D. SHRUM (223), Department of Plant Pathology, University of Minnesota, St. Paul, Minnesota 55108

[1] We regret that an error was made in the address of Dr. Horst Lyr, author of Chapter 13 in Volume I. The correct address is: Horst Lyr, Institute for Plant Protection Research, Kleinmachnow, Academy of Agricultural Sciences of the German Democratic Republic, German Democratic Republic.

PAUL E. WAGGONER (203), The Connecticut Agricultural Experiment Station, New Haven, Connecticut 06504

H. R. WALLACE (181), Waite Agricultural Research Institute, University of Adelaide, South Australia

HEINRICH C. WELTZIEN (339), Institut für Pflanzenkrankheiten der Universitat, D 53 Bonn, Federal Republic of Germany

J. C. ZADOKS (63), Department of Phytopathology, Agricultural University, Wageningen, The Netherlands

Preface

He that publishes a book runs a very great hazard, since nothing can be more impossible than to compose one that may secure the approbation of every reader.

Cervantes

We worry about things like that. Will plant pathologists like our design? Will they like the chapters? Will they read the book? We can only hope they will.

To manage disease, as shown in Volume I, is to prevent it from enveloping a whole population of plants in a farmer's field, a county, a country, or a continent. To study disease in populations is epidemiology —that is the subject of Volume II.

During the last decade, botanical epidemiology has leaped across the space from primitive description to sophisticated quantitation (from state to process). Computers with their giant memories and enormous speed of data manipulation enable us at long last to deal with the rapid sequence of events as one biological process follows another under the variable drive of the fickle and ever-changing weather.

Now we are much better prepared to perform statistical tests, to find holes in our knowledge, to forecast disease for use by farmers, and to reduce the principles to abstraction for teaching and further research.

Volume II tells us all about this new science of epidemiology—from aerodynamics to understanding how the cycle of disease looks like a helix as it moves through time (see Bateman, Volume III).

Sixteen authors from all over the world have come together to produce the volume. We have asked them to go beyond a comfortable exposition of where we have been in plant pathology to tell us where we are going. We have asked them to show us where the prospecting for new knowledge will be most productive and how our profession can grow in usefulness to the society that pays its bills.

Sterling Hendricks was percipient indeed when he said, "The opportunity to inquire into the nature of things is a privilege granted to a few by a permissive society." We have used that privilege in designing this treatise to inquire into the basics of our science and into the useful arts of crop protection. We hope this treatise will stretch the minds of its readers by being comprehensive and timely, provocative and forward looking, practical and theoretical in outlook, and well-balanced in its coverage. We hope you will find it so.

James G. Horsfall
Ellis B. Cowling

Contents of Other Volumes

Tentative Contents of Other Volumes

Chapter 1

Prologue: How Disease Develops in Populations

ELLIS B. COWLING AND JAMES G. HORSFALL

I. INTRODUCTION

"Chemical industry and plant breeders have forged fine tactical weapons, but only epidemiology sets the strategy."

Van der Plank

Yes—epidemiology sets the strategy. We have devoted Volume I essentially to the tactics of disease management. And now we turn to the strategy. In Volume II we deal with disease in populations of plants. This is a rapidly moving subject, as can be seen as one peruses the volume.

After long and earnest debate we decided to put management in Volume I to emphasize the useful side of our science. We chose to put epidemiology in Volume II to provide the strongest possible foundation for understanding both the theory of disease management and the dynamic nature of disease.

It is curious that most general texts on plant pathology have largely ignored the subject of disease in populations. They have concentrated

1

instead on disease in individual plants. Perhaps this preference arises because plant pathology grew out of botany instead of ecology.

The last volume of the original treatise was devoted to the dynamics of disease in populations. It contained van der Plank's * (1960) first analysis of epidemics. Later he told us that his determination to write three more books on plant epidemiology grew from his experience in preparing this chapter (van der Plank, 1963, 1968, 1975).

II. What Is an Epidemic?

This volume is entitled, "How Disease Develops in Populations." We could have called it "Botanical Epidemiology," as suggested by Zadoks and Koster (1976) or simply "Epidemiology."

The word "epidemic," like numerous other words in science, stems from the Greek, *epi* (on) and *demos* (people). Webster's International Dictionary defines epidemic as "affecting or tending to affect many individuals within a population, a community, or a region at the same time—an outbreak, a sudden and rapid growth, spread, or development." The word usually contains a connotation of disease, but we also say "the practice has reached epidemic proportions." Rock-and-roll music has certainly reached epidemic proportions in some countries.

Why not use the term "epiphytotic"—*epi* (on), *phyton* (plant)? Many plant pathologists use the term. For analogous reasons some zoologists use "epizootic." To us, these "refinements" are unnecessary bits of jargon. Epidemic is a well-understood word outside our field as well as within it. Thus, for better or worse. we shall use epidemic.

In recent years the term epidemiology has come to have a broad meaning within plant pathology. The term has been variously defined as: the study of disease in populations; the study of environmental factors that influence the amount and distribution of disease in populations; and the study of rates of change (either increases or decreases) in the amount of disease in time, in space, or both.

III. THE ELEMENTS OF AN EPIDEMIC

Until recently, epidemics have been exposited in terms of the famous disease triangle—host, pathogen, and environment. The host must be susceptible, the pathogen virulent, and the weather favorable. But to understand an epidemic fully, a fourth dimension—time—must be added

---

* He tells us he now prefers the original spelling of his name, Vanderplank, but citations must use the spelling as they were written.

to the triangle to give a disease pyramid or disease cone, as shown by Browning *et al.* in Volume I. The first three factors listed above must not only be at or near their optimum, but they must remain so for a period of time. If not, the epidemic will not flourish. Let us consider each essential element in greater detail.

A. The Role of the Host

The role of the host in an epidemic is not a new element in epidemiology. Man has known the host since the beginning of agriculture itself. He has chosen from the wild the plants that were most useful. He has eaten the host. He has built his house with the host. He has fed the host to his livestock. Over time he learned to collect the seeds and to plant and nurture and harvest the plants that served him best.

The seeds that early man kept for the next crop were selected from the healthiest, most suitable, and best-formed plants. Even as far back as Theophrastus (ca. 300 BC), man recognized that some varieties of plants were more vulnerable to epidemics than others. Selecting those most resistant to disease has narrowed the genetic base, so that today most of man's food is derived from relatively few species and varieties of plants. Through his own volition man has made his crops highly vulnerable to epidemics.

For convenience in planting, tending, and harvesting, man has clustered his plants together on the most fertile land. This is convenient not only for man but also for the pathogens that cause epidemics.

B. The Role of the Weather

The role of the weather is an old story, too. Theophrastus observed that plants in the foggy river bottoms were more liable to disease than those on the drier hillsides. The Irish people learned to their sorrow that the cold, wet summers of 1845, 1846, and 1847 brought on an epidemic of blight in their potatoes.

There is, of course, no one set of weather patterns that favors disease in all crops. The potato blight fungus likes cool, wet weather. The black stem rust of wheat develops best when it is warm and wet.

C. The Role of the Pathogen

Knowledge of the pathogen is hardly more than one century old. During the previous two or three thousand years, man suffered innumerable famines when epidemics devastated his crops. But he did not understand the role of the pathogen until the middle of the nineteenth century.

In natural ecosystems hosts and pathogens exist in a dynamic balance in which they have grown accustomed to each other (see Chapter 14, this volume). The plants that persist are those that adapt to the local climate, soil, and other factors of the environment. When man moves his favorite plants into new environments, he often manages to take the local pathogens along too. He typically clusters the host plants together in artificial monocultures. As a result, the homeostatic balance achieved in the natural ecosystem is lost. The host meets new pathogens and the pathogens meet new hosts; the climate, soil, and other conditions are not the same. The balance shifts toward a new equilibrium—sometimes with catastrophic results—for plants, pathogens, and man.

If man is to maintain stable supplies of food and fiber crops, he must learn to manage his crops so that epidemic disease does not occur. He must change the host or the environment and keep the pathogens under control during all the time necessary to obtain a healthy crop. And that leads us to the last and least understood of the four elements of an epidemic: the element of time.

D. The Role of Time

As indicated above, the disease triangle is not enough to describe an epidemic. The host, weather, and pathogen must be in conjunction, to be sure, but they must remain in conjunction long enough for damage to result from disease. Intuitively, plant pathologists have always recognized that time is necessary for disease. We could see that disease moves more rapidly through a crop in some seasons than in others.

Research leading to a sound understanding of the importance of time in plant disease epidemics is only about 30 years old. Before that, field experimenters who tested plant varieties for resistance or fungicides for efficacy were content to measure disease only once in a season, if at all. They depended on yield as a measure of disease severity. "It's yield that counts," they said. As we shall see in the later chapters of this volume, that perception has limited our thinking and therefore inhibited our progress in understanding plant disease.

Barratt (1945) was the first to sense the importance of rate in epidemics by plotting the amount of disease against time. His powerful idea was published in a short abstract with a very long title: "Intraseasonal Advance of Disease to Evaluate Fungicides or Genetic Differences." His abstract describes the "rate of progress of disease" and his measurements of that rate.

It is regrettable that Barratt's abstract got lost in the literature. Large (1952) missed it and had to rediscover the concept of rate. Van der

Plank (1960, 1963) missed it, too, when he was calculating his famous *r*—the rate of advance of disease. Barratt (1945) had used an *r* nearly two decades earlier to differentiate treatments and varieties.

Priorities aside, the point is that we must measure rates if we are to understand the kinetics of epidemics well enough to manage them efficiently.

Barratt left his original data at The Connecticut Agricultural Experiment Station, where some of it was used more than 20 years later to test the performance of EPIDEM—the first mathematical simulator of plant disease (Waggoner and Horsfall, 1969).

Merrill (1968) has made practical use of *r* in the case of Dutch elm disease. He calculated the rate of spread of the disease through the elms of cities. He showed that *r* was smaller if the city used sanitation measures than if it did not. He also used his calculated *r* values to predict how long the elms would remain alive in cities that used different methods of management. Given these estimates, city authorities could calculate the costs and decide what action to take, if any, to deal with the disease in their city.

IV. THREE PHASES OF THINKING ABOUT PLANT DISEASE

Evolution in our understanding of disease in plants is a continuously changing process. Scientists develop new ideas and patterns of thinking to replace those proving to be incomplete or inadequate under the tests of new evidence and improved insight. We suggest it is possible to recognize at least three major and several lesser phases of thinking about plant disease: the descriptive phase, the dynamic–quantitative phase, and the theoretical–synthetic phase.

A. The Descriptive Phase

All sciences begin with a descriptive phase. Plant pathology is no exception. The pathogen, the host, the weather, and the physiology of disease all have undergone descriptive analysis.

1. The Pathogen Was First

During the middle of the nineteenth century we slowly and painfully unraveled the role of the pathogen in the complex web of causality. It was DeBary who finally broke through the strangling cords of mysticism, scholasticism, and authoritarianism to establish firmly that fungi are the cause and not the result of disease (see Chapter 2, Vol. I). This breakthrough initiated a prodigious outpouring of knowledge about a myriad of new pathogens and the diseases they cause. At first, tens and then thousands of plant pathologists described hundreds and then thousands of fungi, bacteria, viruses, nematodes, and seed plants that cause disease. Many historians of our science call this the mycological era because fungi were the chief pathogens to be discovered and described. Our understanding of disease caused by recently discovered pathogens (e.g., air pollutants and mycoplasma) is still largely in the descriptive phase.

2. The Host Comes into Vogue

During the second part of the descriptive phase, the host came into its own. Plant pathologists (who were still called botanists or mycologists) emphasized the anatomy of plant diseases. Pathological histology became the rule of the day. At first the naked eye and the microscope were used almost exclusively. Later the more sophisticated tools of scanning electron microscopy, physiology, histochemistry, biochemistry, and biophysics were added.

At this stage in the evolution of our thinking, the study of plant disease had two major dimensions—host and pathogen—and the interactions between the two dominated the way plant pathologists thought, wrote, and talked at their meetings in the botanical and agricultural societies of those days. Ernst Gäumann was the first and most outstanding of all the men who promoted plant pathology during this stage. He also wrote the first comprehensive treatise on botanical epidemiology (Gäumann, 1946).

3. We Rediscover the Weather

About midway into the descriptive phase, plant pathologists suddenly rediscovered the weather and the role of the environment in disease processes. This added a third dimension to the study of plant disease and the term "disease triangle" became a household word in the rapidly expanding departments of plant pathology.

Plant pathologists soon learned to use the tools of the physical sciences. Temperature, rainfall, soil moisture, humidity, wind velocity, turbulent flow, boundary layers, and other physical factors of the environment were added to the lexicon of the descriptions of host–pathogen interactions. L. R. Jones, J. C. Walker, and others of the Wisconsin School of Plant Pathology called all the pathogens "incitants" to emphasize the important role of the weather. This term helped our thinking about the environment but muddied the water of our thinking about the processes of infection and pathogenesis. While a riot in the street will continue without the original incitant, we know of only one plant disease (crown gall) which will continue without the pathogen. Disease is induced by pathogens, not incited by them.

B. The Dynamic–Quantitative Phase

When plant pathologists expanded their thinking to embrace the dynamics of disease in populations, quantitative measurements and patterns of thinking came into vogue. How many spores does it take? How viable are they? How far can they fly? How rapidly can they germinate, invade, and colonize the host? How soon will the pathogen sporulate? These and many other quantitative questions stimulated the development of new concepts—inoculum potential, inoculum dose, inoculum density, disease gradients, probits, single- and multiple regressions, critical-point models, and disease progress curves. Gradations of virulence and avirulence and of tolerance and resistance were measured in quantitative terms.

Our first quantitative thinking was about the foliar diseases induced by fungi. But eventually the same quantitative questions were asked about bacteria, viruses, and pathogenic seed plants. Despite the complexities involved, they are now being asked about pathogens that operate below ground.

Quantitative thinking about disease in populations led to discussions about dispersal phenomena and the concepts of rate. Barratt (1945) was the first to emphasize the rate of progress of disease, but it remained for van der Plank (1960, 1963) to make such measurements commonplace. The concept of rate requires the measurement of disease over time, Thus, time has become a fourth dimension for thinking about plant disease. As a result, "disease progress curves," "disease pyramids," and "disease cones" (see Chapter 11, Vol. I) are replacing the "disease triangle" as a framework for thinking, and quantitative assessments of disease loss are becoming more common (James, 1974; Carlson and Main, 1976; see also Chapter 4, Vol. I).

More than any other single scientist, van der Plank (1960, 1963, 1968, 1975) is the man who "pulled the bung on the epidemiological barrel." At present new knowledge is spilling out rapidly in ever-widening pools of understanding. With his stimulation, quantitative thinking has spread like an epidemic across the field of plant pathology and it appears that the field will never again be the same.

C. The Theoretical–Synthetic Phase

Today, two new trends are emerging in the continuing evolution of thought about plant disease. Both trends represent a departure from our traditional dependence on empiricism. Truly theoretical phases are now beginning to develop in the genetics of host–parasite interactions and in botanical epidemiology. Also a new extent of integration is developing in the arts of disease management.

1. Theoretical Plant Pathology

Looking back now, it is clear that theoretical plant pathology had one of its origins in Flor's (1942) gene-for-gene hypothesis. Today, his brilliant idea is being coupled with the emerging concepts of quantitative and theoretical genetics. Using a combination of systems-analytic and simulation models, these ideas are beginning to provide powerful new insights into the dynamic genetic interactions in host–parasite systems. At present, these new tools of analysis are being applied to crop systems with a single pathogen in highly controlled agroecosystems. Eventually these concepts may be extended to include diseases of multiple etiology and the more complex host populations in natural ecosystems.

Theoretical developments are also emerging in the population ecology of pathogens and in predictive epidemiology. Certain aspects of the population dynamics of nematodes are approaching a theoretical phase. In addition, as Zadoks and Koster (1976) stated in their recent history of botanical epidemiology. "The characteristics of this new science, which permit it to be ranked among the modern natural sciences, are its tendency towards abstraction, . . . its quantification leading to statistical tests and predictions, its leaning towards theoretical concepts and models . . . [and] its manifold interdisciplinary relations . . ."

By contrast, despite the impressive investments and progress in research on the physiology of pathogenesis, we expect this aspect of our field to remain in a descriptive phase for some time to come. Our conceptual knowledge of disease physiology is yet too weak to bridge the gaps with the approximations that theory requires.

2. Integrated Management

Stimulated by the rise of quantitative plant pathology, Waggoner and Horsfall (1969) developed EPIDEM, a mathematical simulator of plant disease. It was the first attempt to combine available knowledge of the multiple factors that determine the course of a plant disease epidemic. Since then, new and improved simulation models have been created and tested for their capacity to predict disease development under real-world conditions. Gradually the gaps in knowledge that account for lack of predictive efficiency are being filled.

Today, crop simulation models are being built which incorporate an increasingly large array of stress factors including diseases, insects, weeds, and such physiological factors as drought, heat, nutrient imbalances, etc. These models are being used to develop integrated systems for management of both agricultural and timber crops.

Predictive modeling is placing unprecedented demands on our knowledge of disease processes. It requires plant pathologists to expand their traditional frames of reference to include new factors such as: (1) the interactions among multiple stress factors as they influence the same host population; (2) the structures of interdisciplinary teams that can embrace the whole agricultural or forest ecosystem in which crops grow; (3) the influence of crop production practices on plant disease; (4) sustained yield management and cost-benefit relationships; (5) the changing costs of energy, transportation, and labor as well as local, regional, national, and even international economics; and (6) environmental protection and a multitude of other social, political, and legal constraints on crop and disease management.

The age of synthesis is at hand. Farming has always been an integrating art with the integrating left to the farmer. As the pressures on world food and fiber supplies continue to mount, however, plant pathologists and other scientists will have to join and assist farmers and foresters as never before. Our daily bread requires a renewed commitment to synthesis in our thinking about crop production and crop protection.

As this volume goes to press, we dare not venture further in predicting the evolution of our science. And so we pause to remember the prophet's prayer: "Lord, help our words to be gracious and tender today, for tomorrow we may have to eat them."

V. AN OVERVIEW OF THE TREATISE

We chose to call this treatise "Plant Disease," not "Plant Pathology." The term "plant pathology" means the study of disease and thus the term has a certain anthropocentric ring to it—it's man-centered. Study is something man does. Man may suffer when disease hits his crops, but his suffering is secondhand. It is the plant that is sick—not the man. We would prefer to understand plant disease as plants experience it and thus make the treatise plant-centered.

To look at disease as plants do, plant pathologists must learn enough to predict the reactions of plants to pathogens. That is a tall order! It is hard enough to understand how healthy plants grow. It is even more difficult to understand how plants behave when they are sick, especially when there are whole fields of them, all in various stages of disease.

Many earlier books about disease in plants were given titles with plural subjects such as "Manual of Plant Diseases," "Diseases of Trees and

Shrubs," "Diseases Caused by Ascomycetes," or just plain "Plant Diseases." Use of the plural is understandable—the total number of plant diseases is as astronomical as the national debt.

It is impossible to learn all the diseases of plants and probably foolish to try. Even to learn the diseases of one plant is a tremendous challenge. We designed this treatise to emphasize the unifying principles and concepts that will integrate our thinking about plant disease. This method of organization wa salso used in the original treatise (Horsfall and Dimond, 1959, 1960). It emphasizes the common features of disease rather than the diversifying factors that tend to fragment our thinking. As a symbol of our desire for synthesis and unification, we left the "s" off diseases and called this treatise just "Plant Disease." *

Given five volumes in which to set out the art and science of plant disease, we found the subject could be divided readily into the required five parts:

 I. How Disease Is Managed
 II. How Disease Develops in Populations
 III. How Plants Suffer from Disease
 IV. How Pathogens Induce Disease
 V. How Plants Defend Themselves

When the original treatise was designed in the late 1950's, plant pathology was reaching for maturity as a science with basic research coming into its own. For that reason the original treatise was organized around the scientific foundations of disease processes.

But during the last decade, society has called on plant pathology to demonstrate its usefulness in a world of worsening hunger. This treatise was conceived in 1975 when the world passed another great milestone along the road to global starvation. In that year the world population reached four billion people, enough to form a column marching 30 people wide and 1 m apart around the equator. It scared the "wits" out of us!

The growing urgency of the world food problem made us decide that it was timely, even urgent, that this treatise begin with the arts of disease management and go on to the science of plant disease. In this way it also would relate the basics of our science to its usefulness to society.

Volume I is not a cookbook on how to control specific diseases. Rather, it is a theoretical and philosophical treatment of the principles of man-

---

* We later found that we are not the first to see value in this title. Russell Stevens (1974) used it for his recent introductory text.

aging disease by altering the genes of plants, by changing the associated microbiota, by selecting or altering the environment, or by using chemicals. Since the first volume sets the stage for the others, Volume I also contains chapters on the profession of plant pathology, its sociology, and how it works to benefit society.

Botanical epidemiology is the subject of Volume II: How Disease Develops in Populations. Since 1960, explosive progress has been made in understanding epidemics of plant disease. The latest explosion has come in the mathematical analysis of factors that make epidemics increase and subside (Waggoner and Horsfall, 1969). This provides the foundation for the emerging new field of theoretical plant epidemiology (see Zadoks and Koster, 1976). Volume II will also include analyses of the genetic basis of epidemics; the methodology and technology of epidemiological analysis and forecasting; the concepts of inoculum potential and dispersal; the climatology and geography of plant disease; and the use of quarantines as a defense against epidemics of introduced disease.

In Volume III we move from disease in populations to disease in individual plants: How Plants Suffer from Disease. The early chapters set the stage for all the later chapters in Volume III plus those in IV and V. First, they describe how healthy plants grow. Next, a modern conceptual theory of how disease develops in plants is presented. Here disease is presented as the end result of all the positive and negative influences of hosts on pathogens and pathogens on hosts. The later chapters describe the many different kinds of impairments that can occur when plants are diseased. This volume considers various dysfunctions: in the capture and utilization of energy; in the flow of food; in the regulation of growth; in the processes of reproduction; in intermediary metabolism and mineral nutrition; in the integrity of membranes; and even in the biological rhythms of plants.

Having set out in Volume III the potential for dysfunction due to disease, Volume IV considers how pathogens induce these various dysfunctions: How Pathogens Induce Disease. This volume describes the concepts of single, multiple, and sequential causality and their relationship to stress; the evolution and energetics of parasitism and pathogenism; the concepts of allelopathy and iatrogenic disease; the structure and function of toxins; and the role of pathogen enzymes in disease processes. Next we compare and contrast the unique features of all the various pathogens of plants—fungi, bacteria, insects, Mycoplasma, Rickettsia, parasitic seed plants, nematodes, viruses, viroids, air pollutants, etc. How are the effects of pathogens similar? How are they different? What offensive weapons does each type of pathogen use to be successful? Finally, we consider the effects of diseased plants and pathogens on livestock and man.

Volume IV, in turn, sets the stage for Volume V: How Plants Defend Themselves. The chapters of this final volume are closely linked with the analogous chapters in Volumes III and IV. Volume IV deals with how pathogens thwart the defenses of the host. Volume V describes how plants thwart pathogens.

Plants have many natural enemies and they have evolved a magnificent array of armaments to keep their enemies out or to minimize the damage they cause once they get in. Some plants escape from disease. Others defend against it with great success. Still others have evolved mechanisms to tolerate disease and grow well in spite of their sickness. Volume V describes defense by analogy to the defense of a medieval castle. There are defenses at the perimeter—the outer walls and gates— but if they do not hold, internal defenses come into play. There are dynamic defenses triggered by the invaders and even defenses triggered by previous invaders. A final chapter describes how the metabolic resources of the plant are allocated to maintain and repair its defenses.

The dynamic competition between the offensive weapons of the pathogen described in Volume IV and the defensive weapons of the host described in Volume V should read like "a battle royal."

Since the treatise begins with management in Volume I, it is fitting that it should end with defense in Volume V—a major goal of integrated management being the enhancement of natural defenses against disease.

VI. SOME HIGHLIGHTS OF VOLUME II

We continue this volume with a chapter entitled "Some Epidemics Man Has Known." Here we have considered the sociology of epidemics just as we dealt with the sociology of plant pathology in Chapter 2 of Vol. I. How have plant disease epidemics influenced man's food, his social customs, his economics, and his ability to wage war? How has man encouraged epidemics in his own crops? We shall soon see.

After Chapter 2 was written we were astonished to note that textbooks by professional plant pathologists simply do not deal with the social impact of epidemic disease; or they pass over it lightly with a paragraph or two on the dreadful Irish famine. The books of Stakman and Harrar (1957), Walker (1957, 1969), Agrios (1969), and Stevens (1974) are examples of this curious phenomenon.

By contrast, however, amateur plant pathologists have tackled the sociology of epidemics, and generally have done it well. Large (1940), at that time a novelist, approached it first. Then came the two magnificent volumes: one by Woodham-Smith (1953, 1962) on the impact of the Irish famine, and the exciting book by Carefoot and Sprott (1967) "Famine on the Wind." Fuller's book (1968), "The Day of St. Anthony's Fire," is a dramatic treatment of the ignorance, stupidity, and greed that triggered a horrible epidemic of ergotism in France in 1951. Thanks be to the amateurs!

In Chapter 3 Kranz develops the comparative anatomy of epidemics. He slices up epidemics so they can be compared in terms of their structure, their patterns of development, and their dynamics.

Chapters 4–6 deal with the methods of epidemiological research. In Chapter 4 Zadoks describes the rational processes of epidemiological research and shows how they differ from the processes used to investigate disease in individual plants. In Chapter 5 Pennypacker follows up with a discussion of the instrumentation for measuring the weather component. Measuring the amount of disease is covered by Horsfall and Cowling in Chapter 6.

No subject is more central to the development of epidemics than the abundance of inoculum and its capacity to cause disease. Baker discusses the theory and the measurement of inoculum potential in Chapter 7.

The importance of time in the analysis of epidemics has been emphasized earlier in this Prologue and is developed further in Chapters 8, 9, and 12. In Chapters 8 and 9, Aylor and Wallace discuss the dispersal of pathogens in both time and space. Aylor deals with these processes in above-ground disease while Wallace discusses the movement and maintenance of infectivity by pathogens that operate below ground. Many earlier analyses of epidemics were based on the simplifying (but false) premise that plants do not change in susceptibility with time. In Chapter 12, Populer sets this subject squarely before us by showing how the same plant may change in susceptibility from one time to another.

In matters of great complexity like flying to the moon and synthesizing knowledge of plant disease epidemics, computers are necessary to cope with all the variables. Waggoner discusses computer simulators of plant disease in Chapter 10. In Chapter 11 Shrum shows how to use predictive models to forecast epidemics so that management decisions can be made in time to head them off.

Man keeps changing the genes for resistance in his major crops and the pathogens keep changing to keep up. In Chapter 13 Day discusses the problems of genetic uniformity and susceptibility and the breeding and deployment strategies that are needed to cope with these problems.

Most plant pathologists study disease in agricultural monocultures.

But we need to learn more about the natural homeostatic mechanisms that regulate disease in natural ecosystems. Using the natural forest for most of his examples, Schmidt discusses in Chapter 14 the role of functional stability in limiting disease epidemics.

The influences of climate and weather on epidemics are discussed by Rotem in Chapter 15. Here he describes the climatic habitat of airborne and soil-borne pathogens and then puts forth a series of hypotheses concerning compensating factors within hosts and pathogens that limit or increase the likelihood of disease under specific climatic conditions. In Chapter 16, Weltzien provides a new analysis of the geographical and climatic distribution of plant diseases in various parts of the world.

Farmers and foresters aim to produce healthy crops, but sometimes their best-intentioned efforts make epidemics worse. In Chapter 17, Cowling provides a chronicle of hazardous practices that have favored epidemics in the past and offers some conclusions about how they might be avoided in the future.

Volume II concludes with a discussion by McGregor about the efficacy of plant quarantines. This subject has always been a source of controversy in plant pathology. McGregor provides stimulus for our thinking by focusing attention on the different probabilities of success that may be expected for quarantines against diseases of various types.

The chapters of Volume II have all been designed to stimulate our thinking. They certainly have aroused our interest. We hope they stimulate yours as well.

References

Agrios, G. N. (1969). "Plant Pathology." Academic Press, New York.

Barratt, R. W. (1945). Intraseasonal advance of disease to evaluate fungicides or genetical differences. *Phytopathology* **35,** 654.

Carefoot, G. L., and Sprott, E. R. (1967). "Famine on the Wind." Rand McNally, Chicago, Illinois.

Carlson, G. A., and Main, C. E. (1976). Economics of disease-loss management. *Annu. Rev. Phytopathol.* **14,** 381–403.

Flor, H. H. (1942). Inheritance of pathogenicity in *Melampsora lini. Phytopathology* **32,** 653–669.

Fuller, J. G. (1968). "The Day of St. Anthony's Fire." Macmillan, New York.

Gäumann, E .A. (1946). "Pflanzliche Infectionslehre." Birkhaeuser, Basel.

Horsfall, J. G., and Dimond, A. E. (Eds.) (1959, 1960). Plant Pathology: An Advanced Treatise. 3 Vols. Academic Press, New York.

James, W. C. (1974). Assessment of plant diseases and losses. *Annu. Rev. Phytopathol.* **12,**27–48.

Large, E. C. (1952). The interpretation of progress curves for potato blight and other plant diseases. *Plant Pathology* **1,**109–117.

Large, E. C. (1940). "Advance of the Fungi." Holt, New York.

Merrill, W. (1968). Effect of control programs on development of epidemics of Dutch elm disease. *Phytopathology* **58**, 1060.

Stakman, E. C., and Harrar, J. G. (1957). "Principles of Plant Pathology." Ronald Press, New York.

Stevens, R. B. (1974). "Plant Disease." Ronald Press, New York.

van der Plank, J. E. (1960). Analysis of epidemics. *In* "Plant Pathology" (J. G. Horsfall and A. E. Dimond, eds.), Vol. 3, pp. 229–289. Academic Press, New York.

van der Plank, J. E. (1963). "Plant Diseases: Epidemics and Controls." Academic Press, New York.

van der Plank, J. E. (1968). "Disease Resistance in Plants." Academic Press, New York.

van der Plank, J. E. (1975). "Principles of Plant Infection." Academic Press, New York.

Waggoner, P. E., and Horsfall, J. G. (1969). EPIDEM: A simulator of plant disease written for a computer. *Conn., Agric. Exp. Stn., Bull.* **698**, 1–80.

Walker, J. C. (1957). "Plant Pathology." McGraw-Hill, New York.

Walker, J. C. (1969). "Plant Pathology." McGraw-Hill, New York.

Woodham-Smith, C. (1953). "The Reason Why." McGraw-Hill, New York.

Woodham-Smith, C. (1962). "The Great Hunger: Ireland 1845–1849." Harper, New York.

Zadoks, J. C., and Koster, L. M. (1976). A historical survey of botanical epidemiology. A sketch of the development of ideas in ecological phytopathology. *Mededelingen Landbouwhogeschool.* Wageningen, The Netherlands. **76**, 1–56.

Chapter 2

Some Epidemics Man Has Known

JAMES G. HORSFALL AND ELLIS B. COWLING

Man suffers from epidemics. His crops suffer from epidemics. Whole nations suffer from epidemics. In fact, human suffering and epidemics of plant disease have gone hand in hand since the earliest history of man.

Epidemics of plant disease have influenced man's food, his health, his social customs, his economics, and even his ability to wage war. Let us look at a few examples, beginning with effects on war since that is the most dramatic.

I. IMPACT OF PLANT DISEASE EPIDEMICS ON WAR

In human medicine the classic case of the influence of epidemics on war is that of lice and typhus. Lice carry the typhus pathogen. The pathogen kills soldiers, or, if it does not kill them, it weakens them and destroys their ability to fight.

17

The classic case of a plant disease that has influenced war is that of ergotism at Astrakhan in Russia in 1772. Carefoot and Sprott (1967) relate the dramatic tale. First, let us examine the disease of rye called ergot and the disease of man called ergotism (Barger, 1931).

A. Ergotism at Astrakhan

Ergot is a disease of rye. The fungus invades the grain and produces a bloated, obscene, purple-black body that resembles a rooster's spur. The mycologists call it a sclerotium. The French call it an ergot. The English call it a purple cockspur. Whatever the name, it contains large amounts of alkaloids—alkaloids of many types—including LSD, the well-known hallucinogenic drug used by some modern young people to "open their minds." When this purple cockspur is ground into flour along with the grain, the alkaloids are baked into the bread.

When contaminated bread is eaten, the alkaloids produce a fantastic array of distressing symptoms. If small amounts of the contaminated bread are eaten by pregnant women or contaminated grain is eaten by pregnant cattle, the alkaloids induce abortion. If larger amounts are ingested, there is first a tingling in the fingers and toes. This is followed by a raging high and burning fever. The fever can lead to mental derangement and often to death. Sometimes gangrene sets in and toes fall from the foot, feet from the ankle, fingers from the hand, hands from the arm. In other cases the disease causes unbelievable hallucinations and the victim may die of convulsions.

The terrible scourge of ergotism flourished in the Middle Ages. The first recorded epidemic was AD 857 when thousands died in the Rhine Valley. On the basis of one of its major symptoms, the doctors gave it a Latin name, *Sacer ignis*, the Holy Fire.

Another devastating outbreak hit France in AD 1089. The monks of the order of St. Anthony were able to relieve the symptoms, presumably by feeding ergot-free bread. The monks were so successful that the disease became known as St. Anthony's Fire, and the name Holy Fire died out.

The fire raged on in the rye-growing regions of France and Germany during the eleventh, twelfth, and thirteenth centuries.

Ergotism had an influence on war. According to Carefoot and Sprott (1967) Peter the Great in Russia set forth for the south in the spring of 1772 to force the Turks to give him control of certain warmwater ports on the Black Sea. When his soldiers moved into the delta of the Volga River at Astrakhan, they stumbled into an epidemic of ergotism. Being ignorant of the disease, they bought rye hay for their horses and rye bread for themselves. Almost immediately 100 horses were down with the "blind staggers" and many soldiers were doubled up with convulsions. It was St. Anthony's Fire. Peter's dream turned sour and his army turned back. Twenty thousand people died of ergotism at Astrakhan that year. Peter could not win the Dardanelles because a disease of rye destroyed the ability of his army to fight. One wonders what would have happened if the disease had not been there.

B. "The Charge of the Light Brigade"

In his famous poem, Alfred Lord Tennyson wrote:

Forward the light brigade!
Charge for the guns, he said.
Into the valley of death
Rode the six hundred. . . .

The charge of the light brigade during the battle of Balaclava in the Crimean War of 1854 is one of those cases in which history repeats itself. It was George Bingham, better known as Lord Lucan, a Lieutenant General in the British army, who gave that fateful command, "charge for the guns." It was he who sent the Light Brigade charging into the valley of death, into what he knew was almost certain death and bloody destruction. Only one-third of his men and horses came stumbling and crawling back. Why did he do it?

Who would have dreamed 10 years earlier that the bitter experience of the Irish would be repeated at Balaclava in the charge of the Light Brigade. Cecil Woodham-Smith tells the story in her two books, "The Reason Why" (1953) and "The Great Hunger" (1962).

In "The Reason Why" (p. 130) she says, "In the opinion of Kinglake, the historian of the Crimean campaign, it was Lord Lucan's conduct in Ireland, his ruthlessness, energy, disregard for human sentiment, and contempt for public opinion, which decided the government to select him for command in the Crimea." How did this come about?

Lord Lucan was an Englishman who, in 1837, had inherited a manorial estate at Castlebar in the County of Mayo in Ireland. At the same time he also inherited a host of Irish tenants who evidently shared the contempt of the Roman Catholic Irish for the Anglican English, especially Anglican English noblemen. He responded to their prejudice with a vengeful prejudice of his own. He saw the Irish tenants who swarmed over his land as ignorant, shiftless, and, to boot, Roman Catholics.

In the middle of the nineteenth century, the people of Ireland displayed the same symptoms we see in many hungry nations in the twentieth century. Its population had expanded to the very limits of its food supply. Early in the eighteenth century the population of Ireland had come into a reasonable balance with the food available from oats, wheat, and barley. With the introduction of the potato, however, they could produce nearly three times as much food per hectare as before. The population promptly exploded—it doubled and doubled again. Farms were divided, subdivided, and divided again. By the time Lord Lucan took over his estate, the population of Ireland had exceeded the capacity of the land to feed it. A confrontation over food supplies was inevitable. The onset of the blight triggered it.

In 1844 the potato blight had mysteriously invaded the potato fields of the United States; but nobody worried about it 5000 km away in Europe. The next year, however, the blight swept over Ireland like a hurricane. Starvation stalked the land the next winter, and the two that followed. In Ireland there was no alternative to the potato. The people suffered with malnutrition, starved, and died. Some emigrated from Ireland if they could.

Before the blight-induced famine struck, Lord Lucan had already seen the dreadful overcrowding of his land and had begun the process of evicting some of his tenants to reduce their numbers to a density the land could support. When the "potato cholera" struck the potato, however, the situation became desperate. Lord Lucan's innate, blind ruthlessness came to the fore and his evictions became more merciless than ever.

In one area of his land, 10,000 people were evicted and Lord Lucan brought in a Scotsman, another hated "foreigner," and put him in charge of farming operations on 7000 hectares of his land. In order to make sure that the evictions were permanent, Lucan tore down the huts of the evicted. The ignorant, starving, and terrified people took refuge with their neighbors. However, the neighbors were forbidden to receive the evicted families or they themselves would be evicted. Lifeboat ethics became a reality a century before Hardin. No wonder the people called Lord Lucan "The Exterminator." He regarded his tenants as a threat, and so he barged ruthlessly ahead. He displaced the people he considered to be vermin infesting his land.

It is no wonder that, a few years later, he could just as blindly order the Light Brigade to "charge for the guns"—directly into the mouths of the waiting Russian cannons.

As we all know, the blight that blackened the potato vines and rotted the tubers was caused by a fungus. We call it *Phytophthora infestans*. The word *Phytophthora* comes from two Greek words: *phyton* (plant) and *phthora* (destruction). The fungus destroyed the potatoes, then the Irish, and then the Light Brigade. What a gruesome chain of events!

Phytophthora infestans also destroyed some previously sacred social and political institutions. For example, Lord Lucan paid £25,000 for his original commission in the army, as was the custom of that day. After the disastrous battle at Balaclava, no one could buy a commission. The whole military system was overhauled in Britain, on the continent, and in many other parts of the world.

By destroying the potato, *Phytophthora* also helped to destroy the ill-famed corn laws which had been in effect for years in Britain to protect grain farmers from foreign competition. With the defeat of the corn laws, Britain became a free-trade nation and her economy boomed as a consequence.

Carefoot and Sprott (1967) cite another striking repercussion of the potato famine. The blight swept all across Europe and on into Russia. There was less starvation than in Ireland because European agriculture was more diversified than in Ireland. Nevertheless, the hunger was widespread.

This gave great ammunition to radicals like Kossuth, Engels, and Marx who urged the people to insist on new forms of government. Several governments fell in 1848 when an economic depression added to the misery of hunger. Louis Philippe fell in France and government reforms were instituted in Italy, Austria, and Hungary.

The starvation induced by *Phytophthora infestans* frightened and worried many people—both citizens and scientists alike. This gave great impetus to plant pathology and the development of Agricultural Experiment Stations both in Europe and in North America (see Chapter 2, Vol. I). Roland Thaxter was perhaps the tenth professional plant pathologist to be hired in the United States. His first assignment in 1888 was to control potato rot, as he called the disease, caused by *P. infestans.*

II. IMPACT OF WARS ON PLANT DISEASE EPIDEMICS

Not only do plant disease epidemics affect wars, but wars also affect plant disease epidemics. The most often cited case of the effect of an epidemic on food crops caused by war is that of the Hessian fly insect on wheat. When King George III of England hired Hessian mercenaries from Germany to fight for him during the American Revolution, they brought wheat straw as bedding for their horses. Along with the straw came the Hessian fly which killed off American wheat. This was unwitting biological warfare.

A. Late Blight of Potatoes and World War I

Carefoot and Sprott (1967) describe how late blight of potatoes helped the British defeat the Germans in World War I.

Bordeaux mixture, that holy water of plant pathology, was discovered in France in 1882. This complex mixture of copper sulfate and lime was just the tool needed to save the potato from disease and the people who depended on the potato from starvation. When potatoes are anointed with Bordeaux mixture, they are protected from blight and the people are saved from hunger. Potato farmers were inhibited from doing so in Germany in 1916 during World War I when an epidemic of late blight enveloped their potatoes.

The weather in Germany during the growing season of 1916 resembled that in Ireland in 1846—cold rain, cold rain, cold rain. The Bordeaux mixture that was so specific for potato blight was missing, the military leaders would not release the copper needed to make it. They were as arrogant as Lord Lucan had been in the previous century. They had to have it for shell casings, electric wire, and, we suspect, for brass buttons, too. The farmers, being low in the Prussian-dominated pecking order, lost the argument and the nation went hungry the next year. According to Carefoot and Sprott (1967) all the grain and most of the potatoes went to the army in 1916 and 1917. The soldiers were not hungry but their families at home were. Many of them starved to death. The morale of the soldiers was weakened. The military might weakened and it collapsed in 1918.

Phytophthora infestans played a role. It reminds us of the old saying that "For want of a nail the shoe was lost; for want of a shoe the horse was lost; and for want of a horse the battle was lost."

B. Dutch Elm Disease

At the end of World War I, this infamous disease swept through the elms of Holland killing them as it went. As a result the disease was dubbed "Dutch elm disease." The Dutch do not like the epithet. We can hardly blame them for that, but the name has been retained.

Since the Asiatic elms are resistant, the disease presumably originated in eastern Asia. If so, how did it move 10,000 miles into Holland? Apparently, the war did it. During the war the Allies imported hundreds of Chinese laborers to dig trenches in Flanders. They moved their meager belongings in wooden wicker baskets made from the tough fibrous wood of the Chinese elm. Some of the pieces carried bark and the vector beetles. Presumably the fungus escaped into the low countries.

In any event the ensuing epidemic killed millions of elms in Europe. Despite a quarantine against importation of diseased seedlings, veneer manufacturers in Ohio and other states purchased the dead logs and thereby imported the fungus into the heartland of their own country where it has done untold damage to the elms that formerly lined the streets of hundreds of American cities. Although the disease is still spreading farther and farther west, it has now come into a generally stable equilibrium with the American elm in the eastern states where it has now existed for up to four decades. It is a significant case of a war-induced epidemic of plant disease.

C. The Great Bengal Famine

Almost 100 years after the potato famine that killed an estimated 1.5 million Irishmen, another famine struck in Bengal halfway around the world and killed an estimated 2 million of its citizens (Padmanabhan, 1973).

Unlike the Irish famine it was exacerbated by war. Otherwise the two situations were astonishingly alike. Like Ireland, Bengal was dreadfully overpopulated. And, like Ireland, it was dependent on a single crop—rice in this case. The Bengal weather in 1942, like that in Ireland in 1845 and 1846 was unusually rainy and unusually cloudy. *Helmintho-sporium oryzae* struck and the rice was destroyed. The rains were a little too late for the fungus to kill off the early maturing varieties. Their yield was cut by only 50%. The disease hit the later varieties harder and reduced their yields by 75 to 90%. Fortunately for the Bengalese, and in contrast to the hapless Irish, the weather did not favor *Helmintho-sporium oryzae* in the subsequent years so that their travail was limited chiefly to 1943, the year after the disease epidemic.

Dr. Padmanabhan, in his fine monograph on the disease (1973), relates that he was sent to Bengal to investigate the disease. He says, "The author was appointed as mycologist in Bengal when the famine was at its height. When he traveled to join his new assignment on the 18th of October 1943, he could see dead bodies and starving and dying persons all along the way. This horrendous situation of thousands of men, women, and children dying of starvation continued through October, November, and December." The dying continued until the new crop arrived in the summer of 1944.

The war worsened the famine because the Japanese army had occupied neighboring Burma and thereby shut off the normal flow of rice from that important source. The British were fighting with their backs to the wall and could not possibly provide enough rice from elsewhere.

Thus you have the complex relation between plant disease epidemics and war. A plant disease helped the Turks defeat the Russians in the eighteenth century. Another helped the Russians defeat the British in the nineteenth century, and a third helped the British defeat the Germans in the twentieth century.

III. IMPACT OF PLANT DISEASE EPIDEMICS ON HUMAN CULTURE

We have already seen how the famine in Ireland had incredible effects on human culture, sometimes far from Ireland. Plant pathologists have rarely written about the sociological impact of plant disease epidemics. Perhaps they have been too busy researching the diseases themselves, or perhaps they are too timid to venture out of their field into the history of human affairs. Fortunately, a few epidemics have been treated in this way by historians like Woodham-Smith (1962), Large (1940), and Carefoot and Sprott (1967).

A. Coffee Rust

The history of coffee and its rust has been treated by Large (1940) and by Carefoot and Sprott (1967). From them we gather that the Arabs brought coffee from Ethiopia about AD 1000. They found it stimulating; it helped them to outlast the interminable rituals of Islam. The Islamic Turks had established coffee houses by 1550 and the roasted (not fresh) coffee beans began to move into Europe to whet the appetites of the wealthy.

The Arabs were careful not to let live beans move out because they wanted to maintain a monopoly. Coffee provided foreign gold just as their oil does now 400 years later.

Eventually, however, a Muslim from South India on his holy pilgrimage to Mecca was able to rise above his religion and steal seven live coffee beans. When he planted them in India, they flourished. From India, coffee was taken to Ceylon (now Sri Lanka). Coffee was as well adapted to Ceylon as the potato had been to Ireland. In 1835 the British in Ceylon were growing only 200 ha of coffee. By 1870 they had about 200,000 ha in coffee and they exported 50 million kg of beans per year. Ceylon flourished and so did the Oriental bank that handled the golden stream of money that flowed back.

In the meantime the citizens of the British Isles had learned empirically to boil their drinking water to avoid intestinal "fluxes." They found that the boiled water was more fun to drink if they boiled coffee with it. By 1870, in the last quarter of Queen Victoria's reign, they had become a nation of coffee drinkers.

Then the devastating rust disease struck Ceylon. It killed the coffee trees faster than they could be replaced. In the 19 years after 1870 the yield dropped from 50 million kg to essentially zero. Four hundred and seventeen planters went broke as did the Oriental bank. The life savings of its depositors were swept away and Ceylon was essentially bankrupt. The thousands of south Indian laborers who had been imported to pick the coffee were sent home without funds and the local Singhalese went hungry because they had no money with which to buy rice. No wonder Berkeley gave the fungus the name of *Hemileia vastatrix*. It surely was devastating.

To endure in their cold climate, the British needed a stimulant to replace coffee. So what did they do? They planted tea, of course. By the end of the coffee boom Ceylon had only about 500 ha of tea. By dint of a herculean effort by 1880 the planters had covered about 140,000 ha with tea bushes. England put tea in its boiled water and thereupon became a nation of tea drinkers, as they are today. They established the pleasant custom called elevenses—tea at eleven o'clock. Americans call it the "coffee break," but their coffee comes from South America, not Ceylon.

The disease has now jumped the Atlantic Ocean to Brazil and Columbia where it is spreading. What will it do there? What effect will it have on the American coffee break? Time will tell.

B. Peach Yellows

Peach yellows is another dramatic disease. Since it attacked a luxury crop, the peach, it did not induce starvation among the citizens but it seriously affected peach growers along with the villagers that depended on them.

The disease broke out, or at least it was first reported, on a farm in what is now Fairmont Park in Philadelphia about 1791 (Smith, 1888).

Since Philadelphia is an ocean port, it is tempting to assume that the virus entered this city on imported seedlings. This theory is difficult to defend, however, because in 1888, one century later, Erwin F. Smith was unable even then to find records of the disease abroad. We must conclude, then, that the virus probably arose *de novo* as did X disease of the peach in 1933.

By 1810 the disease had spread eastward to central New Jersey, by 1814 to Long Island, and by 1820 to Connecticut. It reached the northeastern limit of peach culture in Massachusetts by about 1840.

Similarly, it spread north up the Hudson River to Newburgh by 1828, on into western New York by about 1840, into Ohio by 1849, and onward to its western terminus in Michigan by 1866.

It differed from all other epidemics on trees that we know of in that it waxed, waned, and waxed again. In New Jersey it ravaged the peaches between 1806 and 1814, declined to an almost negligible amount by 1830, and irrupted again about 1846. Smith (1888, p. 28) says, "It is apparent that the ups and downs of peach growing in New Jersey have been many." He lists the general outbreaks as 1791, 1806–1807, 1817–1821, 1845–1848, 1874–1878, 1886–1888. By the end of the nineteenth century the disease disappeared just as mysteriously as it had appeared a century before. Today it is essentially impossible to find a diseased peach tree. One cannot help but wonder why. The best hypothesis seems to be that the rise in insect control about the turn of the century killed off the vector or vectors.

Smith (1888) provided us with some interesting grains of understanding about the epidemiology of peach yellows. He sorted the grains from the chaff of observations made by "dozens of farmers, judges, and nurserymen and others." These observations, however, were mainly the jottings of amateurs made before there was a significant number of plant pathologists in the United States. Most were made before the germ theory of disease was established. (For further discussion about the role of amateurs in the study of plant disease, see Chapter 2, Vol. I.)

In 1806 one year ahead of Prévost, a judge named Peters wrote: "I find that sickly trees often infect those in vigor near them by some morbid effluvia . . . The disorder being generally prevalent would, among animals, have been called an epidemic." In 1817, Mr. Prince, a nurseryman says:

> . . . (the disease) is evidently contagious . . . A decisive proof of its being contagious is that a healthy tree inoculated (i.e. grafted) from a branch of a diseased one, instead of restoring the graft to vigor and health immediately becomes itself infected with the disease. This disease is spread at the time when the trees are in bloom, and is disseminated by the pollen or farina blowing from the flowers of the diseased trees and impregnating those which are healthy and which is quickly circulated by the sap through the branches.

A farmer wrote in 1847 in the middle of the Irish famine that he could transmit the disease by budding and "that the disease, however it may have originated, has not its origin in either soil or climate is pretty evident." At the same time Lindley, editor of the Gardner's Chronicle in Britain, was still arguing that the *Phytophthora* was the result and not the cause of the potato disease.

In 1855 a Connecticut resident wrote that the disease ". . . is propagated . . . from blossom to blossom by insects."

In these astute observations you have most of modern plant pathology —enunciated by amateurs working with peach yellows. This disease was not due to spontaneous generation, not caused by the weather, was graft transmitted, was insect transmitted, and possibly pollen transmitted. It even moved in the sap stream. Marvelous!

What were the social and economic consequences of peach yellows? Perhaps the most dramatic evidence is expressed in the curve shown in Fig. 1 and constructed from data in Smith (1888). It shows the rise in production of a 45 ha peach orchard and its decline after infection. As the young orchard grew in the absence of yellows the yield grew geometrically but when the outbreak of peach yellows in the 1870's hit the field, the yields of the orchard plummeted essentially to zero—the whole cycle occurring in 10 years.

Hedrick (1917) says, "The best peach lands are seldom fit for other

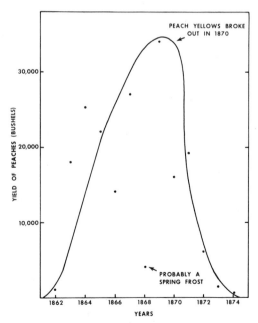

Fig. 1.   Rise in peach production and decline in yield following infection with peach yellows.

crops, so that in Delaware, New Jersey, and Michigan the whole community including railroad and steamboat lines suffers to the verge of bankruptcy when yellows exterminates the orchards."

Referring to Michigan he says that "The peach industry was literally swept out of Berrien County in one decade. The county had 6000 acres (2700 ha) of peaches in 1874. By 1877 it had only 503 acres (203 ha). The depreciation of peach lands at this time due to yellows was so great as to threaten the community with bankruptcy. Pitiful was the case of the growers of Berrien County."

Peach yellows bankrupted a few counties. Potato blight and coffee rust bankrupted whole nations.

C. Chestnut Blight

The destruction of the American chestnut by *Endothia parasitica* is perhaps the best-known epidemic in the United States. The trees were so plentiful and popular that the natives named many places as Chestnut Hill, Chestnut Lane, and Chestnut Street. The trees produced delicious nuts for the natives to pick up in the woods. The dramatic biological and human events of this epidemic and the attempts to stop it have been documented recently in an excellent paper by George Hepting (1974).

The chestnut filled two important niches in the American economy at the time, decay-resistant timber and tannin for the preservation of leather. The wood is extremely resistant to decay and so it was used for ties (sleepers) for building the railroads and for poles for building the telephone and telegraph system of the United States.

The bark that was removed to make ties or poles was widely used in the tanning of leather for the shoes the Americans walked on. The tree was also unique in providing a highly dependable supply of food for wildlife.

When the blight swept essentially all the chestnuts from the land in the first quarter of this century, the pole, tie, and tanning industries died and two new industries arose to replace them. The engineers built huge creosoting plants to inject creosote into rot-susceptible pine and the chemists devised synthetic tanning agents for leather. The latter are fuel intensive. Fossil fuel must be substituted for solar energy.

IV. MAN ENCOURAGES HIS OWN EPIDEMICS

In Chapters 2 and 6 of Vol. I, we discussed the sociology of plant pathology. There is a sociology of epidemics, too. We can find several cases where man has set the stage for epidemics in his own crops. The agricultural and forest practices by which this is achieved are discussed more fully in Chapter 17 of this volume. Here we will mention a few classic cases that are largely of historical interest.

A. Barberry/Wheat Rust

Wheat originated in the Near East and so did the barberry (*Berberis vulgaris*), the nemesis of wheat. As all pathologists know, *Puccinia graminis*, the cause of wheat rust, goes through the sexual stage on barberries. Therefore, the barberries provide sources of inoculum to start the epidemic anew each season. Barberries and wheat rust have produced epidemics and subsequent famine in the Near East since prehistoric times.

In Roman times, traders carried wheat by ship from the Mediterranean Sea through the Pillars of Hercules to northern and western Europe, but they did not take the barberry. As a result wheat flourished there for centuries, free of rust (Carefoot and Sprott, 1967).

Elsewhere, movements were afoot that would one day introduce rust into the western wheat. Mohammed encouraged his followers to spread the faith of Islam. Some did so by the sword, when necessary. For example, the Saracens rode north, east, south, and west. They spread all across northern Africa and on into Spain at Gibraltar. Conquering Spain, they flowed into France, where they were met and finally defeated by Charles Martel at the battle of Tours. They fled back across the Pyrenees to Spain.

Sometime during this great push of the Saracens, one of them in Italy, homesick for his homeland, imported a barberry bush, "and the fat was in the fire." Birds ate the delicious red berries and spread the seeds north and west. Peasants liked the pretty bush and planted it in their door yards. The missing link was now provided and rust for the first time began to destroy European wheat.

The European farmers discovered the connection between barberries and wheat rust by 1660. By then the idea was so firmly established that the French lawmakers passed a law requiring the eradication of barberries. We think it can safely be assumed that English farmers, at least the observant ones, knew of the relationship. For that reason the English colonists who moved to the United States in the early part of that century had a golden opportunity to move wheat once again across the water without moving its deadly enemy, the barberry. But they bungled it. Whether out of ignorance or stupidity, someone brought barberries to New England and the chance was lost. Wheat rust now raged on in the new land. Man had encouraged his own epidemic. The New Englanders caught up a century later. In 1726 Connecticut passed a barberry eradication law as did Massachusetts a decade later.

It is strange that, despite the knowledge of barberries displayed in the colonial laws, the New Englanders took their barberries along with their wheat seed when they moved west—another bungled opportunity.

Man in his incredible ignorance displays astonishing ingenuity in messing up the great ecological system of which he is inescapably a part.

B. Maize Leaf Blight

The classic recent case, where man in his ignorance set the stage for an epidemic, is the recently "lamented" corn leaf blight in the United States in 1970. This event generated a detailed study by the National Academy of Sciences (Horsfall, 1972).

To produce hybrid corn, seedsmen must make controlled crosses. At first they did this by removing the male tassels by hand from the female line. To save the high cost of hand detasseling, the seedsmen altered procedures in the 1950's to using a female line with sterile pollen. The sterility is cytoplasmically inherited. By the end of the 1960's essentially all the corn from Miami to Maine and Minnesota had been produced from so-called male-sterile cytoplasm. Suddenly there appeared a new strain of *Helminthosporium maydis* that flourished on sterile cytoplasm.

The epidemic started in Florida in the late winter of 1970 and spread northward with the greening wave of corn, reaching the northern limits of the crop in September. Some 15% of the nation's corn crop was destroyed. This was enough corn that, if fed to cattle and converted to beef, would have produced over 30 billion (3×10^{10}) quarter-pound hamburgers (Horsfall, 1975).

No one can be blamed when nature produces a new strain of a fungus. We simply forget the often-repeated lessons from previous epidemics— "Don't put all your eggs in one basket." We put all our eggs in the basket of sterile cytoplasm and paid the price. Such is the influence of man on an epidemic.

V. ONE EPIDEMIC LEADS TO ANOTHER

We have discussed above three epidemics: wheat rust, ergot on rye, and potato blight. These three follow a pattern in which one leads into another (Horsfall, 1958). Wheat rust likes a warm, wet spring. In northern Europe the spring weather is wet enough but too chilly for wheat rust. In Italy, the springs are warm enough, but a little dry for maximum rust. This leaves the central part of Europe where the weather tends to be optimal. In those central European countries where rust is a serious threat to wheat, the citizens turned to rye as a source of flour for bread. You can see this in the local names for bread. In Britain bread means wheat bread. In Germany brot means rye bread. In Italy, of course, bread is made from wheat but called macaroni and spaghetti as well as pan.

In Europe, the distribution of ergotism was the mirror image of that for wheat rust. Ergotism occurred in central Europe where rye is grown. It seldom or never occurs in northern and southern Europe. To say it another way, in northern Europe there is relatively little rust on wheat, no rye, and no ergotism. In central Europe there is much rust on wheat, much rye, and much ergotism. In Italy there is little rust on wheat, little rye, and no ergotism.

A similar pattern occurs in the United States. When the English colonists brought wheat, barberry, and rust to North America, they encountered a similar ecological situation to that found in Europe. In New England the weather was similar to that in England, wet enough but a little cool for maximum rust. Farther south in Virginia, the weather was both wet enough and warm enough for rust. Finding wheat difficult to grow in Virginia, the colonists there turned not to rye but to maize (they called it corn), thereby escaping ergotism. To this day in New England bread means wheat bread, but in the South, bread means corn bread. Diseases do change eating habits, difficult as that may be.

Ergotism in Europe began to fade in the eighteenth century with the arrival of the potato. The people were less dependent on rye and hence experienced less ergotism. The pattern about ergotism in central Europe is clear. Wheat rust led to rye and ergotism. The potato freed them from ergotism, but the potato then became widely planted. The population of Europe and the British Isles became dependent on it. When the blight struck the potato, we had more human misery.

With the cautions amply stated and the uncertain hope that enlightened understanding of plant disease epidemics may help man learn to minimize the vulnerability of his major crops, we now turn to the chapters that follow. They contain the principles and the theory that supports the arts of disease management as outlined in Volume I.

References

Barger, G. (1931). "Ergot and Ergotism." Gurney & Jackson, London.
Carefoot, G. L., and Sprott, E. R. (1967). "Famine on the Wind." Rand McNally, Chicago, Illinois.
Hedrick, U. P. (1917). The peaches of New York. *N.Y. State Dep. Agric. & Markets. Rep.* **24**, Vol. 2, Part II, 119–130.
Hepting, G. H. (1974). Death of the American chestnut. *Forest Hist.* **18**, 60–67.
Horsfall, J. G. (1958). The fight with the fungi. The rusts and the rots that rob us,

the blasts and the blights that beset us. *In* "Fifty Years of Botany" (W. C.
    Steere, ed.), pp. 50–60. McGraw-Hill, New York.
Horsfall, J. G., chm. (1972). "Genetic Vulnerability of Major Crops." Natl. Acad. Sci.,
    Washington, D.C.
Horsfall, J. G. (1975). The fire brigade stops a raging corn epidemic. *U.S., Dep.
    Agric., Yearbook* pp. 105–114.
Large, E. C. (1940). "The Advance of the Fungi." H. Holt & Co., New York.
Padmanabhan, S. Y. (1973). The great Bengal famine. *Annu. Rev. Phytopathol.* **11,**
    11–26.
Smith, E. F. (1888). Peach yellows. *U.S., Dep. Agric., Bot. Div. Bull.* **9,** 1–254.
Woodham-Smith, C. (1953). "The Reason Why." McGraw-Hill, New York.
Woodham-Smith, C. (1962). "The Great Hunger, Ireland 1845–1849." Harper &
    Row, New York.

Chapter 3

Comparative Anatomy of Epidemics

J. KRANZ

Anatomy, derived from the Greek for "cutting up," is the science which, by dissection, examines the structures of plants and animals (Henderson and Henderson, 1963). In this chapter epidemics will first be dissected into their component parts and then compared.

An epidemic may be defined as any change (increase or decrease) in a plant disease in a host population in time and space (Kranz, 1974b). Behind such ups and downs are interlocking complexes of processes determined by the many reciprocal cause-and-effect relationships that are characteristic of biological systems (Watt, 1966). We thus may regard epidemics as systems—open, coupled, and dynamic systems (Kranz, 1974b; Robinson, 1976). Epidemics belong to the population level in the hierarchial systems of biology (with populations of pathogens, hosts, and lesions as manifestations of disease).

Every biological system has a structure and a more or less specific behavior. So have epidemics. The structures (see Section II below) are

33

composed of important rate-determining elements. The behavior of these elements can be analyzed by describing the patterns of the epidemics (e.g., the shape of the disease progress curve—see Section III) and the rate of progress of the epidemic (Section IV).

This chapter will outline and compare the structures and behaviors of epidemics in relation to their external rate-determining factors. A record of past achievements in comparative epidemiology will not be attempted here. Instead, concepts and methods of analysis that may lead to new conclusions will be emphasized.

I. COMPARATIVE EPIDEMIOLOGY

Comparison is a powerful tool in science. Our comprehension of complex phenomena is enhanced when we can identify elements of a system or, for that matter, whole systems that are different, similar, or identical and why they are so. Thus comparative analysis is a major technique for the derivation of principles and models of epidemics.

The objectives of comparative epidemiology are: (1) to reduce to their fundamentals the numerous phenomena involved in a wide array of plant diseases caused by different etiological agents; and (2) to boil down the large number of apparently dissimilar epidemic phenomena to a convenient number of basic systems or types to which individual epidemics may be assigned, and with which future disease management can prosper.

Ideally, comparative epidemiology should be both quantitative and experimental. Comparative analyses can be made of an entire epidemic system (e.g., progress curves), or to a limited number of components (e.g., infection or spore production). Theories derived from comparative observations in turn should be amenable to experimental testing and verification. Systems analysis appears to be the method of choice for comparative epidemiology.

A. Description of Epidemics

Every science begins with the description of phenomena and then proceeds by deductive comparison to the development of generalizations. When a generalization proves to have predictive value, it can be raised to the rank of a principle. For example, a statement like: "An endemic pathogen is less virulent on native than on introduced host species" is now recognized as a valuable principle of epidemiology (Yarwood, 1973).

Descriptive terminology like "simple interest diseases," "compound interest diseases" (van der Plank, 1963), and "endemicity" (van der Plank, (1975) refers to structures or patterns of epidemics. This is also true of Gäumann's (1951) distinction between annual versus secular (or perennial) epidemics, and endemy versus pandemy. The terms explosive and tardive are used to describe differences in the rates of development of epidemics. Similarly, terms describe extremes in disease intensity, as within $0 < y \leqslant 1$ (Kranz, 1974b; van der Plank, 1975).

Patterns of epidemics may also be compared by reference to some famous or well-documented epidemics, e.g., the "Heines VII–yellow wheat rust epidemic" (Zadoks, 1961) or the northern and southern Japan types of rice blast epidemics (Kato, 1974). The effect of an epidemic on the host may also be considered (Kranz, 1968a). For example, in relation to loss in yield, terms like short and long epidemics with medium or severe attack as well as "early" and "late predictive disease" (James, 1974) have been used.

Comparisons of epidemics also can be made in glasshouses versus open fields, diseases in forests versus in fields, foliage versus root diseases, introduced versus endemic diseases, etc. Such comparisons provide inferences that can be fruitful for teaching and even for policy making. Chapters 2, 16, and 18 in this volume contain examples of such references. But comparisons of this type tend to violate the rule that experimental results are valid only for the specific conditions under which they have been obtained. Furthermore, many such comparisons are either not amenable to experimental proof, or if they are, only with difficulty (Kranz, 1974a). For this reason, their usefulness in research is limited.

B. Experimental Comparison of Epidemics

At its best, comparative epidemiology involves experimentation in which hypotheses can be tested by simultaneous measurement of structures and behaviors, within one or between two or more epidemics or their components. So far, very few studies satisfy these requirements. The next best method for comparison of epidemics involves the analysis of experimental data developed by various authors who have used similar methods to study different epidemics under essentially the same conditions or the same epidemics under different conditions.

Methods for comparison of results obtained in such experiments range from simple plotting, mapping, or tabulation of quantitative data to elaborate statistical, mathematical, and computer methods. These methods have been summarized and discussed elsewhere (Kranz, 1974a).

It is important to distinguish between "immediate comparison" and "classification" of epidemics. An immediate comparison between epidemics, or their components, has its greatest utility in the evaluation of effects of different treatments, environmental factors, or genetic differences in the pathogens or hosts involved in epidemics. Classification, on the other hand, involves the assignment of epidemics or their components to certain well-defined patterns or models. The latter implies a higher degree of abstraction, which is sometimes very advantageous. Progress curves that look different may belong to the same class when stripped of their chance variation. A classification of various types of progress curves for fungal diseases is given by Gregory (1968); similarly Thresh (1976) has compared progress curves for various types of viral diseases.

A major criterion for comparison are the similarities among epidemics. But what are these similarities? If we consider epidemics as open, coupled, and dynamic systems, the first thing we must do is look for correlations between state variables (Table I). Such correlations exist between the function of the system's components (see Table I), or external rate-determining factors and components and their outputs. For instance, pycnidia as sporulating structures are highly correlated with splash-borne dispersal, and this in turn with steep gradients; infection is commonly correlated with leaf wetness, etc. Correlation coefficients developed by regression analysis are well-known indicators of correlations among factors from the disease quadrangle. But since such coefficients often are highly variable, even within the same host–pathogen system, the only correlations that are useful for purposes of comparisons (provided they are not redundant) are those with large correlation coefficients. Henceforth in this chapter we shall refer to such highly correlated factors as invariant correlations. The more invariant correlations that pathogens and hosts, as well as their epidemics, have in common, the more similar one can expect their epidemics to behave.

Invariance stands for low variability and/or a high degree of correlation (consistency). Epidemiological structures with consistent (invariant) outputs, i.e., invariant patterns and invariant rates, have a high degree of similarity. This consistency should be defined quantitatively, possibly with confidence intervals or similar limits to allow for reasonable variation. In some epidemics, certain correlations may be rigidly fixed or even absolute. For example, obligate parasitism is an invariant characteristic of rust diseases. So too are the sequence of events involved in spore germination, the requirement of a water film for infection, the need for virulence in a pathogen, the existence of vertical resistance, S-shaped progress curves in some diseases, etc. Invariance may be very narrowly limiting or, in some cases, it may be modifiable by long-lasting changes in environmental factors, control measures, changes in host or pathogen genetics, etc. Adaption of pathogens to previously unfavorable conditions or genetic vulnerability due to new breeding methods indicates that such modifications can occur.

For a more static comparison of epidemic structures homologue and analogue components can be distinguished. Homologue components are phylogenetically alike, e.g., infection by means of conidia. If components of the epidemic structure are only similar in function, e.g., infection taking place by means of rhizomorphs, they should be regarded as analogous.

Although direct experimentation provides the strongest tests in comparative epidemiology, additional evidence should also be evaluated. Examples of this approach are given by Populer (1972) with regard to powdery mildew of rubber (*Oidium heveae* Steinm.) and by Zadoks (1961) for stripe rust of wheat (*Puccinia striiformis* West.). Both authors have supplemented their own data with other records. In this way they were able to identify traits that epidemics of these diseases have in common, and to explain some of the sources of variation in the behavior of these epidemics.

To a certain extent, records for epidemics of different diseases can be utilized for some cautious deductions; this has been done by Gäumann (1951) and by van der Plank (1960, 1963, 1968, 1975).

II. STRUCTURES OF EPIDEMICS

Systems are organized sets of components, subsystems, and elements, each with their specific relations and interactions. Each of these entities has a characteristic pattern of behavior both as a unit and as a whole. But unless we understand the total organization of these components, the kind, strength, and flexibility of their relations, as well as their contributions to the behavior of the system, our knowledge of the total structure will remain incomplete and the system will remain a "black box."

Improved knowledge of plant disease epidemics will help identify where man can interfere with the ongoing population dynamics. Improved knowledge will permit the development of more refined, more efficient, and more dependable methods for disease management.

Structural elements of epidemic systems stem from pathogen–host interactions, or, more strictly, from those of their populations. Environmental and human interference become part of the system by triggering or inhibiting the action of these elements (e.g., leaf wetness triggers in-

TABLE I

Components of Epidemic Structures

Subsystem	Element	State variable [a]	Variable unit [a]	Operational mechanisms [a]
Pathogen	Pathogenicity	Type of parasitism	Yes-no	Obligate, necrotrophic, etc.
		Mode of infection	Number of appressoria, or haustoria	Direct penetration through stomata, wounds, etc.
		Mode of action	Type of symptom	Vivotoxins, active growth, ethylene, etc.
	Virulence	Host specialization	Formae speciales, race	Yes-no reaction (in this context)
	Sporulation	Inoculum	Number of spores produced	Conidia, ascospores, etc.
	Dispersal	Deposition gradients	Number of spores caught	Growth; wind, water, vectors,[b] etc.
	Survival	Longevity	Time units	Oospores, saprophytic growth, etc.
Host	Plant type	Herb, tree, shrub; annual, deciduous, perennial	Yes-no	(Determines largely whether annual epidemic, endemicity, etc.)
	Growth rhythm	Growth stages	Key designations	Tillering, fruit setting, etc.
	Growth pattern	Abundance	Leaf area index, biomass	Plant stands, vegetative growth, etc.
	Propagation	By seeds	Tons per hectare	Grains (out- or inbred), tubers, cuttings, ratoons, etc.
	Population resistance	Disease incidence	Disease severity (%) or frequency (%)	Vertical or horizontal resistance; oligogenic, polygonic resistance
	Reaction to disease	Type of symptom	Proportion of types(s)	Defense reactions

Disease	Infection	Number of lesions disease severity (%) (per host plant)	By spores, rhizomorphs, systemic, etc.
Pathogenesis	Incubation period	Time units	Active growth, vivotoxins, etc.
Latency	Latent period	Time units	Start of sporulation
Lesion growth	Lesion size	Disease severity (%)	Active growth, vivotoxins, etc.
Infectiousness	Inoculum	Spore number/time unit/area unit	Sporulation, removal
Spread	Infection gradient	Disease intensity in space, spore catches	See pathogen; seed borne,[b] etc. homogeneous or heterogeneous infection chains; esodemics, exodemics
Multiplication	Infection cycle, generation	Duration and/or number of generations (per season)	Mono- or polycyclic
Survival	Longevity	Time units	See pathogens, perennial, in plant refuse, etc.

[a] These are not complete lists; the items included are only examples.
[b] From an epidemiological point of view, it might be desirable to have a subsystem of spread, or a subsystem vector. The latter is obvious because the insects involved have a life cycle of their own.

fection or a protective fungicide inhibits germination of spores). Here we must bear in mind that there is only one incontestable cause-and-effect relationship in epidemics: no disease can develop without both a pathogen and a susceptible host. Beyond this, cause becomes relative and functions prevail. Obviously, every phase in an epidemic is the function of one or more previous functions; for example, there can be no infection without inoculum and no inoculum without previous infection, etc.

To the epidemiologist, the components of an epidemic are known by their functions $g(x)$ or state variables (see Table I). Similarly, the structure of the whole epidemic is known by the sum of all the functions involved; as $Y = f(x_i \ldots x_n)$. Although each $x_i = g(x)$ is indispensable to the total structure, each has a variable quantitative impact or weight determined by its inherent scope, by its previous state (i.e., the condition of the state variable), and by various antagonistic factors. For this reason, each change in an x_i results in a different output.

Consequently, epidemics are not determined by linear causal relationships but by programs which are determined by the components and functions that are characteristic of their structures. These are open programs. Environmental and human interference confine the programs and provide the stimuli for their operation. Together with some features inherent in populations (like random processes, recurrence, limits or dimensions, discontinuities and thresholds) they stimulate and regulate an otherwise latent structure, which then results in behavior, e.g., growth and decline of a population of lesions, their spatial and age distribution, etc. Mathematical or computer simulators of epidemics (Waggoner and Horsfall, 1969) are designed to imitate this kind of systems control through structural elements.

A. Anatomy of Epidemic Structures

When dissecting epidemics to understand their structure, we should be aware that the behavior of any single component cannot represent the behavior of the entire structure (Fig. 1). Invariably the question arises of how far this cutting up should go. Ideally, every component and all its interactions should be known. In practice, however, this is rarely feasible. Very often we do not know, and need not necessarily know, all the components and all of their interactions in detail. For satisfactory understanding and comparison of epidemic systems we need to know the components that characterize the "core dynamics" (Patten, 1971). Even mathematical descriptions of some of the components as "black boxes" may suffice to understand the organization and function of an epidemic. Mathematical compartment analysis—a method akin to systems

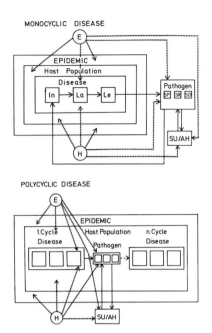

MONOCYCLIC DISEASE

POLYCYCLIC DISEASE

Fig. 1. Schematic model of epidemic structures for monocyclic (top) and poly-cyclic diseases (bottom). The terms mono- and polycyclic indicate that the pathogen produces one or more generations during the vegetation period of the host. Each epidemic structure consists of three subsystems represented by rectangular boxes—host, disease, and pathogens—as well as of elements represented by square boxes. Environmental (E) and human interference factors (H) are shown as circles. Solid lines represent the interaction between elements; dotted lines represent the action of the external factors upon elements. Key to abbreviations: In, infection; La, latency; Le, lesion growth; Sp, sporulation; SR, spore release; SD, spore dispersal; SU, sur-vival; AH, alternate host. See also Table I.

analysis—implies that a "higher" component need not "know" all about an "inferior" component in order to regulate it. The coordination of parameters of only a few components may suffice. This implies a certain hierarchy of components.

Structural components of epidemics were listed in Table I (see also Fig. 1). Obviously, the pathogen, host, and disease are subsystems. Their subsystems in turn are designated elements. For instance, the subsystem, disease, comprises parts of the disease cycle, e.g., infection, pathogenesis, lesion growth, etc. Elements are usually measurable and tractable as state variables or, according to Patten (1971), as "parts of observable attributes, or arbitrary groupings of parts or attributes for particular purposes." The state variables are measurable in specific "variable units." The particular state variables and variable units listed in Table I are

offered only as examples; others could be added. The same applies to
what we call "operational mechanisms," which may clarify the content
of elements. But they also can be utilized as a basis for comparison of
epidemics, e.g., of obligate versus necrotrophic pathogens.

The subsystem, disease (see Table I), can be seen as the interaction
between a single host plant and a pathogen. The end result of this
interaction is a disease lesion in a single host plant. The subsystem,
disease, also can be seen as the interaction between a population of host
plants and a population of pathogens. The end result of this interaction
is an epidemic.

Development of an epidemic requires the interaction of all the elements in subsystems, sporulation, spore dispersal, infection, etc. (see
Table I and Fig. 1). Each infection cycle results in one or more daughter
lesions in the next disease cycle or generation. These new lesions can be
on the same host plant or on previously noninfected host plants. Robinson (1976) has referred to the former as esodemic infection and the
latter as exodemic infection. The inoculum may be supplied from the
same host species (homogeneous infection chain) or from another host
species (heterogeneous infection chain) (Gäumann, 1951).

The host plants also can be seen as subsystems, with a number of
relevant elements (see Table I). Many of these elements are complementary to characteristics of the pathogen; e.g., virulence and susceptibility. Hosts and pathogen, of course, constitute systems in their own
right. This is most evident for the pathogen during phases of survival
outside hosts or, most markedly, with soil-inhabiting pathogens.

B. How to Compare the Structures of Epidemics

Identification of the structure of an epidemic begins with the elucida-
tion of the disease cycle and usually continues with studies of the interactions of pathogens, hosts, and environmental factors.

Epidemiological structures can be compared in two ways: (1) by
comparison of sets of relevant elements and their organization; and (2)
by comparing the epidemiologically relevant functions of elements and
the correlations among their output–input relationships, e.g., between
intensity of infection and length of incubation and periods.

In comparisons of the first type, the main criterion of decision is the
degree of similarity in the sets of elements involved in the epidemics to
be compared. For instance, in infectious diseases, the structure of an
epidemic is determined in large part by the sequence of events involved
in infection. Individual elements may have the same function but differ
in their operational mechanisms (see Table I), e.g., one pathogen may

require wounds for infection while another may require stomata. The different operational mechanisms by which infection is achieved in these two cases would make a difference in the structures of the two diseases under comparison.

In comparisons of the second type, the main criterion of decision is the degree of similarity in the functions and correlations involved in the epidemics to be compared. Such comparisons can be made both between and within host–pathogen systems. The latter usually involves different races of the pathogen, cultivars of the host, sites, climates, or other treatments. In comparisons of this type, it is not so much the influence of a given factor on, for instance, sporulation as such, but rather its effect on the dynamics of the epidemic. For the study of these implications, simulators are an ideal tool. Incidentally, the requirements for development of an adequate simulation model often will indicate what state variables, variable units, and operational mechanisms of an element are biologically relevant for observation. Sometimes the necessary data are academically trivial but epidemiologically essential, for example, the number of sporophores formed per mm² of diseased tissue.

C. Comparative Anatomy of Epidemic Structures

Diseases caused by different etiological agents almost invariably have different epidemiological structures. This is obvious when comparing the sets of elements involved in different host–pathogen systems. For instance, in tobacco the epidemiological structure of weather fleck, an abiotic disorder, will differ radically from that of a vector-borne virus disease, such as tomato spotted wilt, and that for a fungal disease, such as blue mold.

Even among closely related fungal or bacterial diseases, differences in epidemiological structures will exist, although the differences involved may be relatively subtle. Table II contains an analytical comparison of the structural elements involved in three fungal diseases in California. The schematic representation of these diseases in the paper by Ogawa *et al.* (1967) shows three structural elements: survival, infection, and sporulation. Although all infectious diseases have homologous sets of elements, just as nuclei of a given species contain homologous sets of chromosomes, their individual programs are often quite different. For instance, *Sphaerotheca pannosa* (Wallr.) Lév. and *Pseudoperonospora humuli* (Miyabe and Takah.) G. W. Wilson both survive in buds; but the infection chain of the former is heterogeneous while infection by the latter is homogeneous, i.e., with inoculum from the same host species. Apart from hetero- and homogeneous infection chains (Gäumann, 1951)

TABLE II

Comparison of the Structural Elements of Three Host–Pathogen Systems [a]

Structural element	Monilinia laxa on almond	Sphaerotheca pannosa on plum	Pseudoperonospora humuli on hop
Survival on	Blighted blossoms, twigs, mummies and peduncles of previous year	Dormant buds and new shoots of roses	Dormant hop crowns and buds
Infection on	Blossoms; monocyclic and homogeneous (sometimes also fruits)	Fruits of plums; mono- or bicyclic and heterogeneous	Shoots (incl. leaves, stems, stipules) and crown; polycyclic and homogeneous
Sporulation on	Blighted blossoms, twigs, (fruits) and mummies, recurrent	Shoots of roses; recurrent	Shoots (incl. leaves, stems and stipules); recurrent

[a] Adapted from Ogawa et al. 1967.

we must also take into account that epidemics can either be mono- or polycyclic. That is the pathogen produces one or more gernerations during the vegetation period of the host.

The possibility for experimental comparison of separate structural elements in an epidemic are legion (see Table I). However, laboratory studies of this kind, e.g., sporulation of fungus P_1 in comparison to P_2 under condition x, so far usually bear little relation to the epidemics caused by P_1 and P_2. Nevertheless, such results may serve to formulate mathematical functions which may then be compared as submodels of a simulator for their quantitative impact on the course of an epidemic.

Published comparisons of the element "infection" are often very useful in understanding epidemics. An excellent example of such studies is the work of Bashi and Rotem (1974) on the effect of the duration of leaf wetness and temperature on the disease intensity of four pathogens under semiarid conditions (see also Section IV,C,2 below).

III. PATTERN OF EPIDEMICS

Interactions of structural elements (see Table I) in epidemics that are triggered by factors in the environment or by human interference result in systems behavior that is expressed in patterns and rates. Epidemic patterns can be expressed in curves that show the progress of a disease as it spreads over distance or over time. These curves yield rates.

Patterns are important indicators of essential and, in many cases, specific biological facets of an epidemic. The point of origin and shape of disease progress curves may well tell us about changes in host susceptibility during the growing period, the intermittent availability of inoculum, or recurrent weather events, effectiveness of control or cultural measures, age distribution of lesions, or other features. Linearization of entire progress curves sometimes tends to sweep this valuable information under the carpet. This is intolerable for a full comprehension of epidemics. Consequently, transformation of progress curves should either be consistent with patterns (Hau and Kranz, 1977) or be confined to biologically relevant sections of the curves (see also Sections III,B and IV below).

Nearly all biological curves for growth, metabolism, etc., are S-shaped. This is also true for most progress curves, although it apparently is more common for epidemics of some diseases (e.g., late blight of potatoes, stem rust of wheat) and less common for many other diseases. Fracker (1936) many years ago noted in rate-determining factors considerable deviations from the S-shaped, logistic ideal. These deviations appear to be specific enough to distinguish some basic patterns or families of disease progress curves. Similar characteristics also prevail in the progress curves of virus diseases (Thresh, 1974). On the other hand, the same progress curves produced by epidemics of two different diseases need not be the result of the same structural interactions with environment and human interferences.

A. Some Basic Patterns of Epidemics

Disease progress is commonly plotted in cumulative curves over time (Fig. 2, solid lines in graphs 1 to 6), while gradients are plotted in hyperbolic curves over distance. Both types of curves aptly describe the additive amount of disease and thus the course of an epidemic in time and space. But the more informative of the two is the derivative of the cumulative progress curves—the dashed curves in Fig. 2, indicating the rate of growth of the epidemic. These rate curves identify at least three major classes of epidemic patterns: symmetrical (bell-shaped curves Fig. 2, graphs 2a and 2b) and asymmetric curves with either positive skewness (Fig. 2, graphs 1a and 1b) or negative skewness (Fig. 2, graphs 3a and 3b). All of these curves, and Fig. 2, curve 4, depict more or less continuous development of disease until either harvest or death of the host population due, for example, to frost. In this sense, they are incomplete, since a complete, undisturbed progress curve often is bilateral, i.e., it has both progressive and degressive legs

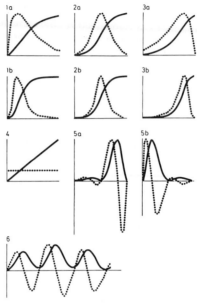

Fig. 2. Schematic diagrams of basic epidemic patterns. The full lines represent the disease progress curves resulting from assessments of disease intensity y at each time t_i; dotted lines depict the rates of change in disease intensity y between t_{i-1} and t_i as derived from the cumulative curve (see text).

(Kranz, 1968a, 1974a, 1975; Kranz and Lörincz, 1970). But progress curves also can be bimodal (Fig. 2, graphs 5a and 5b), multimodal (Kranz, 1968a, 1975; van der Plank, 1975), or oscillating (Fig. 2, graph 6).

In the analysis of progress curves, whether they are complete or incomplete cumulative curves, one can discern various phases which can be quantified. Baker (1971) suggested four consecutive phases: (1) the true logarithmic phase, (2) the synergistic (or, more commonly, exponential) phase, (3) the transitional phase, and (4) the plateau phase. These, and the more detailed curve elements as suggested by Kranz (1968b, 1974a), represent biological facets of epidemics. If we identify each phase of the curve as a definable parameter, then these particular parameters can be used to measure the effects of rate-determining variables during certain phases of epidemics. They can also be used as a basis for quantitative comparisons, providing that their inherent variability is adequately understood.

Progress curves of a given disease have only a limited scope of variation (Kranz, 1968c; Kranz and Lörincz, 1970). Nevertheless, they may overlap with those of other diseases because of variation caused by environment, varieties, changes in pathogen population, or control measures. Consequently the same disease need not always appear in the same family (or cluster) of progress curves. But with improved techniques for defining the limits of variation for the same family of progress curves and/or for measuring and evaluating the effects of various elements of epidemics, it may become easier to correctly relate a given disease progress curve to particular classes or families of epidemics (Kranz, 1974a).

Multimodal (periodic) progress curves (Fig. 2, graphs 5a, 5b, and 6) often reflect disease cycles, discontinuities in infection processes, variation in incubation periods, changes in host susceptibility due to new growth, or other factors. Periodic progress curves may also be recorded in multiple-wave epidemics. A multiple-wave epidemic refers to epidemics developing with somewhat distinct lapses of time on different organs of host plants. Examples are rice blast with seedling blast, leaf blast, neck, and panicle blast; or *Phytophthora* sp. on *Hevea* attacking fruits first, then leaves, and finally tapping panels on trunks. If, however, the pathogen passes on to another host, as when *Phytophthora infestans* (Mont.) de By. passes from potato to tomato it should better be regarded as an independent, heterogeneous epidemic.

B. How to Compare Patterns of Epidemics

As shown in Section II,B, the patterns of epidemics are revealed by plotting untransformed or transformed observations of disease intensity y over time or distance, or both. Untransformed but smoothed progress curves are epidemiologically most meaningful, as was pointed out earlier. But linear progress curves are a prerequisite for statistical or mathematical analysis. If this is required, a transformation formula should be chosen that gives the best fit in terms of χ^2. Strictly speaking, comparisons are possible only when a comparable transformation formula has been used (Hau and Kranz, 1977). On the other hand, curves that require a different type of transformation, if consistent, would be an additional evidence of different patterns of epidemics (see Fig. 2). It should be emphasized, however, that for comparison over weekly or other short intervals, most disease progress curves can be considered linear even when they are not transformed. Wavelike annual and perennial progress curves, or substantial variation of the infection rate r, cannot appropriately be described by straight line transformations (van der Plank, 1975). For these multimodal or serial functions, more sophisticated approaches are needed, such as Fourier analyses or oscillation functions (Kranz, 1974a).

The number of basic bilateral patterns among disease progress curves is limited (Kranz, 1968a, 1974a; Kranz and Lörincz, 1970). But even if we confine ourselves to the progressive leg of the common unilateral progress curves, some distinct shapes are most prevalent (see Fig. 2). As suggested in Fig. 2, all smoothed, unilateral progress curves can be compared to the most prevalent patterns. They also can be compared by the following criteria: starting point or position, initial disease intensity (y_0), slope of the curve after various periods of time, existence and, if applicable, duration of an asymptote, and maximum disease intensity (y_{max}) (Kranz, 1974a).

Additional criteria would be required for analysis of entire bilateral cumulative disease progress curves, as suggested by Kranz (1968b, 1974a, 1975) and Kranz and Lörincz (1970).

When defining basic patterns of progress curves as in Fig. 2 it is necessary to distinguish the curve type itself from the parameter "disease intensity." If these two features are not kept separate when comparing disease progress curves, one might too easily be carried away by apparently individual morphologies of epidemics. For instance, Croxall and Smith (1976) have used the data of Cox and Large (1960) and Hirst and Stedman (1960) to depict a large variety of basically S-shaped progress curves for potato late blight caused by *Phytophthora infestans*. Invariably, the progress curves for this disease fit well into that particular "family" of curves. Apparent deviations can usually be attributed to different ways of defining the specific phases or parameters of the curve in question. In some cases, however, e.g., with a late starting point, curves may shift into a different curve family.

C. Comparison of Epidemic Patterns

There is evidence in the literature for practically all the patterns or types of progress curves shown in Fig. 2. Records of potato late blight commonly show essentially sigmoid progress curves (see Fig. 2, graphs 2a and 2b), though substantial deviation to Fig. 2, graphs 1, 3, and even 4 can be caused by weather, differences in varietal susceptibility, and control measures. Apple scab (*Venturia inaequalis* (Cooke) Aderh.) and powdery mildew of rubber (Analytis, 1973; Populer, 1972) commonly follow Fig. 2, graph 1. Here the hypersusceptibility of young leaves causes very high rates of infection early in the epidemic.

Figure 2, graph 3 applies to the large number of diseases that start slowly but speed up considerably as host susceptibility increases late in the epidemic. The so-called "slow rusting" varieties of wheat fit in this category—the speed-up usually occurring after the time of flowering

(Kranz, 1968a). Figure 2, graph 4 is typical of epidemics in which disease progress is determined mainly by the rate of multiplication of a pathogen which is not inhibited by weather or host resistance. Bimodal (Fig. 2, graphs 5a and 5b) and oscillating progress curves (Fig. 2, graphs 5 and 6) typically occur in certain diseases conditioned by, for instance, a regular sequence of weather conditions and/or of growth flushes. Coffee leaf rust (*Hemileia vastatrix* Berk. and Br.) and blister blight of tea (*Exobasidium vexans* Massee) (Visser *et al.*, 1961) are examples.

But there are also examples of differing patterns within the same host–pathogen system (Kranz, 1968c; Kranz and Lörincz, 1970). In Kenya, for example, the shape of the progress curve for coffee leaf rust may be either uni- or bimodal, depending on the pattern of annual rainfall in a given year. Bimodal curves are retained, even with little rain, up to 1800 m altitude; but above this they flatten out and invariably become unilateral (Bock, 1962). Also, progress curves for epidemics of the coffee berry disease (*Colletotrichum coffeanum* Noack sensu Hindorf) in "early crops" are more or less linear (Fig. 2, graph 4), whereas the ones for "late crop" show a distinct tendency to asymptotes (Fig. 2, graphs 1a and 1b). Gibbs (1971) reports that this is true irrespective of the type of fungicide applied. When certain recommended spray schedules are used, however, the shape of the curve changes to Fig. 2, graphs 3a or 3b often with a small initial peak such as that shown in Fig. 2, graph 5a (Steiner, 1973). A spray schedule based on measurements of amount of inoculum results in a much lower, periodic curve (Fig. 2, graph 6), which, however, stretched from the onset of the epidemic in the unsprayed plots until nearly harvest time (Steiner, 1973). Although cultivars with vertical resistance, as defined by van der Plank (1968), do not affect the shape of the progress curve, changes in shape can be rather pronounced in cultivars with horizontal resistance.

From this apparently conflicting evidence one may be inclined to think that similarity in the shape of disease progress curves is not a dependable indication of the pattern of epidemics. Some variation in disease progress curves certainly can be ascribed to imperfect and/or inconsistent disease assessment. A great deal of variation, however, is due to the influence of environmental and human interference factors. Obviously it is essential to know whether these changes in patterns are consistent or not, and if so, to what extent. Although little evidence is yet available, a few cases are worthy of mention. For example, the progress curves for epidemics of northern leaf blight of corn (*Drechslera turcica* (Pass.) Subram. and Jain) in Florida (Berger, 1974) show a high degree of consistency (invariance) despite the overriding influence of weather. In all seasons of the year epidemics of this disease follow a consistent pattern—with a fairly rapid spread below 1% disease incidence, a moderate spread prior to full silk, and a slow spread for the last 10–14 days before harvest. A second noteworthy case is the consistency with which Fig. 2, graph 5 is observed in progress curves for *Ramularia rubella* (Bon.) Nannf. on *Rumex* (Kranz, 1975). Finally, many soil-borne pathogens give consistently similar shapes when inoculum density is plotted against disease intensity (Baker, 1971).

As was discussed in Section III,B, for effective comparison of epidemics we need to know both the epidemic pattern(s) a disease can attain under normal conditions and the correlations between changes in pattern and the essential components of the epidemic. Knowledge of these two factors will enhance our capacity to predict future disease development and thus to develop better systems for disease management. Finally, it should be emphasized that different rates need not necessarily alter the pattern of disease development.

IV. DYNAMICS OF EPIDEMICS

Both states and rates are essential characteristics of the interactions and the dynamics of a system. States in a mathematical sense give a transient picture of their variables (Patten, 1971). In a disease epidemic, rates express the response of the system to a stimulus in relation to a previous state of the system—usually disease intensity, spore catches, etc. Rates also can be interpreted as ". . . a sequence of state changes in time . . . (either) discrete or continuous" (Patten, 1971). Rates usually are proportional to the size of transient population states. Sometimes, however, disproportionately small quantitative changes (thresholds) can trigger large qualitative and characteristic changes.

If a disease shows a very low proportionality in rates, it eventually will become an endemic disease. Endemicity is usually due to a balance of coexisting host and pathogen population. This means that the average number of progeny per lesion is about one; in such cases, time becomes unimportant (van der Plank, 1975). In addition to structural elements, epidemiologists seek to identify factors that influence the rate of determining factors in the development of epidemics. In many cases these factors can be found in the genetic and phenological variability of host and pathogen populations in space and time. They also can be found in the many environmental factors that influence the processes of infection, pathogenesis, sporulation, dispersal, survival, and other structural elements of epidemics. In these next few paragraphs we shall try to identify and compare some of these rate-determining factors.

A. Rates that Gauge Dynamics

Several different terms have been proposed as infection rates in epidemiology. Each term is derived from some measurement of disease progress. Many of these terms are useful for comparative analysis of epidemics.

Van der Plank's (1963) logarithmic and apparent infection rates, r_1 and r, respectively, have become popular. They react ". . . like a speedometer and nothing else" (van der Plank, 1975). They are very sensitive to both environmental and host factors. Analytis (1973) has proposed that the rate of change in disease intensity, y, should be formulated so as to take into account the asymptote of progress curves. Kiyosawa (1972) has proposed a yearly infection rate λ, which is estimated from the daily infection rate r.

Regression coefficients describing rates are still widely used for comparison of linear progress curves, particularly for diseases in which full experimental data are available. They can be tested for significance, which helps to avoid inferences from variation due to chance alone. So far, this is the only rate used to describe changes in disease gradients (Gregory, 1968).

The basic infection rate, R (van der Plank, 1963), is the average infection rate per unit of inoculum (or infectious tissue) per unit of time (van der Plank, 1975). The amount of diseased host tissue which is not yet sporulating or which is no longer sporulating, is not covered by R, but by r. R reflects inoculum potential, and, in the case of spread the epidemic threshold, iR, is defined as the average number of effective cycles of dispersal from a lesion during its infectious period.

B. How to Compare the Dynamics of Epidemics

The dynamics of an epidemic is best described in terms of patterns and rates. This is also true in the comparison analysis of epidemics.

Since rates can vary during an epidemic (Analytis, 1973; Berger, 1975; Gassert, 1976; Kranz, 1975; Steiner, 1973), transformation of whole progress curves into lines should be handled with care. As was discussed in Section III,B, it often appears advisable and biologically acceptable to linearize and make comparisons only within distinct sections of disease progress curves. An entire progress curve or gradient, once made linear, invariably has a constant rate but it relates to patterns of epidemics only through its transformation formula, if it is properly chosen.

Rates of epidemics are sensitive to environmental or experimentally varied factors. For that reason they can be useful tools for comparison.

But, by nature, they lack consistency and may, therefore, be of limited value in quantitative comparisons. In qualitative comparisons, however, rates may be very valuable since the quality of their response to, or the extent of their correlation with, factors is very likely to be consistent and therefore useful for comparative analysis. Thus, the most useful question to ask is: "How does the epidemic react to a given factor?," or perhaps, "In what order of magnitude does an epidemic react to a given factor?" Both of these questions would be superior to: "Exactly how much does an epidemic react to this factor?"

C. Dynamics of Epidemics Compared

An epidemic begins when the interaction among the components of a given host–pathogen system produces a threshold value of iR that is greater than 1 (van der Plank, 1963). But only in about 16% of 59 epidemics which I selected for study in 1968 (Kranz, 1968b) did the epidemic cause the complete destruction of the host population even though no plant protection efforts were made (Kranz, 1968b). Van der Plank (1963) relates this to a natural balance—a more or less transient stability—that is achieved between new infections and new growth. Decline with negative rates is due either to excessive removal (or disappearance) of affected plant parts, or to the rapid development of new growth.

1. Some Host Factors as Rate Determinants

If we look for and attempt to compare host factors that are rate determining, we must know the kinds of correlations that exist between these host factors and how consistent they are. The more invariant a correlation, the more specific it will be for the epidemics to be compared.

Let us now discuss some apparently specific correlations that have been established by comparative epidemiology between changes in rate of development of epidemics and the following host factors: type of crop, density of host stands, genetics of resistance, and predisposition.

a. *Type of Crop.* The rate of development of epidemics appears to vary with different types of crops. In the United States, the highest rates have been reported for vegetatively propagated crops, next highest with self-pollinated crops, and lowest rates with cross-pollinated crops. Also, epidemics appear to be more frequent in valuable crops (van der Plank, 1960).

In woody plants the rate at which a virus spreads is usually relatively slow or even imperceptible (Thresh, 1974). But this still may suffice for an epidemic, as is shown by the tristeza disease of citrus—about 35,000 aphids flying to or from a single tree per year in California is sufficient to ensure a multiplication rate of about two new infections per year for each diseased tree, even at a very low rate of transmission (Carter, 1973). Many fungal diseases of trees also progress slowly. On the other hand, in annuals such as tobacco, cotton, pepper, cucurbit, maize, vegetable, and other crops, the rates of virus dissemination by aphids, whiteflies, leafhoppers, or beetles are similar to rates of spread of fungal diseases. Their epidemics often develop in a few weeks.

b. *Stand Density.* Dense stands of plants achieved by close planting and abundant rain or fertilizers, etc., may affect the rates of epidemics in various ways. Though more foliage provides more sites for infection, leaf area as such rarely appears to be a limiting factor in continuously growing hosts (Kranz, 1975; Zadoks, 1961). However, spread and infection appear to be facilitated. In spacing trials with *Cercospora apii* Fresen twice as much disease was observed in closely planted celery plots as in widely spaced plots in 8 weeks once a 2% disease intensity level was surpassed (Berger, 1975). This was due to slightly lower r values in plots with the widest spacing during these 8 weeks. Thereafter, however, the situation was reversed. While r slowed down in the most densely planted plots there was a considerable increase in disease in the more widely spaced ones. Consequently, final disease intensity was virtually the same in both the densely and widely spaced plots. In virus diseases increased plant size facilitates rapid spread of disease by increasing the changes of direct contact between infected and noninfected plant parts and by increasing the amounts of both infected and susceptible host tissue that are accessible to vectors. Thus, the actual rate of spread of swollen shoot disease of cacao in a plantation is directly proportional to the number of healthy trees in contact with infected ones at the beginning of the year (Thresh, 1974).

Distance between host plants may impede the rapid buildup of epidemics. On the other hand, Kato (1974) and Kiyosawa (1972) did not find a significant contribution of host density to the apparent infection rate and the yearly infection rate of the rice blast disease. Epidemics of systemic diseases that receive their inoculum from outside the field may even be slowed down by a dense crop, as van der Plank (1960) inferred from the Poisson theorem.

c. *Genetic Resistance.* Genetically controlled resistance has a major impact on the dynamics and rates of epidemics. Rates are affected by the horizontal or race-nonspecific resistance in cultivars whereas vertical resistance primarily affects y_0, i.e., the initial inoculum (Robinson, 1976;

van der Plank, 1968, 1975). The greater the horizontal resistance, the lower the infection rate r; this effect is most prominent in long-lasting epidemics. The decrease in amount of disease by multilines is due partly to reduced rates.

When two grass species were sown in mixed stands, the rate of development of damping off (*Pythium irregulare* Trow.) and its spatial distribution within plots were determined mainly by the abundance of the most susceptible grass in the mixture. It was the insertion of resistant plants rather than the removal of susceptible ones which resulted in a linear reduction of the infection rate r (Burbon and Chilvers, 1976).

Developmental changes in susceptibility are a major cause for varying rates in the progress of an epidemic. Rates can differ widely when *Alternaria longipes* (El. and Ev.) Mason infects different stalk positions of tobacco (from $r = 0.16$ up to 0.29); also, at different times, rates may vary $r = 0.17$ up to $r = 0.48$ at the same position (Norse, 1971).

d. *Predisposition.* Environmentally conditioned predisposition of host plants to increased susceptibility usually is not race specific. Therefore, r is most likely greater with increased predisposition. Predisposition has been reported to increase the infection rate of rice blast (*Pyricularia oryzae*) by about 40% (Kato, 1974). Numerous environmental or human interference factors can change predisposition of hosts. So far, this aspect of disease development has hardly been explored from an epidemiological point of view, but its influence on infection and duration of latent periods should not be underrated.

2. Some Effects of Pathogens on Rates

Changes in structural elements of pathogen populations can be either qualitative (new races or biotypes, aggressiveness, formae speciales, etc.) or quantitative (amount of inoculum, etc.). Some of the correlated effects on rates have already been reviewed extensively by van der Plank (1963, 1975). Some additional aspects will be discussed here.

There is no evidence for important differences between rates of epidemics caused by viruses, bacteria, and fungi. But some differences in rates have been reported to exist between certain taxa (Kranz, 1968b). In the same expreiment *Cercospora* sp., *Phyllachora* sp., and *Puccinia* sp. had mean r values of 0.18, 0.16, and 0.12, respectively. In addition, the epidemics of *Phyllachora* sp. tended to cause relatively more frequent and rapid total breakdowns of their host populations, whereas *Cercospora* sp. attacked their hosts earlier and over longer periods. Also, both *Phyllachora* sp. and *Cercospora* sp. had a greater area under progress curves than did the *Puccinia* epidemics studied. Wheat leaf rust is usually slower in developing and less explosive than wheat stem rust. This means that it has a smaller value of r. Overall damage, nevertheless, may be greater with leaf rust since there is less fluctuation from year to year (van der Plank, 1960).

Mechanisms of survival of pathogens often determine the primary inoculum available early in the cropping season. Seed-borne pathogens, for instance, ensure random distribution of inoculum throughout the field better than any other type of pathogen. This creates a maximum opportunity for infection of young plants, particularly through matching of virulent races to susceptible cultivars. It is obvious that the number of infections and dispersals (or cycles) required to cover a given crop area is less with seed-borne diseases than with diseases that develop from more localized but equally strong primary foci. For this reason, other things being equal, seed-borne pathogens will develop at more rapid rates.

Other favorable conditions such as larger amounts of primary inoculum appear to be consistently correlated with earlier disease incidence and higher initial disease intensity (Kranz, 1968c; van der Plank, 1963). Most experimental evidence seems to bear this out. For instance, the higher the carryover inoculum of *Phytophthora infestans* the more frequent a fast multiplication rate of potato late blight (Croxall and Smith, 1976). Furthermore, this rapid rate is more important for the degree of potato defoliation than is the starting date of the epidemic (see below).

In the case of northern and southern corn leaf blights (caused by *Drechslera turcica* and *D. maydis*, respectively), the amount of initial symptoms appearing in the field is the most obvious factor in determining final disease intensity (Berger, 1974; Summer and Littrell, 1974).

Earliness of start is highly correlated with greater final severity of disease (y_{max}), larger area under progress curves, and a greater frequency of plants killed in 59 fungal pathosystems studied (Kranz, 1968b). This agrees with 42 years of observation on wheat stem rust (*Puccinia graminis* Pers.) in the Mississippi Basin—the earlier the first incidence the higher y_{max} tends to be (Hamilton and Stakman, 1967). Also, for virus diseases the date on which infection first appears within a crop is of crucial significance in their epidemiology (Thresh, 1974). The proximity of older or overwintering crops with abundant inoculum ensures early infection and can have a pronounced effect on rates of various plant disease epidemics.

The delay between emergence of the host plant and onset of the epidemic is a major factor affecting dosage response curves, growth functions, and regression equations. This factor has proved to be consistent enough to be of predictive value (Schrödter and Ullrich, 1965). The amount of initial inoculum and the time of its availability are major determinants of the speed with which a "takeoff" amount of disease, or threshold, is reached within the crop. This amount is critical, because once it is surpassed a general outbreak occurs. The threshold amount differs greatly from one disease to another. In studies extending over five years, Hirst and Stedman (1960) found that the frequency of stem lesions must be of the order of 0.52 to 0.79% before the epidemic started to "take off." Rice blast epidemics show two distinct phases of progress in some areas of Japan. During the first phase the disease intensity gradually reaches the 5% level of y, then the infection rate increases steeply (Kato, 1974). With coffee berry disease this happened after 3–5% of berries had been infected (Steiner, 1973).

According to van der Plank (1975), a rapid infection rate at the start of an epidemic tends to remain rapid—or a slow one slow—throughout the epidemic unless a change of circumstances occurs. Such a rapid rate brings about a low average age of the population of lesions and this tends to favor further development of the epidemic. This relationship may be consistent for linearized progress curves, but there is evidence to the contrary from nontransformed epidemics. However, the effect of large and/or early primary inoculum on the subsequent epidemic may be altered by prolonged changes in r. Such changes could involve the reproductive ability of a pathogen, infection requirements, as well as weather conditions.

Another pathogen or, rather, disease factor that determines the maximum rate of infection and the upper limit of explosiveness of an epidemic is the duration of latent periods. In simulated progress curves the most rapid increase in abundance of stipe rust of wheat occurred in host–pathogen systems with the shortest latent period, the most rapid basic infection rate R, and the longest infectious period (Zadoks, 1971). These three factors are listed in order of their decreasing effect on the daily increase of infections. Varieties with greater horizontal resistance have longer latent periods. The longest latent periods occur in endemic diseases of trees (van der Plank, 1975) and, since the length of latent period and the infection rate r are inversely related, r invariably becomes very small.

3. Environmental Factors Determining Rates

The various environmental factors affect an epidemic through its various structural elements (see Table I). These have differing effects on states, patterns, and subsequent rates. Some of these factors are discussed extensively by Rotem in Chapter 15 of this volume.

Very often environmental factors have relative rather than absolute effects on epidemics. Thus a favorable level of one or more factors may compensate for a certain deficiency in another factor (Rotem et al., 1971). The compensating effects may be achieved through the host, the pathogen, or both. For example, with a given combination of temperature and leaf wetness, the infection rate of late blight on tomatoes (Phytophthora infestans) increases with larger amounts of inoculum. If, however, inoculum and leaf wetness are constant, then temperature becomes the dominant determinant up to a certain point (Rotem et al., 1971). But with Stemphylium botryosum Wallr. f. sp. lycopersici R.,C., and W. on tomato, the duration of leaf wetness has a greater effect on infection than either temperature or number of spores, and it even influences the effect of the two latter factors (Bashi et al., 1973). High temperatures and low levels of soil moisture increase the number of cereal seedlings succumbing to Fusarium culmorum (W. G. Sm.) Sacc. and F. avenaceum (Wr.) Sacc. (Colhoun et al., 1968). However, this in turn is conditional upon a certain minimum amount of inoculum on the seeds. Fusarium avenaceum requires a much higher level of soil moisture in order to cause serious attacks. Large numbers of spores of F. culmorum cause very severe attacks over a wider range of environmental conditions than do small spore loads. It is obvious from these few examples that comparison of rate-determining environmental factors can be rather complicated. Compensating interactions must also be considered. Since this has not always been done in the past, published evidence should be viewed with some caution. Nevertheless, the few examples given above suggest that certain deductions are even now within reach. There may be a "family" of fungi which behaves rather like Phytophthora infestans, another like Stemphylium botryosum, and still another like the Erysiphaceae, which do not necessarily need a water film at all. For still another group of fungal pathogens leaf wetness may have to originate from rain to be conducive to infection.

4. Effects of Human Interference on Rates

We use the notion of "human interferences" to cover intentional control measures as well as any other agricultural practice or policy that may have an unintentional effect on plant diseases and their epidemics. Any measure, intentional or unintentional, eventually affects either y_0, the primary inoculum, or the relative infection rate r, or both. Control measures acting on either factor may show some exclusiveness, as exemplified by vertical and horizontal resistance (van der Plank, 1968) or shading of tea. The latter apparently does not alter the infection rate r so much, but the progress curve in unshaded plots follows at a near-constant rate of change with time (Visser et al., 1961). The early development of southern corn leaf blight is enhanced by minimum tillage practices leaving infected plant residues on the field surface. Plowing reduces inoculum by burying it and thus disease intensity to about one-fifth compared to disc harrowing. This reduces both y_0 and r (Summer and Littrell, 1974).

Irrigation by sprinkling has a more pronounced effect on r than do other techniques (Rotem and Palti, 1969). The intensity of powdery mildew (*Erysiphe betae* Weltzien) in sugar beet crops is increased by large amounts of applied nitrogen (Huber and Watson, 1974). The application of chlorocholine chloride to control lodging of wheat decreases the intensity of *Septoria nodorum* Berk., and this effect is enhanced by increasing doses of chlorocholine chloride and by increasing amounts of nitrogen fertilizers (Brettschneider-Herrmann and Langerfeld, 1971).

According to van der Plank (1963) the efficiency of fungicides depends on iR_c, the product of the length of the infectious period and the corrected basic infection rate. The latter determines the proportion of spores that must be killed in order to keep iR_c below 1. Late blight with $iR_c = 7.2$ is much easier to control than wheat stem rust with $iR_c = 360$. Apple scab has a large i. For this reason it can be controlled only if R_c is kept small. A pathogen like *Alternaria longipes* on tobacco (Norse, 1971) should pose no serious problems since R is usually less than 0.6, and the maximum multiplication rate is 2.1 daughter lesions per cycle. If properly applied, effective fungicides decrease the rate and often also delay the onset of visible disease. In experiments on powdery mildew of roses (*Sphaerotheca pannosa*) (Price, 1969) the r for the untreated control plants was 0.019 on leaves and 0.024 on blooms. The most effective fungicide brought r down to 0.006 and 0.013, respectively. In spray trials against coffee berry disease (Gibbs, 1971) fungicides altered the slope of the curve by decreasing r. In two trials with the same disease Steiner (1973) found infection rates r to be 0.19 and 0.29 (untreated), and 0.17 and 0.13 (treated), respectively.

V. EPILOGUE

In this chapter, I have described how I envision comparative epidemiology as a scientific tool. Basically, epidemics must be understood as systems with all the implications and complications pertaining thereto. This involves knowledge of their structures and behavior. Behavior, however, is not expressed in rates only, but also in the patterns that epidemics tend to follow in their dynamics. The requirements of future disease management are so demanding that adoption of such an approach is desirable.

Let us imagine that some years ahead the objective of crop protection no longer is control of one or two diseases or insects in one crop at one time. Plant protectionists then may be obliged to maintain fields reasonably free from the constraints of all relevant pests. In such cases we will face a variety of very different epidemics of many different diseases, weeds, and insects at one time in one field. Haphazard deductive comparison of epidemic features will not be very helpful. Systems analysis certainly will be required to cope with all the complexities involved. For all these reasons, therefore, we must begin the experimentation necessary to study and compare both the structures and the behaviors of epidemics. This may be complemented by cautious deductions from observational and other evidence.

Some criteria for comparative epidemiology have been proposed in this chapter and in the other references cited. Many of these criteria cannot yet be employed consistently because available evidence is incomplete and too erratic. Also, space did not permit me to elaborate, particularly on methodology. Nevertheless, I hope that these and other ideas presented in this chapter will stimulate new ideas, which, after some forthright discussion, may encourage the development of a promising branch of epidemiology. The usefulness of comparative epidemiology is obvious. Comparison within and between epidemics, of structures and behaviors is a *conditio sine qua non* for modeling, or at least their generalization. Modeling in turn is necessary to achieve the goals of our science: to describe and explain essentials in epidemics, and to enable us to manipulate them in the interest of mankind.

References

Analytis, S. (1973). Methodik der Analyse von Epidemien dargestellt am Apfelschorf (*Venturia inaequalis* (Cooke) Aderh.). *Acta Phytomed.* **1.** 76 pp.

Baker, R. (1971). Analysis involving inoculum density of soil-borne plant pathogens in epidemiology. *Phytopathology* **61,** 1280–1292.

Bashi, E., and Rotem, J. (1974). Adaption of four pathogens to semi-arid habitats as conditioned by penetration rate and germinating spore survival. *Phytopathology* **64,** 1035–1039.

Bashi, E., Rotem, J., and Putter, J. (1973). The effect of wetting duration, and of other environmental factors, on the development of *Stemphylium botryosum* f. sp. *lycopersici* in tomatoes. *Phytoparasitica* **2,** 87–94.

Berger, R. D. (1974). A guide to spraying sweet corn for *Helminthosporium* blight by likely disease spread. Univ. Tl. Agric. Res. Educ. Cent., Belle Glade Res. Rept. EV-1974-15, 3 pp.

Berger, R. D. (1975). Disease incidence and infection rates of *Cercospora apii* in plant spacing trials. *Phytopathology* **65,** 485–487.

Bock, K. R. (1962). Seasonal periodicity of coffee leaf rust and factors affecting the severity of outbreaks in Kenya Colony. *Trans. Br. Mycol. Soc.* 45, 289–300.

Brettschneider-Herrmann, B., and Langerfeld, E. (1971). Untersuchungen über Beziehungen zwischen CCC-Behandlung und *Septoria*-Befall bei Sommerweizen unter klimatisch kontrollierten Bedingungen. *Z. Acker- Pflanzenbau* 133, 137–156.

Burbon, J. J., and Chilvers, G. A. (1976). Epidemiology of *Pythium*-induced damping-off in mixed species seedling stands. *Ann. Appl. Biol.* 82, 233–240.

Carter, W. (1973). "Insects in Relation to Plant Disease," 2nd ed. Wiley, New York.

Colhoun, J., Taylor, G. S., and Tomlinson, R. (1968). *Fusarium* diseases of cereals. II. Infection on seedlings by *F. culmorum* and *F. avenaceum* in relation to environmental factors. *Trans. Br. Mycol. Soc.* 51, 397–404.

Cox, A. E., and Large, E. C. (1960). Potato blight epidemics throughout the world. *U.S. Dept. Agric. Handbook* 174, 230 pp.

Croxall, H. E., and Smith, L. P. (1976). The epidemiology of potato blight in the East Midland. *Ann. Appl. Biol.* 82, 451–466.

Fracker, S. B. (1936). Progressive intensification of uncontrolled plant-disease outbreaks. *J. Econ. Entomol.* 29, 923–940.

Gassert, W. (1976). Zur Epidemiologie der Kaffeekirschen-Krankheit (*Colletotrichum coffeanum* Noack sensu Hindorf) in Äthiopien. Ph.D. Dissertation, University of Giessen.

Gäumann, E. (1951). "Pflanzliche Infektionslehre," 2nd ed. Birkhaeuser, Basel.

Gibbs, J. N. (1971). Some factors influencing the performance of spray programmes for the control of Coffee Berry Disease. *Ann. Appl. Biol.* 67, 343–356.

Gregory, P. H. (1968). Interpreting plant dispersal gradients. *Annu. Rev. Phytopathol.* 6, 189–212.

Hamilton, L. M., and Stakman, E. C. (1967). Time of stem rust appearance on wheat in the Western Mississippi Basin in relation to the development of epidemics. *Phytopathology* 57, 609–614.

Hau, B., and Kranz, J. (1977). Ein Vergleich verschiedener Transformationen von Befallskurven. *Phytopathol. Z.* 88, 53–68.

Henderson, I. F., and Henderson, W. D. (1963). "A Dictionary of Biological Terms" (J. H. Kenneth, ed.), 8th ed., Oliver & Boyd, Edinburgh.

Hirst, J. M., and Stedman, D. J. (1960). Epidemiology of *Phytophthora infestans*. I. Climate, ecoclimate and phenology of disease outbreak. *Ann. Appl. Biol.* 48, 471–488.

Huber, D. M., and Watson, R. D. (1974). Nitrogen form and plant disease. *Annu. Rev. Phytopathol.* 12, 139–165.

James, W. C. (1974). Assessment of plant diseases and losses. *Annu. Rev. Phytopathol.* 12, 27–48.

Kato, H. (1974). Epidemiology of rice blast. *Rev. Plant Protect. Res.* 7, 1–20.

Kiyosawa, S. (1972). Theoretical comparison between mixture and rotation cultivations of disease resistant varieties. *Ann. Phytopathol. Soc. Jpn.* 38, 52–59.

Kranz, J. (1968a). Eine Analyse von annuellen Epidemien pilzlicher Parasiten. I. Die Befallskurven und ihre Abhängigkeit von einigen Umweltfaktoren. *Phytopathol. Z.* 61, 59–86.

Kranz, J. (1968b). Eine Analyse von annuellen Epidemien pilzlicher Parasiten. II. Qualitative und quantitative Merkmale der Befallskurven. *Phytopathol. Z.* 61, 171–190.

Kranz, J. (1968c). Eine Analyse von annuellen Epidemien pilzlicher Parasiten. III. Über Korrelationen zwischen quantitativen Merkmalen von Befallskurven und Ähnlichkeiten von Epidemien. *Phytopathol. Z.* 61, 205–217.

Kranz, J. (1974a). Comparison of epidemics. *Annu. Rev. Phytopathol.* **12**, 355–374.

Kranz, J. (1974b). The role and scope of mathematical analysis and modeling in epidemiology. *In* "Epidemics of Plant Diseases: Mathematical Analysis and Modeling" (J. Kranz, ed.), pp. 7–54. Springer-Verlag, Berlin and New York.

Kranz, J. (1975). Beziehungen zwischen Blattmasse und Befallsentwicklung bei Blattkrankheiten. Z. *Pflanzenkr. Pflanzenschutz* **82**, 641–654.

Kranz, J., and Lörincz, D. (1970). Methoden zum automatischen Vergleich epidemischer Abläufe bei Pflanzenkrankheiten. *Phytopathol. Z.* **67**, 225–233.

Norse, D. (1971). Lesion and epidemic development of *Alternaria longipes* (Ell. & Ev.) Mason on tobacco. *Ann. Appl. Biol.* **69**, 105–123.

Ogawa, J. M., Hall, D. H., and Koepsel, P. A. (1967). Spread of pathogens within crops as affected by life cycles and environment. *In* "Airborne Microbes" (P. H. Gregory and J. L. Monteith, eds.), pp. 247–266. Cambridge Univ. Press, London and New York.

Patten, B. C. (1971). A primer for ecological modeling and simulation with analog and digital computers. *In* "Systems Analysis and Simulation in Ecology" (B. C. Patten, ed.), Vol. 1, pp. 4–121. Academic Press, New York.

Populer, C. (1972). Les épidémics de l'oidium de l'hévéa et la phénologie de son hôte dans le monde. *INEAC, Ser. Sci.* No. 115.

Price, T. V. (1969). Epidemiology and control of powdery mildew (*Sphaerotheca pannosa*) on roses. *Ann. Appl. Biol.* **65**, 231–248.

Robinson, R. A. (1976). "Plant Pathosystems." Springer-Verlag, Berlin and New York. **7**, 267–288.

Rotem, J., Cohen, Y., and Putter, J. (1971). Relativity of limiting and optimum inoculum loads wetting duration and temperature by *Phytophthora infestans*. *Phytopathology* **61**, 275–278.

Schrödter, H., and Ullrich, J. (1965). Untersuchungen zur Biometeorologie und Epidemiologie von *Phytophtora infestans* (Mont.) de By. auf mathematisch-statistischer Grundlage. *Phytopathol. Z.* **54**, 87–103.

Steiner, K. G. (1973). Der Einfluß der Fungizidbehandlung auf den Verlauf der Kaffeekirschen-Krankheit (*Colletotrichum coffeanum* Noack). *Z. Pflanzenkr. Phytopathology* **64**, 168–173.

Sumner, D. R., and Littrell, R. H. (1974). Influence of tillage, planting date, inoculum survival and mixed populations on epidemiology of southern corn leaf blight. *Phytopathology* **64**, 168–173.

Thresh, J. M. (1974). Temporal patterns of virus spread. *Annu. Rev. Phytopathol.* **12**, 111–128.

Thresh, J. M. (1976). Gradients of plant virus diseases. *Ann. Appl. Biol.* **82**, 381–406.

van der Plank, J. E. (1960). Analysis of epidemics. *In* "Plant Pathology: An Advanced Treatise" (J. G. Horsfall and A. E. Dimond, eds.), Vol. 3, pp. 229–289. Academic Press, New York.

van der Plank, J. E. (1963). "Plant Diseases: Epidemics and Control." Academic Press, New York.

van der Plank, J. E. (1968). "Disease Resistance in Plants." Academic Press, New York.

van der Plank, J. E. (1975). "Principles of Plant Infection." Academic Press, New York.

Visser, D. R., Shamuganathan, N., and Sabanayagam, J. V. (1961). The influence of sunshine and rain on tea blister blight, *Exobasidium vexans* Massee, in Ceylon. *Ann. Appl. Biol.* **49**, 306–315.

Waggoner, P. E., and Horsfall, J. G. (1969). EPIDEM. A simulator of plant disease written for a computer. *Conn., Agric. Exp. Stn., New Haven, Bull.* 698.

Watt, K. E. F. (1966). The nature of systems analysis. *In* "Systems Analysis in Ecology" (K. E. F. Watt, ed.), pp. 1–14. Academic Press, New York.

Yarwood, C. E. (1973). Some principles of plant pathology. II. *Phytopathology* 63, 1324–1325.

Zadoks, J. C. (1961). Yellow rust on wheat. Studies in epidemiology and physiological specialization. *Tijdschr. Plantenziekten* 67, 69–256.

Zadoks, J. C. (1971). Systems analysis and the dynamics of epidemics. *Phytopathology* 61, 600–610.

Chapter 4

Methodology of Epidemiological Research

J. C. ZADOKS

I. INTRODUCTION

Epidemiology is moving rapidly and so are the methods for dealing with it. This chapter is about those methods, their history, their philosophy, and their use. A method is a special form of procedure in any branch of mental activity; it serves to achieve a certain objective. Methodology is the science of methods. Methodology in its broadest sense is a branch of philosophy; in its narrowest sense it is the description in the "materials and methods" section of a scientific paper. The methodology of epidemiologic research to be discussed here is about midway between the two extremes, the wide and the narrow, the abstract and the concrete. It aims at the generalities of how to approach an epidemiologic problem; more specifically, it aims at that part of epidemiologic research that is between problem identification and statistical design.

Methodology refers to the conceptual side of a science, here epidemiology; technology refers to the material side. The tools of the mind differ from the tools of the hand; the former are concepts, the latter are objects made of steel, copper, or glass. Of course, technology influences methodology and vice versa. Methodological requirements stimulate the development of new techniques; new techniques lead to changes of method.

II. THE METHODOLOGY–TECHNOLOGY INTERACTION IN HISTORICAL PERSPECTIVE

Experiments are the bread and butter of modern science, but they did not exist in the Middle Ages. They are a fruit of the Renaissance. But what did an experiment signify in 1767 to the then famous Florentine scientist, Felipe Fontana? He explains his experiment in some detail (see translation, 1932). It consisted of looking at a fungus through the microscope after he had transferred some fungal material to an object holder. Such was the meaning of the word experiment in plant pathology until about 1850.

Certainly, there were a few enlightened scientists who performed phytopathological experiments before 1850. The experiments on wheat bunt (*Tilletia tritici*) by Tillet, first published in 1755, were classic (see translation, 1937). Similarly, Duhamel's (Duhamel de Monceau, 1728) experiments on the pathogenicity of *Rhizoctonia violacea* to the saffron crocus are up to modern standards. Though Duhamel was awarded a membership of the Royal Academy of Sciences of France, his methodological innovation was not recognized and accepted by the scientific world of that day. The experiment, not as an observation but as an intervention in a natural process, slumbered during the period of fights between pathogenetists and autogenetists, conflicts which culminated around 1850. It seems that Duhamel himself did not recognize the value of his own work, however, because in later writings on wheat rust he made the usual remarks on atmospheric conditions instead of performing an experiment (from Tull and Duhamel de Monceau, 1764). Evidently, the rare experimenters before De Bary were outside the main stream of phytopathological lore.

A. De Bary's Innovation

In 1853, de Bary (1969) introduced a methodological innovation. He described the complete infection cycle from spore through germination, penetration, colonization, and back to spore again. Later he conducted experiments in the modern sense(de Bary,1861).Why did the world's self-respecting phytopathologists in the second half of the nineteenth century visit de Bary? It was because he had established a new methodology: the experiment in the modern sense, used to reconnoiter the infection cycle and/or the life cycle of a fungus, and to prove that the fungus is the cause and not the consequence of the disease. To modern eyes the new technology needed to implement the new methodology is simple: (1) a good microscope; (2) a greenhouse to grow plants under some protection; and (3) the modest but then new tools used in mycological and phytopathological practice.

B. Koch's Innovation

The next methodological innovation was the set of prescriptions known as Koch's postulates, first clearly pronounced in 1890. The postulates are a set of prescriptions to prove that one contagious disease is caused by one and only one pathogen, disease and pathogen both being identifiable by following the rules. Koch's postulates can be tested only by means of an appropriate technology. Koch's work was made possible, or at least facilitated, by the innovation of using solid media as a substrate for culturing bacteria and fungi: Koch's innovation.

A methodological innovation of older date but of more recent effect was proclaimed by the French Royal Academy of Sciences, which published annually a series of meteorological observations: "The purpose of that work was not simple curiosity; people knew at that time how the variations of the air temperature could affect the organized bodies." "The botanico–meteorological observations that M. Duhamel began to publish in 1741 are among the first fruits obtained thereby; they demonstrated how much the different temperatures could influence the crops" (Anonymous, 1751). H. L. Duhamel de Monceau began to report on the weather and its effects on the condition of the crops. It is not known whether an analysis of an eighteenth-century sequence of annual reports has yielded any tangible result.

In 1767 Targioni Tozzetti (1952) was probably the first to make a detailed analysis of the weather preceeding and during an epidemic of *Puccinia graminis*, mainly using data from his private weather station. Though his descriptions contain some interesting clues for the modern epidemiologist, he did not come to clear conclusions. Lutman (1911) in Vermont, was perhaps the first to demonstrate convincingly that weather had a serious impact on *Phytophthora infestans* epidemics on potatoes, using a still modern analysis of a 20-year series of weather and disease records. Of course, the relation had been known intuitively (or "felt") for a long time, but intuitive "knowledge" and knowledge based on scientific proof are two different things. Work on *Plasmopara viticola* of vine paved the way to disease forecasting, which reached a milestone with van Everdingen's (1926) rules for outbreaks of potato late blight. The concept of the critical period, a period of some hours to days during which the weather is conducive to infection, was born. The concomitant technology consisted of the synoptic weather stations, where standardized meteorologic observations were made continuously.

III. THE NEED FOR METHODOLOGY

Modern science is analytic in that it looks for cause-and-effect relations. There are, however, several snags. For example, it often assumes that every effect has one and only one cause, which can be identified by induction and experimentation. The cause-and-effect relation may be misinterpreted, as in the case of the signalman's daughter. A train always appeared after she heard the bell ring. She naturally concluded that the ringing of the bell caused the train to appear. Hers is a common error, that of confounding a temporal with a causal relation.

There may be more than one cause that contributes to the effect. The one pointed out may not necessarily be the main cause; it may even be the wrong one. This error was common in the eighteenth century, when outbreaks of cereal rust were attributed to weather or even to "cosmic" constellations (Unger, 1833).

The problem may be resolved by calling the fungus the primary cause and the weather the secondary cause, but room for confusion still remains. That the pathogen is the cause of a disease is readily accepted, but what about the pathogen as the cause of an epidemic? If we take the pathogen for granted, we might as well call the weather the cause of the epidemic if the weather was exceptionally favorable to the pathogen. In a way, the eighteenth-century scientists, up to Unger, were right.

In present-day terminology, the fungus is considered to be the only cause, and the weather is ranked among the constraints, the limiting conditions that allow or inhibit the cause to exert its effect. This way of looking at causality is a matter of convention, be it an appropriate one among plant pathologists, a convention resulting from the victory of the pathogenetists. Other conventions are possible, and indeed the convention used in dynamic simulation is decidedly different from the elementary cause-and-effect type of thinking discussed here.

IV. THE TOOLS OF THE TRADE

The carpenter needs a hammer, pincers, a screwdriver as the tools for his job. If the chisel does not fit, he takes another one; if it is blunt, he sharpens it.

In science the tools of the mind are concepts, that is, ideas or notions of classes of objects or phenomena. Concepts are the units of thought, clearly differentiated from each other. Concepts need definition but no proof. A definition is a set of attributes characterizing the concept, usually in a single sentence. The concept to be defined cannot appear in that sentence of course. "The latent period is the span of time from inoculation until the first appearance of new spores." This is a general definition. It covers many cases (although not that of pathogenic *mycelia sterilia*), but it is not precise.

Definitions are rooted in common parlance, where there is an intuitive understanding of the words used. Some words are given a specific meaning within a certain scientific context by defining them, thus creating basic concepts. From the basic concepts, more complex concepts can be constructed. Mathematics and physics have gone the farthest in this process, thus becoming very abstract sciences. In epidemiology the process of constructing concepts of gradually increasing complexity is experienced by any reader of Gäumann's "Pflanzliche Infektionslehre"

(1946) or van der Plank's "Plant Diseases, Epidemics and Control" (1963). These books are still exceptional; many epidemiological concepts are intuitively understood by both reader and author rather than being accurately defined.

An improvement is to make an operational definition of a concept, a definition with the character of a recipe written for a specific purpose. Zadoks (1961) defined the latent period of *Puccinia striiformis* on wheat as "the period in days from the day of inoculation until the day at which the first open pustule was observed." This definition expressed the latent period in whole days, which is incorrect but convenient. It also allows for errors of observation as the overlooking of the first open pustule, either because of bad visibility due to the weather or to poor vision as on the "day after the night before." Operational definitions are transferable by speech and writing within and between institutions.

Many concepts have a physical dimension, that is, they can be expressed in units of length, mass, time, and so on, or simply in numbers of countable entities. Few epidemiologists realize that van der Plank's (1963) apparent infection rate r has the dimension (entities per entity per unit of time) or $[NN^{-1}T^{-1}] = [T^{-1}]$, which is the same as the dimension of the relative growth rate in animal ecology, usually expressed in inverted days: d^{-1}. Neglect of the dimensionality of concepts has led to incredible confusion in ecology; let us avoid such a nuisance in epidemiology.

Definitions of concepts must be consistent, that is, mutually compatible and not contradictory. It must be clear whether they refer to a recognizable state, as the appressorium, or to a process, as appressorium formation. If the subject is the infection cycle, the phases must be defined consistently, so that the subsequent phase begins where the antecedent phase ends. The observation of the first open pustule, which marks the end of the latent period, *ipse quo* marks the beginning of the sporulation period. Consistency among definitions, with respect to dimensionality and sequential alignment, is the usual result of the development of a school of thought. It is an asset in teaching a course, and it is essential for any research project.

Misunderstandings between scientists originate when no definitions or only general ones are given, so that the reader can think of more than one operational definition. Another source of misunderstanding appears when in one school of thought the concepts are well defined and consistent whereas they are not so in the eyes of colleagues not adhering to that school, or when the concepts are applied outside their usual context. A third source of misunderstanding is the gradual and imperceptible change in meaning of concepts, which are not precisely defined. Few present-day phytopathologists will know that, at the turn of this century, the word epidemic was also widely used for outbreaks of insect pests.

V. FORMALIZED INQUISITIVENESS—THE HYPOTHESIS

A hypothesis is a supposition made as a basis for reasoning; it is a starting point for investigation. A hypothesis is an assumption that, though it can be used tentatively to explain a phenomenon, has to be proved. Herein lies the difference from a concept: a concept does not need proof, a hypothesis does.

We have been indoctrinated with black-and-white thinking, the either-or type of reasoning. In statistics we are taught that a problem of testing a hypothesis has four ingredients (Lindgren, 1975): "(1) A hypothesis to be tested, called the null hypothesis, H_0; (2) an alternative hypothesis, H_A; (3) data, and a test statistic computed from the data, as an aid to choosing between H_0 and H_A; (4) a decision rule, . . ." The principle applied is the same as in ancient Roman law, *tertium non datur*: there is no third possibility; right or wrong, false or true. Such straightforward logic, only slightly clouded by the uncertainties called type I and type II (α and β) error, is not necessarily suited to the life sciences. The limited scope of our knowledge and the complexities of the phenomena may necessitate a more subtle procedure. The dilemma pathogenic organism—weather illustrates the point.

The hypothesis in the statistical sense is the highest level in formalization of inquisitiveness. Often we feel satisfied by lower levels of formalization, as in many problems given to students receiving their first research training: "Is there an effect of . . . on . . . ?," or "What happens after such and such intervention in the process of . . . ?." These are open-end questions. They do not lead to the answer in the sense of "leading questions," and they represent a healthy form of inquisitiveness, unless they are used to camouflage the questioner's intellectual laziness (a not uncommon practice). Open-end questions are good in the exploratory phases of a research project, and they will lead to better-phrased questions eventually.

In the applications phase of a research project there is no room for a third answer. Treatment A is better than B, or it is not; a multiple correlation equation P predicts disease outbreaks better than Q, or not. Open-end questions lead to general notions; for specific knowledge the statistical test is needed: either H_0 or H_A is true. There is social and economic relevance in the last remark. Small differences, which are difficult to ascertain, for example, between cultivars or treatments, become extremely important when they lead to decisions affecting thousands of people and millions of hectares. Because of their importance the formal sequence must be followed: hypothesis, experiment, statistical test, and evaluation. This sequence is part of the empirical cycle.

VI. THE EMPIRICAL CYCLE

It is a long way from general notions to specific knowledge. The genius may see truth in a flash and thus arrive at a hypothesis at a high level of abstraction, but the normal scientist works his way up gradually. The pursuance of truth in the sciences follows a certain pattern that has been described in the behavioral sciences as the empirical cycle (de Groot, 1969) (Fig. 1). The cycle consists of phases, beginning with observation and ending with conclusion.

A. Phase 1: Observation

Observation, including collection and classification of data, is the first systematic and purposeful activity of the researcher. Study of the literature is part of this phase. A genius is a keener and more original observer than the average scientist. Intuitively, or on theoretical grounds, he may have an expectation of the observations that are really important, so that he is sensitive to important facts. "No one can be a good observer unless he is a good theorizer," said Charles Darwin.

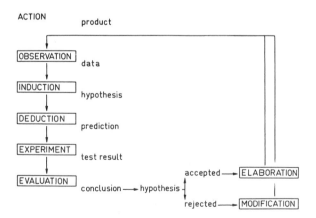

Fig. 1.   The empirical cycle.

B. Phase 2: Induction

Assume that a conclusion of supposedly general validity is drawn from a limited set of observations. This process of the mind is called induction. Induction, or the generalization of observations digested earlier, leads to a (general) hypothesis, such as: *"Puccinia striiformis* of wheat over-winters in the Netherlands" (Zadoks, 1961). When this hypothesis was first put forward, it was contrary to current opinion.

C. Phase 3: Deduction

The generalization obtained by induction must be applied to a new situation, new in the sense that this situation does not belong to the set of situations producing the observations used in induction. If a hypothesis is true in general, specific predictions can be derived from it in special instances, according to the hypothetico–deductive approach. This is the process of deduction. If the general hypothesis stated above about *P. striiformis* is valid, it leads to three verifiable predictions or specific hypotheses; (1) *"P. striiformis* can be found in the Netherlands throughout the winter;" (2) *"P. striiformis* does not overwinter in countries south of the Netherlands;" and (3) "the amount of overwintering inoculum is enough to explain a severe epidemic later in the season."

D. Phase 4: Testing

A hypothesis is said to be valid when it accurately predicts the condition of fresh materials, of new elements not used in the design of the hypothesis. This testing is done by observation or experimentation. Prediction 1 was verified by means of field observations. During a typical, cool and freezing wintry week 40 hours of walking through the fields yielded a harvest of a single wheat leaf with sporulating *P. striiformis* (found on Friday afternoon!). Prediction 2 was verified by travels far and wide through Europe by car and concomitant field observations. Prediction 3 was verified by observational, experimental, and theoretical means, whereby the logistic growth rate r of *P. striiformis* epidemics in the Netherlands was established. A check was run to find out whether overwintering inoculum x_0 and r together could explain terminal severities of up to $x = 1$ before harvest. The result of this complicated testing procedure was positive.

Note that the empirical cycle does not prescribe experiments. A qualitative test by means of observations in the field can be sufficient. It is advisable, however, to proceed to the highest level of formalization,

H_0 versus H_A, and to test H_0 by means of a quantitative experiment. Application of the rules of statistics leads to a test result: H_0 is either rejected or accepted.

E. Phase 5: Evaluation

When the prediction has to be rejected, there comes a painful period of debugging, fault finding. Was there an error in the original observations, the induction, the deduction, the test, or the statistical analysis? If no error can be found in procedure or calculation, there is only one conclusion left: the hypothesis obtained by induction, and/or the prediction obtained by deduction, is wrong. If so, the observation(s) must be renewed or reexamined; hypothesis and prediction need modification.

If the hypothesis cannot be rejected, it is accepted as valid, at least for the time being. The hypothesis "all swans are white" is accepted until the first black swan is observed. The following questions remain: (1) What are the consequences of the decision with respect to other related hypotheses? (2) What is the generalization value of the decision? (3) What is the contribution of the result to general theory?

The hypothesis on the overwintering of *P. striiformis* in the Netherlands was confirmed, but it became clear that the hypothesis was too simple so that the generalization value was limited. The hypothesis was true for some physiological races only, and had to be extended to some other areas where climate favored overwintering and oversummering of the rust. In Fig. 1 the mental process of reflection on an accepted hypothesis was called elaboration.

The result of the evaluation is a conclusion or a set of conclusions. A conclusion can serve as a new observation, which initiates a succeeding empirical cycle. The evaluation leads to renewed and additional hypotheses: (1) "physiological race A is the only race that overwinters in the Netherlands" (a renewed and more specific hypothesis); and (2) "the oversummering process of the rust is relevant to subsequent overwintering" (an entirely new hypothesis). The new or renewed hypotheses close the empirical cycle.

Cycle by cycle the scientist spirals up to more general truths, to higher levels of abstraction. The procedure is tedious but secure. The alternation of induction and deduction, which takes place in the mind of the scientist, is symbolized in Fig. 2. This mental process often takes place simultaneously at various levels of abstraction. At the bottom of the pyramid are large numbers of small and concrete facts, at the top is a single all-embracing and abstract conception—a theory.

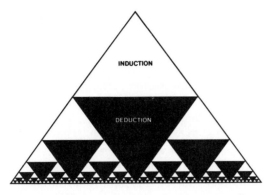

Fig. 2. The alternation of induction and deduction at various levels of abstraction (from Gregory, 1967).

VII. FACTS: HOW OBJECTIVE ARE THEY?

Facts are based on observation, but observations are colored by anticipation. Imagination or a keen intuition may condition the observer so that he actually sees a rare phenomenon when it happens. Misdirected anticipation conditions the observer to see only what he anticipated and to disregard whatever is contrary to his anticipation; unconsciously, he commits the sin of selecting evidence. Conditioning may be the result of anticipation, a hypothesis obtained by induction, local working conditions, training, past experience, or acceptance of authority.

Local working conditions induced J. W. Hendrix (personal communication) to hypothesize that in Washington State, uredospores of *P. striiformis* could oversummer in the dry dust and dirt; the hypothesis was shown to be correct in that semiarid area without summer rain. The idea could hardly occur to a northwest European worker living under conditions characterized by ample summer rainfall.

People have been trained to think in terms of physiological races identified by means of a standard differential set of cultivars. The imprint of teaching is so strong, that it is often forgotten that physiological races are mere artefacts, handy in some conversations but terribly misleading when the genetics of epidemics have to be understood.

The weight of authority can make itself felt so heavily that it obscures or even obliterates an established fact. According to de Bary (1865), a certain Von Bönninghausen in 1818 had provoked rust on marked plants of rye (*Secale cereale*) by transfering to them the rusty powder taken from aecidia on *Berberis*, but the proponents of the insight that the rust on rye came from the barberry were silenced by the damnation ("Verdammungsurtheil," writes de Bary) of the botanical microscopists who led scientific opinion at the time. This amazing matter is discussed in further detail in Chapter 2, Vol. I.

In the trial-and-error phase, the researcher often follows a "hunch," tries over and over again the various arrangements, combinations, and permutations, to find what he intuitively expects. Then there are no conscious, explicit induction and deduction. The facts are more or less "constructed" according to the anticipation. The resulting final observation becomes a fact, maybe even a true fact. Nevertheless, it is "colored" by anticipation, a dangerous but common situation. The scientist is an accurate observer, but at the same time he shapes his research and therewith determines its results (Geurts, 1976).

VIII. THEORY

The dictionary gives two major meanings of the word "theory." The first is an "exposition of the principles of a science." A typical expository account of the principles of epidemiology is found in Gäumann's book (1946). In this masterpiece of epidemiological thinking many consistent concepts were formulated and placed in context. Gäumann's reasoning was qualitative and led to classifications rather than to calculations. The second meaning is a "system of ideas or hypotheses explaining observed phenomena." An additional requirement can be that the explanation is based on principles independent of the phenomena to be explained; this requirement allows for the aprioristic nature of such principles, in contrast to an empirical theory based on experience.

Van der Plank's theory on the genetics of resistance, vertical and horizontal, so penetratingly described in his 1963 book and elaborated in later books, is simple in design with strong aprioristic traits. Robinson (1976) has remodeled and extended the theory. He considered the interacting host and pathogen populations as pathosystems, applying a mode of thinking derived from general systems theory. The approaches of van der Plank and Robinson are essentially qualitative; they classify rather than calculate. An attempt was made by Parlevliet and Zadoks (1977) to quantify the genetic interaction between host and pathogen by applying principles from (quantitative) population genetics.

Another epidemiologic theory has been developed by van der Plank (1963), a "quantitative theory of epidemic growth." The theory, based on the logistic equation, is typically aprioristic, but it describes a number of observed phenomena surprisingly well. It also makes explicit epidemiological concepts hitherto only intuitively understood, as the equivalence of the effects of resistance, environment, and fungicides symbolized by the "fungicide square" (Paragraph 21.1, van der Plank, 1963), and the "threshold theorem" (Paragraph 8.6, van der Plank, 1963). It takes care of the dimensionality of the concepts used. It is a quantitative theory leading to the calculation of infection rates, sanitation quotients, etc.

The power of a good theory, true or false, is impressive. Where Gäumann laid the theoretical foundation of epidemiology, the science acquired real impetus only after van der Plank's book (1963). This book changed epidemiology and plant disease management and revolutionized resistance breeding. As in many natural sciences, a good theory can have great practical consequences. Here, we may also remember Mac-Fadyen's admonition (1975): "If we do not develop a strong theoretical core that will bring all parts of ecology back together we shall be washed out to sea in an immense tide of unrelated information."

IX. VERIFICATION AND FALSIFICATON

A theory is a hypothesis at a high level of abstraction, covering a wide range of phenomena. In terms of Fig. 2, it is a large triangle encompassing a large number of small triangles. A theory is a hypothesis, but not every hypothesis is a theory! As a hypothesis, a theory is but a means toward an end; when it is not good enough, it must be discarded.

The testing of a theory is not fundamentally different from the testing of a hypothesis—it is just more complicated. Let us go back to the evaluation phase of the empirical cycle in Fig. 1 and discuss three steps of evaluation.

A. The Hard Test

Step 1 is a "hard" test. A prediction is made by deduction, leading to a "critical" experiment, which can be analyzed by means of the usual statistical techniques. If the result is negative, the hypothesis (theory) is rejected; if positive, it is accepted for the time being.

B. The Mild Test

Step 2 is a "mild" test. The observation resulting from the experiment is compared with other observations at the same level of abstraction. This comparison necessitates extensive study of the literature on epidemiology and the allied sciences. When the observation fits with comparable observations and is in accordance with current theories, the observation is accepted. Note that "observation" is used here in the methodological sense, so that it can be a simple fact, an accepted hypothesis, or an accepted theory. If, in the latter two cases, there is no fit, the new hypothesis or theory should explain known phenomena better than older ones and solve hitherto unsolved contradictions; otherwise, it must be rejected, or at least be reexamined.

C. The Soft Test

Step 3 is a "soft" test. The observation (again in the methodological sense) should have a function, a biological meaning. This is the teleological interpretation of phenomena, an interpretation that assumes the phenomena to serve a purpose useful to the organism. For example, the release of ascospores by *Venturia inaequalis* is stimulated by far red light (Brook, 1969). What is the function of the red light? A teleological interpretation is that the old apple leaves with perithecia lie on the soil; ascospores are ejected up to 2 cm high; red light implies daytime, and thus turbulence, which transports ascospores from the air layer just above the soil to the newly formed leaves, 1 m or more above the soil. This teleological interpretation, which is not necessarily true, has the merit of placing an isolated observation in a wider ecological context. Teleological interpretation has heuristic value—it leads to new hypotheses. Teleological reasoning is rejected in several natural sciences, but is currently accepted in biology. A frequently mentioned function or desirable purpose is "the perpetuation of the species."

The three steps of evaluation are complementary. The hypothesis or the theory becomes more acceptable when more tests, going from "soft" to "hard," give positive results.

Absolute proof of the truth derived from hypotheses, including theories, cannot be given. Verification, the production of evidence to demonstrate the truth, is sometimes done piecemeal. Van der Plank tried to verify his theory on resistance by providing selected evidence, disregarding all evidence not in accordance with his views. The procedure opposite to verification is falsification, formulated very early by Pascal in 1647. If only one observation does not agree with the hypothesis (the black swan), the hypothesis is false. Remember that in statistical tests, the H_0 is usually formulated so that it will be falsified, rejection of H_0 being the expected test result.

D. Falsifiability

According to Popper (1960) a scientific statement is acceptable only when it can be falsified and is accepted as long as it has not yet been falsified. Falsifiability is the criterion for acceptance of a scientific statement. Van der Plank's theories certainly comply with this criterion; indeed, he is quoted more often by scientists trying to falsify his theory of resistance, and some of the underlying hypotheses, than by those who support them. Especially the hypothesis of homeostasis of stabilizing and directional selection (van der Plank, 1968, Paragraph 4.7) is under fire. Of course, such attacks are not personal; only falsification leads to the progress of science, according to the Popper doctrine.

Whether this doctrine holds remains to be seen. Present-day dynamic simulation models can be regarded as hypotheses of a complex quantitative nature. It is difficult to falsify them because (1) many checks and balances are built into them; (2) assessment of the real-life situation at the same degree of accuracy as the calculated results is often impossible; and (3) a misfit can often be corrected simply by adding another variable or adjusting a parameter. Though for the same reasons verification of the model as a whole is next to impossible, verification of segments of the model is accepted practice. Simulation models as "scientific statements" are hardly "falsifiable" in the sense of Popper; nevertheless they are "acceptable."

X. METHODOLOGICAL HANDHOLDS

It is time now to return from esoteric spheres to problems encountered in daily practice, where we may need some methodological handholds or traffic regulations, as they were called earlier. Real handholds, in the meaning of "do's and don'ts," are rare in epidemiology. Few efforts have been made to construct methodological rules; the empirical cycle is actually one of them. The following is an attempt to bring together other rules or "methodological instruments."

A. Disease Cycle

The disease cycle is a long-used methodological instrument. Everybody knows the pictures with arrows in a circular pattern, pointing from one spore type to the next. *Puccinia graminis* is often given as the example, with one sexual and many asexual cycles per year; host alternation complicates the picture here.

B. Infection Cycle and Infection Chain

Over the years, the notion gradually emerged that a pathogen has to increase its numbers from generation to generation, and that the pathogen has to repeat a cycle of events and morphological appearances in every generation. Duhamel de Monceau (1728) recognized the necessity of multiplication for a parasite as a biological entity in its own right. In 1767, Fontana (1932) regarded multiplication as a criterion for identifying a parasite; in the same year, Targioni Tozzetti (1952) realized that the wheat rust fungus had to multiply during a series of generations before reaching harmful quantities. However, these findings were ignored by later authors. In 1807, Prévost (1939) clearly described an infection cycle of a cereal rust (his point 70), but he did it casually and not as a principle. In 1853, de Bary (1969), interested in the "structure and developmental history" of fungi, speaks of "generation cycles." De Bary's theme, "general laws of development" for the purpose of taxonomy and etiology, was taken up again, and more fundamentally, by Fischer and Gäumann (1929) who spoke of the "Aufeinanderfolge der verschiedenen Fruktifikationsformen" (succession of different fructification forms). Gäumann (1946) finally introduced the "Infektkette" (infection chain), as a fundamental epidemiologic concept; it is a never-ending concatenation of infection cycles. This idea is at the root of present-day computer models for dynamic simulation of epidemics.

C. Infection Cycle and Components Analysis

The infection cycle has three major component phases: sporulation, dissemination, and infection (Hirst and Schein, 1965). In each phase the fungus appears with distinct morphological characteristics. Each phase is susceptible to different environmental influences, and each is thus epidemiologically distinct. The phases can be subdivided into subphases, according to expediency. Each phase can be studied quantitatively, so that a complete quantitative analysis of the infection cycle is possible in principle. This components analysis can serve different purposes if the same technique is always applied. (1) Components of resistance to cultivars can be compared when tested with the same fungal isolate. (2) Components of virulence can be compared between isolates tested on one host cultivar. (3) The effect of environment on various components can be tested when one cultivar–isolate combination is subjected to different environmental factors. Approach (1) is now used in the study of horizontal resistance (Heitefuss and Dehne, 1976; Russell, 1972; Zadoks, 1972b). The combination of approaches (1) and (2) led to the description of characteristics of physiological races other than virulence (Johnson, 1972; Katsuya and Green, 1967; Schröder and Hassebrauk, 1964; Zadoks, 1961) but quite relevant to epidemiology. The combination of approaches (2) and (3) led to the distinction of "ecologic races" (Hill and Nelson, 1976). The results of the components analysis can be reassembled, using life table statistics as in animal ecology, to determine the mean length of a generation, the net reproduction rate per generation, and the intrinsic growth rate of the population (Zadoks, 1977). The results of the components analysis can be applied in simulation models (Kranz, 1974).

D. Koch's Postulates

Koch's postulates, formulated in 1890, are still used (Koch, 1891). Their normative value is great; no organism is accepted as a pathogen unless these postulates are satisfied. The postulates are versatile and can be applied to bacteria, fungi, viruses, nematodes, and even insects. Sometimes, adjustments have to be made, as with viruses which cannot be grown in pure culture but may grow in cell cultures on artificial media outside the intact host. Though the logic of Koch's postulates seems inescapable, they do not fit all cases; for example, they do not allow for symptomless carriers.

E. Postulates of Quantitative Epidemiology

Zadoks (1972a) formulated a set of rules for testing the validity of quantitative evidence on epidemics (Table I), which are, in a way, the

TABLE I

Postulates of Quantitative Epidemiology, or "Rules for Testing the Validity of Quantitative Evidence on Epidemics" [a]

1. The source(s) of inoculum at the onset of the epidemic must be known, and the amount of the pathogen in the source(s) must be expressed quantitatively.
2. The effect of environmental conditions on the development of an epidemic of the pathogen must be known in terms of quantitative relations between independent (usually abiotic) variables and dependent (biotic) variables.
3. The rate of development of the epidemic under the prevailing conditions must be calculated.
4. The successive levels of the epidemic must be calculated from the known amount of pathogen in the source(s) and the calculated rates of development of the epidemic.
5. The calculated terminal level and the calculated intermediate levels of the epidemic must be equal to the observed terminal and intermediate levels.

[a] Zadoks, 1972a.

quantitative counterparts of Koch's purely qualitative postulates. The rules, and especially rule 5, provide checks for errors. Furthermore, they have heuristic value, for they can help to discover unexpected relations. At the time the rules were formulated, there was no disease yet known to which all five rules had been applied. It seems that today, with the present sophistication in simulation models, such checks are essential. The best example is probably that of Shrum (1975) for yellow rust on wheat.

F. Robinson's Rules

Van der Plank (1963), distinguishing vertical and horizontal resistance, initiated a new branch of epidemiology, the genetics of pathosystems. Robinson (1971) elaborated the theme and devised rules to determine when and where one or the other type of resistance would be adequate. These rules are designed primarily for practical breeding purposes, but at the same time they form a methodological instrument for the analysis of epidemics.

G. Auto- and Alloinfection

Autoinfection is infection of plant parts by inoculum coming from other parts of the same plant; alloinfection is infection of a plant by inoculum coming from other plants. Auto- and alloinfection (Robinson, 1976) are terms that primarily refer to individual plants, but the concepts can also be used for individual plots or stands. It is the proportion of auto- to alloinfection that matters; when alloinfection is high in comparison to autoinfection, on a single-plant basis, a multiline variety will give good protection against disease (e.g., Mackenzie, 1978).

When alloinfection is high in comparison to autoinfection in field plots, any comparative experiments on the merits of various fungicides, of various types of resistance, and experiments on crop losses will be biased. This is the "cryptic error of field experiments" of van der Plank (1963, Chapter 23), a representational error that tends to underestimate the value of horizontal resistance and of fungicidal protection by overinfection due to influx of spores from the susceptible or the untreated control. It also underestimates crop loss in the untreated check because of spore losses, spores being blown away by the wind (James et al., 1976). Though in such comparative field trials the ranking of the treatments is correct, the absolute values are not representative for the same treatments applied in commercial fields. The consequences can be serious, because though the errors are always conservative (that is, on the safe side) they may lead to withholding registration of good fungicides or good cultivars, to incorrect economic threshold values for disease control, and to inadequate methods in resistance breeding (Zadoks, 1972c; Parlevliet and van Ommeren, 1975).

H. Crop and Disease Assessment

The Food and Agriculture Organization has published a "Manual on Disease and Loss Assessment," a compilation of observation scales to assess phenologic or growth stages of crops, severity levels of diseases, and so on (Chiarappa, 1971). These scales provide valuable tools for quantitative epidemiology. They become a methodological instrument when they evolve to a certain level of generalization. Indeed there is a slow but steady development. What started as a wheat phenology scale in 1941, and was used in several different forms, became a two-digit decimal code for the observation of growth stages of cereals, including wheat, barley, oats, rye, rice, maize, sorghum, etc. (Zadoks *et al.*, 1974). What began as the Cobb scale in ca. 1890 was developed into a "universal" cereal disease recording system (McNeal *et al.*, 1971). Horsfall and Barratt (1945) provided a background for disease recording scales from perception theory, recalling the Weber–Fechner law. By developing observation scales to higher levels of abstraction, with larger areas of application, and providing them with a theoretical background, they became valuable methodological instruments.

The methodological "instruments" indicated above are not all of the same quality or level of abstraction. Time will tell whether they are accepted or not. They all serve more or less as "check lists." When the researcher also uses the empirical cycle as a check list, he may avoid the errors of incomplete argumentation, overargumentation, and overgeneralization. The scientist applying the methodological instruments can be confident that he is understood by his colleagues. There is no science when there is no adequate communication.

XI. QUANTITATIVE ASPECTS

Any self-respecting scientific discipline eventually turns to quantitative methods. In this respect botanical epidemiology is young. When Newton went to the countryside during the annus mirabilis 1665–1666, because Cambridge University had been closed because of the Great Plague, he laid the foundations for his later work on gravitation theory, a beginning of quantitative physics. In 1811 when Avogadro pronounced his law:

"equal volumes of gases contain—at equal pressure and temperature—an equal amount of molecules," this was a beginning of quantitative chemistry. Quantitative epidemiology began in 1963, when van der Plank's first book appeared and the first international meeting on botanical epidemiology was held (Zadoks and Koster, 1976). Epidemiologists can perform quantitative experiments, and, if they cannot, nature often does so by providing severe epidemics. We will refrain from discussing here the theory of measurement, but will go on to discuss the art of designing experiments.

XII. THE ART OF DESIGNING AN EXPERIMENT

What Fontana in 1767 called an "experiment" (Section II) is no longer regarded so. Experiments—in the present sense—have been done incidentally, however, in the laboratory and in the field, from the first known epidemiologic experiments by Duhamel de Monceau (1728) up to de Bary's time. Except for those by Duhamel de Monceau (which were generally ignored by the cognoscente) and Tillet, they were too clumsy to be convincing. They were mainly infection experiments, transfers of spores to plants in the hope of infecting them. In 1853, de Bary (1969), summarizing all available evidence for pathogenicity of fungi, said: "But it is also subject to no doubt that in cases of this kind a single positive result, as soon as it is established with certainty, has much greater value than many negative ones, since these can be produced by all sorts of untoward accidents over which the observer has no control."

In the 1853 book quoted here, de Bary did not record a single experiment of his own. He was, nevertheless, acutely aware of the importance of experimental evidence and he won his international fame in part through his careful experimentation, for example, his publications on *Phytophthora infestans* (1861) and *Puccinia graminis* (1865). His experiments were mainly qualitative, demonstrating infection cycles and life cycles; he was, however, aware of the importance of comparative experiments. In 1853, he disregarded Duhamel's experiments but mentions comparative experiments by Tillet (first published in 1755) and a few others.

A. Comparative Experiments

Comparative experiments were developed first in physics and later in the other sciences. If all things are kept equal and only one independent variable is changed at a time, the dependent variable can be measured accurately, as is the case when a bar of copper is exposed to various temperatures and its length measured. Much scientific insight has been obtained in relating the reaction of the dependent to the independent variable by mathematical means. The results actually describe a regression of the dependent to the independent variable, or (to use another terminology) of the response to the stimulus, or (and this is written more hesitantly) of the effect to the cause. Whereas in physics the stimulus usually is quantitative, a continuous variable like temperature being given selected discrete levels at predetermined intervals, in epidemiology the stimulus is often qualitative, for example, cultivars, physiologic races, or fungicides.

B. The Problem of Variances

Whereas in physics variance in the response can often be reduced to below the nuisance level, biologists have learned to live with large variances. In good yield trials a coefficient of variation of 3% can be obtained. Such a value would be high for the thermal expansion coefficient of a copper bar but it must be accepted in agriculture; the coefficient of variation of disease records taken in the field may be up to 30%.

Statistics come to our aid, namely, regression analysis and analysis of variance (ANOVA); the latter not only permits us to estimate and compare qualitative differences when all responses are expressed in the same measure, but it also permits us to sort out the main effects and their interactions when different stimuli are varied independently. Multiple regression analysis can render similar services for other types of data, especially for field data (Kranz, 1974; Chapter 3 of this volume). In both cases, there are two or more independent variables, experimental factors, or stimuli, each administered at different levels. When the experiment is completely under the control of the scientist, the intervals between various levels are usually equal or log equal; when the scientist chooses to monitor what nature does for him the levels are never equidistant.

Statistical analysis permits us to find out whether two independent stimuli are additive in effect or not. Additivity means that equidistant portions of a stimulus, e.g., temperature intervals of 3°C, have equidistant effects, and that when portions of the stimulus are added, corresponding portions of the response are added too. Additivity further implies that with two independent stimuli the effects of both on the response are added; two equidistant portions of one stimulus plus two of the other have twice the effect of one plus one portion. A look at the general

J. C. ZADOKS

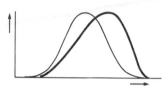

Fig. 3. The general shape of the optimum curve (heavy line) compared to that of the normal curve (thin line). Abscissa—stimulus, ordinate—response.

shape of the optimum curve (Fig. 3), irrespective of the stimulus, shows that in biology additivity must be the exception and its alternative, interaction, must be the rule. Interaction is measurable and it may vary from so small that the main effects in the ANOVA are clearly distinguishable, to so large that there is pure interaction without main effects; the latter happens in experiments on physiologic specialization. In epidemiological experimentation interaction is often the objective of the experimenter, main effects and, more rarely, additivity being only by-products. Interactions between two or three variables, described in the form of two-way or three-way tables with experimental data, can easily be handled in simulation models by entering the tables into the program.

C. Multiple Stimuli

Because in the outside world so many stimuli occur and interact simultaneously, experiments with a single stimulus have less explanatory value than experiments with two or three independent stimuli, in which interaction is assessed. An analogous reasoning can be developed for the responses. Consider an experiment on resistance in which we measure several responses (latent period, sporulation period, and sporulation intensity, for example) of different cultivars inoculated with the same isolate. By means of a set of consistent operational definitions we can produce independent measurements of interdependent responses. The responses are interdependent, as all three reflect resistance, though one perhaps more than the other. Modern statistics permits us to estimate resistance from its components, just as intelligence is "measured" by interrelating independent measurements of interdependent responses of the mind, assessed by means of psychological tests. It may be desirable to measure as many different responses as can be defined and assessed. There has been a development from one to many stimuli, and from one to many responses, or, to change terminology once more, from "single input–single output" to "multiple input–multiple output" experiments. Experiments may also be of the "single input–multiple output" or "multiple input–single output" type.

Economic considerations have stimulated this development. Climatization in growth chambers reduces variance, so that the number of plants and repetitions can be diminished and much of the unrewarding labor of planting and nursing can be eliminated. Many stimuli can be controlled by automation. The emphasis can be laid on measuring responses, which now becomes the most time consuming and the most rewarding work. Here, automation helps. Spore counting devices are an example, but instrumentation to support epidemiologic research is still underdeveloped. Inventive "gadgeteers" could render great service to epidemiology. The numerical results can be many and complex, but their analysis is no longer problematic thanks to modern customer-designed computer software. From the economic point of view, capital replaces labor. In sophisticated field experimentation, electronic monitoring of environmental variables as described by Pennypacker in Chapter 5 of this volume replaces climatization. The statistical techniques change, but the essence of the foregoing remarks remains the same.

D. Who Judges?

The development outlined here has its obvious merits, but it also has some disadvantages. In making experiments so explicit, the scientist has to choose his stimuli and their levels. He may choose the wrong ones. Who judges? As the stimuli are chosen on the basis of induction and/or conditioning, results are practically guaranteed: the H_0, implying no main effects, no interaction, or no correlation, will be rejected; the H_A will be accepted, and the size of the effect estimated. But is the observation relevant? Does it explain epidemiological phenomena in nature? The scientist producing the observation (in the methodological sense) is not always in the best position to judge; he may even be biased by his own result.

Now that we know the importance of leaf wetness periods, the relevance of most of the literature, say up to 1960, on relative humidity of the air is minor. The observations published before 1960 are practically obsolete, particularly because it is still not possible to predict leaf wetness periods from measured or calculated water vapor and energy fluxes (Goudriaan, 1977).

In addition, the more we condition the environment and the more we select stimuli and their levels, the more we run the risk of introducing new but unsuspected variables. For example, most growth chambers have day and night, without twilight. But the largest part of the world has a gradual change from day to night, and night to day. Sudden transitions from day to night or vice versa may affect the responses more than the chosen stimuli do, but who knows? Certainly, constant temperature during day and night, so often applied in older growth chamber work, causes abnormal conditions, with disproportionately high dissimilation during the night. Who knows the effect on disease responses? Anybody who has seen *Puccinia recondita* producing stripes with uredosori not unlike those of *P. striiformis* on adult wheat plants in a growth chamber would mistrust observations from growth chambers unverified by parallel observations in the field.

E. Situational Factors

By emphasizing stimuli or experimental factors, we run the risk of neglecting situational factors. These are all the factors, known and un-known, that surround the experiment but are not varied during the experiment. They may determine the response as much as the experimental factors applied at different levels. Situational factors operate at one level only. Therefore they are not interesting, it seems. Papers, which are very specific about experimental factors in their materials-and-methods section, may be practically silent on situational factors. Nevertheless, these determine the range within which the response varies. In the early phase of the European collaboration on race identification of *P. striiformis*, work was done in Brunswick, Germany, and in Wageningen, the Netherlands, both at about 52° northern latitude and about 5° longitude apart. Both stations used the same seeds and the same isolates but sometimes obtained, nevertheless, different results. This was due in part to a neglect of situational factors, such as light intensity, which happened to be markedly disparate. In addition, the individuals doing the work, their backgrounds, their training, their manual and observational skills were, of course, different. By mutual visits and intensive discussion many, but not all, of the differences could be eliminated. If friendly cooperation and excellent communication cannot succeed in getting around all differences in situational factors, then the present state of poor communication on situational factors must lead to chronic misunderstanding among scientists. It would help if every epidemiologist would use in his publications on research performed in growth chambers the extensive list of circumstantial factors recommended by the American Society of Horticultural Science (ASHS Committee, 1972).

XIII. PLAYING WITH TIME

Pathogens hurry to reproduce when conditions are favorable; they bide their time when the environment is adverse. Therefore, we must understand the importance of the time factor.

TABLE II

Levels of Integration

Level	Epidemiologic processes	τ
Molecular conversions		Seconds
Cellular phenomena		Minutes
Plant physiology		Hours
Recurrent infection cycle	Monocyclic	Days
Development of annual crops	Polycyclic	Weeks
Development of perennial crops	Polyetic	Years
Development of agroecosystems		Decades
Development of natural ecosystems		Centuries

The time scale of a process is characterized by the relaxation time, the time needed by a balanced system to recover from a sudden disturbance. More precisely, the relaxation time is the time τ needed to reduce the disturbance d to de^{-1} ($e = 2.71$). Table II gives a rough classification of biological processes. Epidemiology traditionally covers the area from cellular to population level, and measures time in hours, days, or weeks, with a few exceptions on both sides. For the forced ejection of ascospores τ due to air drag is a few milliseconds. In perennial crops attacked by soil-borne diseases the vegetation period can be interpreted as the rotation period, which varies from 10 to 100 years. Also, there is a new development: the study of epidemics in annual crops over a sequence of years.

A. Time Is Ambiguous

Time is an ambiguous word. If we follow primary school teaching devices and represent time as a time beam, the word time has three meanings: (1) the time beam itself, but of infinite length in the directions of both past and future; (2) a segment of the time beam with a finite length, also indicated by the word "period"; and (3) an infinitely short period, also indicated as instant or moment. The trend toward more precision in epidemiology is demonstrated by two well-known terms, the older "incubation time" and the younger "latent period." Though misunderstanding is rare, more precision and efficiency in communication could be obtained when the words time, period, and instant were used for the three meanings, respectively.

Until recently, epidemiologists have shown relatively little interest in

Fig. 4. Curve of spore germination with time (heavy line) compared to normal curve (thin line). Abscissa—time, ordinate—spore germination.

accurate timing (= determination of specific instants). Exceptions are, for example, the timing of infection days in disease warning systems against *Venturia inaequalis, Phytophthora infestans,* and others. Recent interest was focused rather on the determination of periods: germination, latent, and sporulation. When the experiment is well done, an interesting phenomenon appears.

The instant of germination of an uredospore (operationally defined as the instant at which the length of the germ tube equals the shortest diameter of the spore) is a stochastic variable. At first approximation, it shows a normal distribution around a mean value. This is not quite correct, however. The distribution is really truncated at the lower end and slightly skewed to the left so that it tapers out at the upper end (Fig. 4). If this is so and the standard deviation of the instant of germination is large relative to the period from inoculation to the mean instant of germination, what then is the germination period? And if not all pustules open simultaneously, what is the latent period? Obviously, germination and other periods are stochastic variables, with a mean and a standard deviation. But the common assumption with respect to sto-chastic variables, that values far from the mean occur with low fre-quencies—the further they are the less frequently they occur—does not hold without truncation. In epidemiological periods truncation, usually at the lower end, is obvious, but the instant of truncation is difficult to determine. Again it is a stochastic variable, its value depending on the frequency and number of observations. The wider implications of the fact that instants and periods of epidemiological interest are stochastic variables, with relatively high variance, have hardly been understood. Simulation studies demonstrate that the first open rust pustule can have far more effect than the hundreds of pustules that open 3 days later (Zadoks, 1971). Such is the relevance of studies of time in epidemiology. Present-day simulation techniques can deal adequately with stochastic periods, characterized by mean and variance, under the heading of dispersion (in time; de Wit and Goudriaan, 1974; see also Chapter 10, this volume).

B. Prolonged Sporulation

In seeming contradiction to the statement on the effect of the very first open pustule is the observation that sporulation periods can be so long. The teleological explanation has again been substantiated by means of simulation studies. A prolonged sporulation period is a survival mechanism under conditions adverse to infection. To put it more broadly, when adverse conditions are expected (that is, when the chance of infection is low either because of the absence of the host or of the necessary environmental conditions for infection), the fungus has two options: (1) to produce resting spores whose germination is triggered by the host plant within the limits of certain environmental constraints, or (2) to produce short-lived spores during a long sporulation period at a necessarily low sporulation rate. In the course of evolution, many fungi have opted for one of these alternatives. Rusts in the uredo phase developed toward option (2); interestingly, several tropical *Peronosporales* on cereals evolved toward option (1) (Safeeulla, 1976).

Epidemiology deals with processes at three levels. The first level is that of the single infection cycle, its phases, and subphases, dealing with the monocyclic process (Section X). The second level is that of the infection chain and deals with the polycyclic process. The third level is the gradual buildup and decline of epidemics through the years; this is the polyetic process (Zadoks, 1974).

C. Sequential Analysis

In studies of both mono- and polycyclic processes sequential analysis is useful; it is the quantitative analysis of a sequence of events by means of longitudinal and cross-sectional studies (Fig. 5). Longitudinal refers

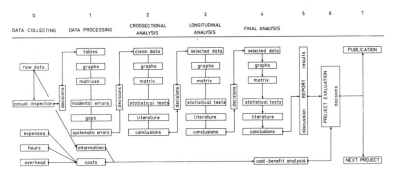

Fig. 5. Procedure in sequential analysis.

to observations at successive instants in time, preferably at equal intervals; cross-sectional refers to comparison of different levels (and replications) of an experimental factor at any instant. The sampling can be destructive or not. Destructive sampling has the advantage of examining the samples at ease in the laboratory and the disadvantage that large populations are needed to avoid disturbance of the process by taking samples. Nondestructive sampling using the same plants at every instant is often inconvenient, but it does permit sampling of relatively small populations and it reduces between-plant variance.

In the case of destructive sampling, samples taken at successive instants from the same population can be regarded as independent samples; this allows for ANOVA with time as an experimental factor and as many levels as there are sampling instants. ANOVA can be a good tool in sequential analysis, especially when the number of instants is low (say two to five). When nondestructive sampling is applied and observations are made at each instant from the same plots, plants, or leaves, successive samples cannot be regarded as independent, and ANOVA with time as an experimental factor is not permitted. Other suitable techniques are available. If the response versus time curve is monotonous and only two treatments are to be compared, a simple technique for longitudinal analysis is to determine the mean difference between the two sets of observations at every instant. These mean differences can be plotted against time, and a regression line or curve can be calculated. There is a significant difference between the two treatments if either the slope of the regression line deviates significantly from zero and/or the intercept of the regression line with the ordinate differs significantly from the origin (van der Wal et al., 1975). When there are more than two treatments, this technique can be used pairwise between treatments, an unsatisfactory procedure.

There are better means of reducing the number of data so that: (1) the individual observations are given their due weight, and (2) the large number of sequentially dependent observations is reduced to a small number of independent statistics. This can be done by adding to the existing information the assumption that the process follows some simple mathematical model. The procedure is easily visualized in the case where disease progress follows a sigmoid curve. Then the disease progress is often adequately described by the logistic equation. Other processes are better described by the cumulative normal curve, again a sigmoid curve. Each of these mathematical models permits the reduction of a large number of sequential data approximately following a sigmoid curve to two independent statistics, which can be visualized as the slope of the logit, respectively probit line, and the instant at which this line reaches the 50% value of the response. These statistics, of which the first is more typical for a longitudinal approach than the second, can be subjected to ANOVA.

When goodness-of-fit tests indicate that the model chosen does not really fit the observed data, more complex models can be tried to correct for the terminal level of the process and for asymmetry. For each item a new and independent statistic is added (Kranz, 1974; see also Chapter 3, this volume). The four statistics mentioned can still be interpreted biologically, but other statistics which have been used cause difficulties in interpretation. Statistical science provides techniques to estimate independent parameters from a multitude of sequential observations without a preconceived model. Such techniques as polynomial analysis have not yet been used in epidemiology and their discussion is beyond the scope of this paper.

The excursion into the realm of statistics gives rise to a warning message: epidemiologists can do a much better job in sequential analysis; they have plenty of good data, but have not put these to their best use. The only exception is the calculation of van der Plank's r values. The area indicated here seems to be very rewarding, because sequential data, showing so much scatter that they yield few significant differences when analyzed cross-sectionally, may show good interpretable differences when analyzed longitudinally (as unpublished experience has taught).

The foregoing techniques are analytical; they try to see significant differences between groups of responses that could be related to differences between stimuli. Dynamic simulation is synthetic because it permits the study of polycyclic processes when enough detail on the monocyclic process is available. In dynamic simulation, biological variables can be made dependent on driving functions representing the outside world, as the temperature of the ambient air, which affects the germination period, the latent period, and the sporulation rate, each in its own way. Simulation techniques are extensively described by Waggoner in Chapter 10 of this volume; it is enough to say that they add immensely to the understanding of the polycyclic process, but they have not yet given good forecasts of disease severities (Zadoks, 1978).

D. Polyetic Processes

Polyetic processes are processes of buildup and decline of epidemics through the years. They are particularly challenging as an object of study in annual crops. An epidemic rarely comes out of the blue; normally it builds up gradually from beginnings that escape observation. Typical situations have been described for *P. striiformis*, where a new race appears at least 1 and often 2 or more years before the actual destructive epidemic. What happens during these years of buildup, when the inoculum may already vary a millionfold or more within a year, but has not yet been observed? What happens when the new genotype of the pathogen is still subject to genetic drift? We do not know.

What happens when an agricultural system adapts to a new and introduced pathogen like *Peronospora tabacina*, which recently had nearly burnt out the European tobacco culture? Changes in cultural methods took place, hygienic measures were advocated, resistance was introduced, and fungicides were applied. But what were the contributions of individual measures to the final safeguarding of part of Europe's tobacco culture? If we knew, we might be able to utilize this knowledge in the future.

Phytopathologists concerned with soil-borne diseases have given much attention to this type of problem, though usually without the degree of quantification which the epidemiologist would like to see. It is time to undertake similar studies in foliar pathogens. A few hesitant starts have been made. The carryover of inoculum of *P. striiformis* has been simulated by Rijsdijk (1975). The increase in frequency of new virulences in *Pyricularia oryzae* has been simulated by Kiyosawa and Shiyomi (1976). Here lies a promising field for future research, where epidemiologists can probably learn much more from entomologists and nematologists.

XIV. ANALYSIS AND SYNTHESIS

Since the Renaissance the natural sciences have tried to solve their problems by reducing them to smaller problems, often at lower levels of integration, that is, with smaller τ. The approach has been called reductionism. The procedure followed is analysis; from epidemic to infection cycle, from this, over phases, to subphases, and so on. There is no end to analysis, since every answer leads to new questions. The procedure can be so fascinating that the entranced researcher may deviate from real-life problems and forget the original question he began with. However, analysis is meaningful only when the small bits and pieces of knowledge can be tied together in an orderly way; this is synthesis. Synthesis has, of course, always been pursued, but until recently it had to be a qualitative, verbal type of activity. Quantitative synthesis became possible only with the advent of simulation techniques. Epidemiology must be brought into effect by means of disease management. To this end, analysis and synthesis should go hand in hand (Fig. 6).

History has seen several periods when philosophical trends accentu-

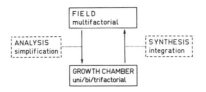

Fig. 6. Analysis and synthesis.

ated a feeling of wholeness in science, holism, implying that the whole was far more than the sum of the parts. Zadoks and Koster (1976) recognized three such periods, that of romanticism in the first half of the nineteenth century, another one of dawning internationalism about 1900, and a third from the late 1950's until today with a nearly super-stitious reverence for ecology. Some people go so far as to claim that the whole cannot be known from its parts; science follows the wrong track, they say. As a scientific approach this attitude is called expansionism, the opposite the reductionism. Present expansionistic theories tend to get entangled in words without becoming operational. It is hard to tell whether this is an inherent trait or only a temporary situation.

The very thesis that the only good scientific statement is a falsifiable statement implies that scientific knowledge is short lived. A fact, once well established, may fade in the light of later knowledge, in part because of the unavoidable subjectivity of the fact finder. Knowledge rapidly becomes obsolete, but present-day knowledge, itself based on earlier knowledge, will be a foundation of future knowledge. Empirical knowledge is based on experiments. Methodology, as streamlined experience, is derived or second-degree empirical knowledge. It is like the road sign, which becomes meaningful only when there is a road. When methodology has collected and streamlined enough experience, it becomes prescriptive, normative. The prescriptions are not needed when things go well, but the scientist will examine the methodological foundations of his activities when things have gone awry. Then he will feel strengthened by methodology, and, reporting his considerations, he will strengthen methodology.

Acknowledgment

The author thanks Mrs. F. Daendels-Wilson for her critical reading of the English text.

References

Anonymous. (1751). Sur les maladies épidémiques observées à Paris en 1746, en même temps que les différentes températures de l'air. *Hist. Acad. R. Sci., Paris, 1746*, pp. 22–23.

ASHS Committee on Growth Chamber Environments. (1972). Guidelines for reporting studies conducted in controlled environment chambers. *HortScience* **7**, 239.

Brook, P. J. (1969). Stimulation of ascospore release in *Venturia inaequalis* by far red light. *Nature (London)* **222**, 390–392.

Chiarappa, L. (ed.) (1971). "Crop Loss Assessment Methods. FAO Manual on the Evaluation and Prevention of Losses by Pests, Diseases and Weeds." Comm. Agric. Bur., Farnham, England (loose leaved).

de Bary, A. (1861). "Die gegenwärtig herrschende Kartoffelkrankheit, ihre Ursache und ihre Verhütung. Eine pflanzenphysiologische Untersuchung in allgemein verständlicher Form dargestellt." Felix, Leipzig.

de Bary, A. (1865). Neue Untersuchungen über die Uredineen, insbesondere die Entwicklung des *Puccinia graminis* und den Zusammenhang derselben mit *Aecidium berberidis*. *Monatsber. K. Preuss. Acad. Wiss.* pp. 15–49.

de Bary, A. (1969). Investigations of the brand fungi and the diseases of plants caused by them with reference to grain and other useful plants. *Phytopathol. Classics* **11**, 1–93.

de Groot, A. D. (1969). "Methodology, Foundation of Inference and Research in the Behavioural Sciences." Mouton, The Hague.

de Wit, C. T., and Goudriaan, J. (1974). "Simulation of Ecological Processes." PUDOC, Wageningen.

Duhamel de Monceau, H. L. (1728). Explication physique d'une maladie qui fait périr plusieurs plantes dans le Gastinois, et particulièrement le safran. *Mém. Acad. Sci. Math. Phys., Paris*, pp. 100–112.

Fischer, E., and Gäumann, E. (1929). "Biologie der Pflanzenbewohnenden parasitischen Pilze." Fischer, Jena.

Fontana, F. (1932). Observations on the rust of grain. *Phytopathol. Classics* **2**, 1–40.

Gäumann, E. (1946). "Pflanzliche Infektionslehre." Birkhaeuser, Basel.

Geurts, J. P. M. (1976). "Feit en theorie. Inleiding tot de wetenschapsleer." Van Gorcum, Assen, Amsterdam.

Goudriaan, J. (1977). "Crop Micrometeorology: A Simulation Study." PUDOC, Wageningen.

Gregory, C. E. (1967). "The Management of Intelligence. Scientific Problem Solving and Creativity." McGraw-Hill, New York.

Heitefuss, R., and Dehne, D. (1976). Investigation on components of resistance of wheat against *Puccinia striiformis*. *Proc. Eur. Medit. Cereal Rusts Conf., 4th, 1976* p. 48.

Hill, J. P., and Nelson, R. R. (1976). Ecological races of *Helminthosporium maydis* Race T. *Phytopathology* **66**, 873–876.

Hirst, J. M., and Schein, R. D. (1965). Terminology of infection processes. *Phytopathology* **55**, 1157.

Horsfall, J. G., and Barratt, R. W. (1945). An improved grading system for measuring plant diseases. *Phytopathology* **35**, 655.

James, W. C., Shih, C. S., Hodgson, W. A., and Callbeck, L. C. (1976). Representational errors due to interplot interference in field experiments with late blight of potato. *Phytopathology* **66**, 695–700.

Johnson, R. (1972). Minor genetic variations for virulence in isolates of *Puccinia striiformis*. *Proc. Eur. Medit. Cereal Rusts Conf., 3rd, 1972* pp. 141–144.

Katsuya, K., and Green, G. J. (1967). Reproductive potentials of races 15B and 56 of wheat stem rust. *Can. J. Bot.* **45**, 1077–1091.

Kiyosawa, S., and Shiyomi, M. (1976). Simulation of the process of breakdown of disease-resistant varieties. *Jpn. J. Breed.* **26**, 339–352.

Koch, R. (1891). Ueber bakteriologische Forschung. *Verh. Int. Med. Kongr., 10th, 1890* pp. 35–47.

Kranz, J., ed. (1974). "Epidemics of Plant Diseases. Mathematical Analysis and Modeling." *Ecol. Stud.* No. 13. Springer-Verlag, Berlin and New York.

Lindgren, B. W. (1975). "Basic Ideas of Statistics." Macmillan, New York.

Lutman, B. F. (1911). Twenty years' spraying for potato diseases. Potato diseases and the weather. *Vt., Agric. Exp. Stn., Bull.* **159**, 215–296.

MacFadyen, A. (1975). Some thoughts on the behaviour of ecologists. *J. Ecol.* **63**, 379–391.

Mackenzie, D. R. (1978). The multiline approach in controlling some cereal diseases. *Proc. Rice Blast Workshop, Int. Rice Res. Inst.* (in press).

McNeal, F. H., Konzak, C. F., Smith, E. P., Tate, W. S., and Russell, T. S. (1971). A uniform system for recording and processing cereal research data. *U.S., Dep. Agric., Agric. Res. Serv.* ARS 34–121, 1–42.

Parlevliet, J. E., and van Ommeren, A. (1975). Partial resistance of barley to leaf rust, *Puccinia hordei.* II. Relationship between field trials, microplot tests and latent period. *Euphytica* **24**, 293–303.

Parlevliet, J. E., and Zadoks, J. C. (1977). The integrated concept of disease resistance; a new view including horizontal and vertical resistance in plants. *Euphytica* **26**, 5–21.

Popper, K. R. (1960). "The Logic of Scientific Discovery," 2nd imp. Hutchinson, London.

Prévost, B. (1939). Memoir on the immediate cause of bunt or smut of wheat, and of several other diseases of plants, and on preventives of bunt. *Phytopathol. Classics* **6**, 1–95.

Rijsdijk, F. H. (1975). A simulator of yellow rust on wheat. *Bull. Rech. Agron. Gembloux* pp. 409–417.

Robinson, R. A. (1971). Vertical resistance. *Rev. Plant Pathol.* **50**, 233–239.

Robinson, R. A. (1976). "Plant Pathosystems." Springer-Verlag, Berlin and New York.

Russell, G. E. (1972). Components of resistance to diseases in sugar beet. *In* "The Way Ahead in Plant Breeding" (F. G. H. Lupton *et al.*, eds.), Cambridge pp. 99–107.

Safeeulla, K. M. (1976). "Biology and Control of the Downy Mildews of Pearl Millet, Sorghum and Finger Millet." Downy Mildew Res. Lab., Mysore.

Schröder, J., and Hassebrauk, K. (1964). Untersuchungen über die Keimung der Uredosporen des Gelbrostes (*Puccinia striiformis* West.). *Zentralbl. Bakteriol., Parasitenkd., Infektionskr. Hyg., Abt. 2* **118**, 622–657.

Shrum, R. (1975). Simulation of wheat stripe rust (*Puccinia striiformis* West.) using EPIDEMIC, a flexible plant disease simulator. *Pa. State Univ. Coll. Agric., Agric. Exp. Sta., Prog. Rep.* **347**, 1–41.

Targioni Tozzetti, G. (1952). True nature, causes and sad effects of the rust, the bunt, the smut, and other maladies of wheat, and of oats in the field. *Phytopathol. Classics* **9**, 1–159.

Tillet, M. (1937). Dissertation on the cause of the corruption and smutting of the kernels of wheat in the head and on the means of preventing these untoward circumstances. *Phytopathol. Classics* **5**, 1–191.

Tull, J., and Duhamel de Monceau, H. L. (1764). De Roest. *In* "De nieuwe wijze van landbouwen," Vol. III, pp. 2–8. Houttuyn, Amsterdam.

Unger, F. (1833). "Die Exantheme der Pflanzen." Gerold, Wien.

van der Plank, J. E. (1963). "Plant Diseases: Epidemics and Control." Academic Press, New York.

van der Plank, J. E. (1968). "Disease Resistance in Plants." Academic Press, New York.

van der Wal, A. F., Smeitink, H., and Maan, G. C. (1975). An ecophysiological approach to crop losses, exemplified in the system wheat, rust, and glume blotch. III. Effects of soil-water potential on development, growth, transpiration of leaf rust-infected wheat. *Neth. J. Plant Pathol.* **81**, 1–13.

van Everdingen, E. (1926). Het verband tusschen de weersgesteldheid en de aardappelziekte (*Phytophthora infestans*). *Tijdschr. Plantenziekten* **32**, 129–140.

Zadoks, J. C. (1961). Yellow rust on wheat, studies in epidemiology and physiologic specialization. *Neth. J. Plant Pathol.* **67**, 69–256.

Zadoks, J. C. (1971). Systems analysis and the dynamics of epidemics. *Phytopathology* **61**, 600–610.

Zadoks, J. C. (1972a). Methodology of epidemiological research. *Annu. Rev. Phytopathol.* **10**, 253–276.

Zadoks, J. C. (1972b). Modern concepts of disease resistance in cereals. *In* "The Way Ahead in Plant Breeding" (F. G. H. Lupton *et al.*, eds), Cambridge, pp. 89–98.

Zadoks, J. C. (1972c). Reflexions on disease resistance in annual crops. *U.S., Dep. Agric., For. Serv., Misc. Publ.* **1221**, 43–63.

Zadoks, J. C. (1974). The role of epidemiology in modern phytopathology. *Phytopathology* **64**, 918–923.

Zadoks, J. C. (1977). On the epidemiological evaluation of fungicide action. *Neth. J. Plant Pathol.* **83**, *Suppl. 1*, 417–426.

Zadoks, J. C. (1978). Simulation of epidemics: Problems and applications. *Eur. Plant Prot. Org. Bull.* (in press).

Zadoks, J. C., and Koster, L. M. (1976). A historical survey of botanical epidemiology. A sketch of the development of ideas in ecological phytopathology. *Meded Landbouwhogesch. Wageningen V* 76–12, 1–56.

Zadoks, J. C., Chang, T. T., and Konzak, C. F. (1974). A decimal code for the growth stages of cereals. *Eucarpia Bull.* **7**, 1–10.

Chapter 5

Instrumentation for Epidemiology

S. P. PENNYPACKER

I. INTRODUCTION

Lord Kelvin's famous dictum says that our ability to understand a phenomenon is proportional to our ability to measure it. This volume is devoted to understanding epidemiology. This chapter describes the measurement of the weather factors in epidemiology as an aid to understanding epidemiology.

The essence of the art of research is to frame the question we ask of nature in such a way that the reply will come back on a language we can understand. We trust that this chapter will contribute to our ability to frame the questions and understand the answers.

II. THE IMPORTANCE OF MEASUREMENT

Natural environments and their biological components are often described in qualitative terms. For example, these terms may include the causal agents of specific diseases and geographical regions of occurrence. For some time yet there will continue to be a need for qualitative descriptions of certain relationships and phenomena.

On the other hand, to analyze disease development rather than disease occurrence we must express our observations on a quantitative basis. The value of data which characterize an epidemic and its meteorological parameters must be judged on the basis of its accuracy of measurement. The term "measurement" refers to the extraction of physical or chemical signals from each parameter of the system under consideration. The primary purpose of utilizing some type of measuring device to obtain such signals is to gain more and better information about a given process than is possible through human intuition alone. The evaluation of disease control techniques, environmental variables, etc., can never exceed the limitations of the associated measuring devices.

Recent advances in instrument design presently allow us to obtain relatively accurate measurements of parameters in the field as well as in the laboratory. The actual selection of appropriate measuring devices for both basic and applied research should be based on the significance of the measurement, required accuracy, environment to which the device will be subjected, and cost. In selecting sensors and recording systems to provide the data necessary to meet the objectives of a investigation, one should also consider the accuracy, hysteresis, repeatability, reproducibility, resolution, response time, and sensitivity of the measuring system.

The output of the measuring device is relative to a previously defined reference and represents the state of the process being investigated. When making measurements we must be careful to minimize any effect of the sensor on the condition being measured. A well-known example of this principle is to measure temperature in minute locations with a relatively large sensor of high heat capacity or a sensing element which possesses a self-heating effect. The purpose of measurement is to obtain the true value of a quantity. An estimate of this value is obtained by comparing the measured signal to a reference. The actual or true value would be obtained by an instrument having no systematic or random errors. In practice, errors are present, but they may be reduced through proper use of the instrumentation network and by compensation. Absolute accuracy is unobtainable because of the human factor. Therefore, the reference values used in the measurement of most variables are only traceable to those existing at the National Bureau of Standards.

Many reports of biological response to meteorological parameters are of little use because the investigator fails to report the exact conditions under which the research was conducted. In order to help understand the entire system, it is important to have knowledge of all possible factors. The trend to utilize results reported in the literature for the formation of disease forecasting systems and computer simulators of plant disease further reveals the need to know the accuracy and repeatability of the investigator's measurement system. This is of importance when one utilizes deterministic models to describe a response and/or establishes what he believes are the critical limits for a response. For example, the temperature range between the optimum and maximum for a biological process may be as small as 2–8 C°. Similarly, some processes may occur within a narrow range at the upper end of the relative humidity scale or be triggered at rather discrete levels of moisture. Unless the instruments are dependable in this range, the data are undependable.

The accuracy and sensitivity of mathematical models used to describe measured responses should dictate the degree of accuracy and repeatability of measurements needed for future use of the models. Epidemiological studies have seldom been approached in this manner. Temperature relationships derived for cultures of organisms under artificial conditions may be in error by 2–5 C°. Although these values may appear small, a 5 C° error in temperature measurement may cause a 10- to 14-day shift in a disease progress curve produced by a computer simulator of a plant disease utilizing deterministic models.

Therefore, it is imperative that we use sound technical judgment in designing studies so we obtain valid and meaningful measurements of both the response and the environment. The sensors must be placed at locations in the environment that are truly representative of the area or process of concern. This is important in both field and laboratory studies especially where conditions such as stratification and shading can cause significant differences among locations. There are many studies in which we must in fact carefully select a location for measurement and acknowledge that it may be somewhat less than truly representative of the biological process under study.

Meteorological parameters associated with plant disease and response of specific life stages of causal agents include radiation, temperature, moisture, and wind. The following will deal with the measurement of these variables.

III. MEASURING TEMPERATURE

Temperature is the meteorological parameter most often correlated with biological responses. It is the factor most easily estimated and that which may best be controlled under artificial conditions.

Temperature is a measure of the degree of thermal equilibrium that exists between two bodies. It is not a flux measurement, but rather a measure of the intensity of a body's heat. Temperature gradients exist in air, soil, and host material under both natural and artificial growing conditions. Therefore, temperature data depend on the measurement of a representative temperature. The signal from the sensor is actually a function of the sensor temperature itself. Therefore, the sensor must come into thermal equilibrium with the medium of unknown temperature. The time required to approach this condition is dependent upon the desired accuracy and time lag in the sensing device. At equilibrium, the heat transferred to the sensor equals the heat transferred away. An error is introduced if heat transfer by radiation and conduction are not minimized.

The continuously indicating temperature sensors commonly used in epidemiological studies include liquid-filled, Bourdon-tube, and bimetallic thermometers.

A. Liquid-in-Glass Thermometers

Liquid-in-glass thermometers are probably the most widely used temperature sensors. Simplicity and low cost make this sensor attractive where manual readings are sufficient to meet the objectives of the investigation. Manufactured in a wide range of scales with various degrees of readability, these thermometers may be obtained with accuracies of ca. ±0.05 C°. Errors may be introduced into the readings, however, by poor readability of the small liquid column, parallax, and radiation. A significant error may also be introduced by using an improper liquid-in-glass thermometer from among the partial, total, and complete immersion types available. Measurement errors of ½ to 1 C° or greater may be acceptable for characterizing certain environments, but when other parameters, such as relative humidity and heat flux, are estimated as a function of temperature, errors of this magnitude may be totally unacceptable.

B. Bimetallic Thermometers

The sensing element of a bimetallic thermometer is formed by firmly bonding together two metals with different thermal coefficients of expansion. The resulting linear deflection with temperature change provides a rugged, low-cost sensor with an accuracy of ca. ±1 C°. Because of its long lag time, this thermometer is of limited value in monitoring rapidly fluctuating environments. The size and principle of operation of the bimetallic thermometer limits its use to obtaining average temperature measurements within thermograph type recorders and controlled environment chambers.

C. Filled System Thermometers

The need to record temperature continuously and remotely requires the use of sensors to transmit signals that are proportional to the temperature. These signal-carrying devices may be either flexible tubes or electrical circuits. Filled system thermometers are normally all-metal assemblies consisting of a bulb, capillary tube, and a Bourdon tube containing a temperature-sensitive medium. Most sensors of this type are utilized in monitoring controlled environment temperatures and a few are used to monitor soil temperature via a thermograph. Since the sensor responds to pressure changes, the system requires very stiff capillary tubes to avoid distortion in transmission of the signal. Although the system is quite simple in structure and produces a linear reading with an accuracy of ca. ± 1 C°, the rather stiff tubes restrict its applicability to limited environments.

D. Thermocouple Thermometers

Practical consideration fairly well restricts the choice of sensor for remote monitoring to thermocouples, resistance thermometers, and thermistors.

A thermocouple thermometer is a temperature measuring system composed of a sensing element, i.e., the thermocouple, electrical conductors, and a current or voltage measuring device. The sensor is formed by welding together two ends of dissimilar metals. A measurable electrical signal develops when a temperature difference exists between the welded junction and the other two ends. The current flow in the circuit is a function of the temperature difference and therefore provides a relative measure. A direct temperature measurement requires that one junction be maintained at a reference temperature. The standard ice bath reference temperature has been replaced in field and many laboratory installations by an electrical type compensator. This device is well received by investigators who must record data over a long time period.

The signal output from the sensor can be measured by a direct current (dc) galvanometer. This measuring system, which basically measures current, is quite simple and ideal for accurate measurement of temperature differences in the laboratory. However, since current is inversely proportional to the circuit resistance, temperature measurement from sensors with long extensions is best obtained via a null balance potentiometer.

Of the many base wire combinations that could be utilized, copper and constantan possess favorable physical and chemical properties for microclimatic investigations of plant–pathogen environments. Termed "type-T" junctions, they are used in mildly oxidizing and reducing atmospheres and are also suitable for use in moist environments. The homogeneity of these two metals is maintained better than that of other wires and commercially produced uncalibrated wire provides an accuracy of ca. ± 1 C°. Yielding an output of approximately 40 μV/C°, the sensors can be calibrated with a given recording system to provide an overall system accuracy of ca. ± 0.1 C°. These type-T sensors are available in wire diameters as small as 0.0254 mm. For construction of soil temperature sensors and all extension leads, I find 0.5106 mm quite suitable when covered with extruded polyvinyl insulation. The flexibility of this wire is a distinct advantage in almost all installations. We must remember, however, that excessive mechanical flexing of the wire may alter calibration of the system.

In addition to errors that result from radiation, ventilation etc., electrical noise may introduce serious error into thermocouple circuits. This error is primarily caused by adjacent wires carrying alternating current (ac) or rapidly varying dc signals. This problem is likely to be encountered if the same data acquisition system is used to record both ac and dc signals. Serious errors may also be introduced if the thermocouple is installed in areas subjected to inductive electrical fields such as electric motors, electric heaters, and fluorescent lighting. Sensors constructed of ferrous materials are more sensitive to these sources of interference than nonferrous sensors. This is another reason why type-T junctions are the most effective for our work. Proper shielding and grounding will minimize measurement errors arising from electrical noise. Common mode voltage errors of significant magnitude may also affect our measurements if the sensor and recorder are grounded at two or more locations which are at different ground potentials. Proper grounding may also alleviate this source of error.

Air temperature measurements may easily be obtained under field and laboratory conditions by using very fine (ca. 0.2546 mm) bare wire, loopshaped thermocouples. This design minimizes possible radiation errors because the convective heat loss of the fine wire will equal or exceed the radiation load on the sensor. This feature thereby alleviates the need for radiation shielding and forced ventilation which otherwise might disrupt the process under investigation. Almost all installations, except in soil, may be made with bare thermocouple junctions. The elimination of a thermowall produces a small sensor with a very short time lag.

Thermocouple sensors used for soil temperature measurement over any length of time should be coated with a material which provides good electrical insulation and good thermal conduction. It is important that the lead wire from the soil sensor be buried at a depth equal to that of the measuring junction to minimize errors due to heat conduction along the wires. It is also important that a portion of the buried lead wire be deeper than the measuring junction to eliminate water accumulation at the point of measurement.

E. Resistance Thermometers

1. Resistor Type

The operating principle of resistance thermometers is that the electrical resistance of metals varies directly with temperature. It is important for the metals to resist corrosion and to possess a high temperature coefficient of resistance. The most sensitive elements are those that have a very high coefficient of resistance. The calibration of some early manufactured sensors might sometimes shift, but high-purity, stable metals possessing ideal resistance to temperature relationships are now available. The sensing element consists of a thin wire, usually platinum, nickel, or tungsten, coiled around a mandrel composed of an electrical insulating material.

The sensing area of the probe depends on the circumference of the mandrel and length of the winding. Ranging from ca. 0.5 to 6 cm in length, they sense the average temperature of a much larger area and have significantly slower response times than do thermocouples. The metals normally used as resistance devices have sensitivities of 0.2 to 10 ohms/C°. A measuring system with this type of sensor may be designed to provide an output signal in the order of 1 to 6 V, a distinct advantage over the often favored thermocouple.

A factor which must be considered in the use of resistance devices is the generated heat which develops from the current applied to produce an output signal. This "self-heating" effect may in fact significantly change the temperature of the environment being measured. Because they are less expensive, copper extension leads are normally used for signal transmission from the specialized wire sensor to the recorder. The inherent resistance of long-distance transmission in conjunction with ambient temperature effects is a potential source of error.

2. Thermistor Type

Thermistors are thermally sensitive resistors formed from mixtures of metallic oxides. They differ from the previously described resistance temperature devices by being semiconductors with large negative coefficients and exhibiting nonlinear resistance-to-temperature relationships. The large change in resistance provides a resolution which greatly exceeds that of thermocouples or resistance thermometers and allows for measurement of temperature differences of ca. 0.0005 C°.

Manufactured in various sizes and shapes, thermistors can be obtained to conform to most all applications. Their high electrical resistivity is also of great advantage for remote measurement, because the resistance of long extension wires is negligible in comparison to the thermistor. Under field conditions, the repeatability is such that differences in repeated measurements are usually smaller than the overall accuracy of the measuring system. In general, thermistors and resistance thermometers possess relatively high absolute accuracies. Whereas commercial grade thermocouples are normally accurate to ca. ±1 C°, thermistors and resistance thermometers provide accuracies of ±0.01 C° or better.

Fine wire thermocouples, very small thermistors, and resistance bulbs, as well as liquid-in-glass thermometers, have been utilized to estimate media surface temperatures. These sensors have in fact been attached to, and inserted into, plant material in order to measure the temperature to which causal agents of disease are exposed. Such installation actually disrupts the tissue and modifies the environment under which the disease reaction occurs. A more realistic approach to obtain surface measurements is to use noncontact temperature sensors.

F. Radiation Thermometers

Radiation pyrometry allows us to measure temperature without establishing physical contact with the object under investigation. The principle of operation is based on a physical law (Stefan–Boltzmann) which defines the energy emitted by the object when at a given temperature. Accurate temperature measurements require modification of the theoretical law to account for surface emissivity, atmospheric absorption, and spectral response of the detector.

The problem of calibrating the pyrometer for accurate sensing of emittance is eliminated in ratio pyrometers. These systems measure the radiation at two different wavelengths. The ratio of these measurements is a function of temperature. This is the most accurate means of measuring the true surface temperature when the emissivity of the object is constant or only slowly changing.

In general, a radiation thermometer consists of a lens and/or mirror system, a detector which compares the incident flux to a standard, and an indicator or recorder. As such, it is a relatively large, heavy, delicate instrument which is not normally designed for unattended operation. They do, however, commonly possess good accuracy and sensitivity along with fast response which are additional, attractive characteristics for intensive investigations centered on surface temperature-response relationships.

IV. MEASURING HUMIDITY

Temperature measurement with thermometers can be made more accurate than necessary. In contrast, instruments for measuring humidity are seldom as precise or accurate as our needs require. In fact, moisture measurements are usually made by secondary instruments which respond to humidity-related phenomena. The instruments commonly used for measuring water vapor in a gaseous medium may be broadly classified as absolute moisture, relative humidity, and dewpoint type hygrometers.

A. Absolute Moisture Measurement

The most direct means to measure the water content of a medium is to compare its dry weight with its weight after water has been absorbed. This gravimetric method is used by the National Bureau of Standards and is accepted as the most accurate way of determining the amount of water vapor in a gas. In practice, a measured volume of atmosphere is passed through a moisture-absorbing chemical and its increase in weight is measured. Under controlled conditions this method is more accurate than any other system and proper technique will yield an accuracy of $\pm$ 0.2% over the range of 0.1 to 100%.

Applicable primarily to constant laboratory conditions, the system is neither simple nor rapid. Its noncontinuous output does not lend itself to transmission from a remote sensor. Therefore, we must usually relinquish this high accuracy and accept a system that is somewhat less desirable in this respect.

B. Relative Humidity Hygrometers

1. The Thermodynamic Approach

The psychrometric method for estimating atmospheric humidity is often used in micrometeorological investigations. The basic simplicity of operation has made the psychrometer the most frequently used instrument for measuring the moisture content of the atmosphere. The sensing device normally consists of two mercury-in-glass thermometers, one of which possesses a wet, muslin-covered bulb located in a forced air stream of at least 2.5 m/sec. The energy that evaporates water from the muslin causes the wet bulb temperature to decrease until equilibrium is reached.

Although no theory completely describes this action, experimental tables and charts are available to convert the wet bulb and ambient dry bulb temperatures to such expressions as relative humidity, dewpoint temperature, absolute humidity, and vapor pressure.

The psychrometer is a local indicating device that is quite acceptable for many laboratory studies. It can also be used to calibrate other secondary sensors. Modifications may be made to allow somewhat unattended operation of this system. These alterations usually include substitution of electrical temperature sensors for the glass thermometers. The validity of such transmitted data is, however, dependent upon maintenance of a moist, clean, muslin wick in adequate ventilation.

The accuracy of the thermodynamic approach depends upon the temperature sensors and the care utilized in carrying out proper procedures for measurement by this method. It is interesting to note that all errors, except for those inherent in the temperature sensors, result in an increase in observed relative humidity. Errors are commonly introduced by insufficient ventilation, improper wick material, dirty muslin, and impure water. A majority of these causes are extremely difficult to avoid in field studies and the use of very sensitive thermometers does not alleviate the need for forced ventilation unless the system is recalibrated and the investigator does not use the standard reference tables.

If epidemiological studies reveal that a biological response is triggered at a given moisture level or only occurs within a specified range of moisture, it is important to strive for a degree of accuracy well within these limits. Few investigators acknowledge that estimates of relative humidity obtained with thermometers readable to 1 C° can indicate a humidity anywhere between 46 and 88% when the true condition is 66% at 10°C. Therefore, it is important that two well-matched, certified thermometers, readable to 0.1 C° be used in the thermodynamic approach to obtain the most accurate and reliable measurements.

2. Hair Hygrometers

Mechanical sensors operate on the principle of the linear expansion and contraction of hygroscopic materials with change in moisture. Of the many materials which exhibit this dimensional change, human hair is probably the most widely used sensing element for measuring humidity.

The amount of water absorbed and desorbed by hair is a function of relative humidity and therefore does not provide a measure of the actual amount of water vapor in the atmosphere. A given length of human hair increases only about 2.5% in length with relative humidity increasing from 0 to 100%. This small change in length is logarithmic and therefore requires some form of mechanical amplification and linearization for recording purposes. The 0 to 100% scale on hair hygrometers is somewhat deceiving because the recognized elongation is primarily in the range of 15 to 85%. If the system is calibrated within this range, exposure to higher or lower humidities may cause a permanent shift in calibration.

When used in plant disease research, hair hygrometers must be calibrated within the range of primary concern. The calibrations may best be accomplished in controlled conditions under constant temperature and humidity levels. When used in field work it is beneficial to check their accuracy periodically against a portable (Assmann-type) psychrometer. This is best done on cloudy days when the temperature is nearly constant. The repeatability of the hair sensor is ca. 3% if maintained at room temperature until equilibrium is reached. Since it has a time lag of ca. 5 min. and a hysteresis of ca. 3%, it is not very suitable for measurement of rapidly fluctuating environments.

3. Electrical Resistance Hygrometers

Several types of electrical resistance sensors are available for measuring ambient humidity. In general, these sensors utilize a relationship between the variation in electrical properties of a hygroscopic material and the change in water content. The sensors are commonly composed of a thin film of lithium chloride spread over intermeshed electrodes on the surface of a flat or cylindrical polystyrene base. The element absorbs and desorbs moisture until equilibrium is reached with the ambient environment. These moisture changes are reflected in the impedance which can be monitored with an ac bridge circuit. Since the electrical resistance is temperature dependent, the readings must be adjusted. This correction for temperature sensitivity is approximately 0.3% RH/C° from its calibrated reference temperature. (RH = relative humidity.)

Sensors of this type may be calibrated over the range of 5 to 95% RH with an accuracy of ca. ± 3%. In an electrical circuit these elements may be used for continuous measurement of humidity at remote locations. That these sensors do not require forced ventilation is an advantage for epidemiological investigations in which unnatural air movement would influence the process under study.

My primary dissatisfaction with this type of element lies in the apparent instability of its calibration. When used for continuous monitoring in field conditions, I note that a significant change in calibration occurs in a majority of the sensors during the growing season. Since a radiation shield also offers protection from rain, this shift or lack of sensor response may be due to contamination with airborne particulate matter. The required ac circuitry for moisture measurement also mandates careful shielding and grounding of any low-level dc signals which are concurrently being monitored.

C. Dewpoint Type Hygrometers

1. Condensation Type

The humidity of the atmosphere may be directly measured by the dewpoint technique. This is a primary measure of the absolute moisture content which may be used in applications requiring high accuracy over a wide range of humidities. The dewpoint temperature is obtained by simply measuring the surface temperature of a mirror which has been cooled to the degree required for condensation to occur. This temperature corresponds to the saturation partial pressure of the water vapor. The actual dewpoint temperature–vapor pressure relationships may be obtained from standard tables.

Optical sensors, rather than visual inspection, are used to detect dew formation in the more sophisticated devices currently available. Used in conjunction with thermoelectric heating and cooling, the system may now be used remotely in an unattended mode. Although the system possesses several desirable features for epidemiological studies, its present design limits it to monitoring at only one point location from which the air must be sampled and bought to the sensor. Furthermore, high cost almost prohibits use of multiple sensors that would be required to monitor vertical or spatial profiles.

2. Saturated Lithium Chloride

An alternate approach to estimating the dewpoint temperature is to replace the mirror with an electrical sensor whose output directly changes with moisture. The sensor normally consists of bifilar wires wound on a tubelike structure impregnated with lithium chloride. An ac current is conducted through the lithium chloride coating between the two nonconnected wires. The circuitry is so designed that the current heats the lithium chloride to the degree required to maintain vapor pressure equilibrium between the salt and ambient atmosphere. When equilibrium is established, the lithium chloride temperature is linearly related to the ambient dewpoint temperature.

Sensors of this type provide a means for remote monitoring of the dewpoint with accuracies of ca. ± 1 C°. This sensor is less accurate than the condensation "mirror" type, and its range is limited to a moisture content in the air greater than approximately 35% relative humidity. This latter disadvantage may not, however, be a serious limitation in disease investigations. Although sensors of this type possess several attractive features, they still do not provide the truly ideal system researchers desire. These sensors may also be used for remote monitoring in the field but, because they are easily contaminated and become unstable when exposed to rapidly changing moisture, they are more suited for controlled environment studies than field investigations.

V. MEASURING AIR MOVEMENT

Wind is the horizontal motion of air relative to the earth's surface and its rate of motion is termed "wind speed." The other important characteristic of this meteorological parameter is the direction from which it approaches the location of concern.

Measurements of wind speed and direction provide information on two of the variables known to be associated with the takeoff, flight, and landing of airborne pathogens. (See also Chapter 8, this volume, on airborne dissemination.) The horizontal surface shear stress to which attached spores, mycelia, etc., are subjected is a function of the frictional velocity. Deposition of these particles is also a function of wind speed and frictional velocity as well as of the dimensional and surface characteristics of the objects upon which propagules are impacted. The associated relationships are most easily estimated from wind tunnel studies. Detailed field studies which include the measurement of all relevant meteorological parameters, spore trapping data, and documented surface characteristics of the host crop may, in conjunction with wind profile theory, verify relationships derived under laboratory conditions. Atmospheric turbulence, diffusion, and convection, as well as wind direction, are additional variables which must be considered in studies related to the transport of airborne organisms.

A. Wind Speed

1. Cup Anemometers

Horizontal wind speed is most commonly measured with a rotating cup anemometer. The instrument normally consists of three or four conical or spherical cups radially mounted to a freely rotating spindle. The mean rotational speed of the spindle is nearly proportional to the mean horizontal wind speed.

Friction within the anemometer prevents rotation when the wind speed is less than the sensor's threshold value. The sensor also overestimates air travel in gusty conditions because momentum causes the cups to reduce their speed of revolution more slowly than they increase with wind speed.

Study objectives and environmental exposure dictate the required sensitivity of the anemometer. Those of rugged construction with low sensitivity may be desirable for long-term, continuous recording of relative wind speed. A mechanical register or counter which displays the total length of air passing the cups may be a suitable monitoring device for such measurements. In contrast, the precise measurement of air flow at low speeds requires an instrument composed of a very light-weight cup and spindle assembly. These sensitive anemometers are commercially manufactured with threshold values approaching 10 cm/sec and electrical outputs which operate counters or recorders.

The precision necessary for precise spatial and wind profile studies requires that all sensors be compared to each other at least daily and their calibration adjusted accordingly. The investigator must also attempt to minimize any potential effect the support mast may have on the indicated measurements and insure exact vertical sensor placement. Accurate measurement at low wind speeds also requires frequent cleaning and lubrication of the entire assembly.

Cup anemometers are ideal sensors for measuring horizontal air flow above plant canopies. Their placement within and below the canopy of most plant species is somewhat limited by the diameter of the rotating sensor.

2. Hot Wire Anemometers

Hot wire anemometers are thermal sensing devices for measuring air flow. The principle of operation is based on the convective cooling of a heated wire by a moving air stream. The sensing element, commonly made of platinum, nickel, or tungsten, may be maintained electrically either to a constant temperature or to a constant current. Under either operating condition, the speed of the wind that passes the element is a function of the temperature and electrical parameters of the circuit.

Hot wire anemometers are used in micrometeorological studies primarily for measuring very low wind speeds and air flow at essentially point locations. Their small size and fast response times are attractive features for monitoring air flow within wind tunnels and plant canopies and very near the surface of leaves and other objects. The sensors are, however, more ideally suited for well-controlled laboratory studies than for unattended continuous monitoring under field conditions. The fine wire sensing elements are easily damaged by moving plant parts, birds, insects, etc., and contaminated by particulate matter carried in the air.

The design and operating principle provides a sensor which is sensitive to air flow from all directions. This response characteristic must be considered in the analysis of data obtained from hot wire anemometers.

B. Wind Direction

1. Electrical Sensors

Wind direction at a given location is estimated with a sensor which points to the direction from which the air is approaching. The simplest form consists of a streamlined body attached horizontally to a freely rotating spindle. This assembly is commonly mounted to a structure displaying the cardinal directions.

Studies of spore dispersal and patterns of epidemic development within a crop may be interpreted if we analyze continuous records of wind direction. This type of study requires the use of well-balanced, aerodynamically designed vanes that respond to rapid changes in direction of very light winds. The continuous records may be monitored electrically by driving a low-torque potentiometer with the vane. The wind direction is recorded as a function of the circuit resistance. Another design utilizes self-synchronous motors to transmit the directional signal to a recording system.

Frictional resistance created by the mechanism that normally alters the electrical properties can be alleviated by using photosensors to detect the position of an ultralight, free rotating vane.

2. Bivane Sensors

The bivane sensor is a light-weight bidirectional vane that measures the vertical and horizontal angles of air currents. A remote record is made possible by signals from two precision potentiometers. The recorded pattern primarily provides relative information on the horizontal and vertical components of atmospheric turbulence.

3. Dispersion Techniques

The release of visible gas or smoke from a point source provides information on the direction of natural air currents and on turbulent dispersion from the central axis of the gaseous cloud. Although the technique does not provide an accurate measure of wind direction, it provides visual information on air flow patterns over irregular sloping surfaces and within canopies of larger species of vegetation.

The wind direction and turbulent dispersion observations must be manually recorded. A more complete documentation and analysis may be made by photographically recording the dispersing cloud at various time intervals.

C. Atmospheric Turbulence

The irregular and random fluctuation of airflow is caused by both the frictional and convectional exchange of mass. The frictional mixing is caused by variations in the roughness of surfaces projecting into the airstream and by shearing between air layers moving at different speed.

In contrast, convection is the vertical movement of air masses caused by heating. This form of exchange is not exactly random in nature. It is caused by changes in air density and therefore contributes primarily to ascending and descending air flows. This type of mixing usually encompasses larger volumes of air than does frictional exchange. These two processes do not contribute equally to atmospheric turbulence. Early in the day the mixing is almost entirely due to friction, whereas a transition to predominately convectional mixing occurs later from radiational heating of the earth's surface. Turbulence is nearly always present in the lower layer of the atmosphere but it may be numerically distinguished from laminar flow by the analysis of profile measurements of temperature and wind speed.

The combined effect of frictional and convectional turbulence causes vertical exchange of volumes of air. This exchange mechanism is the natural process by which spores and other propagules may rise higher than their point of liberation. The direction and speed of spore transport, once the upper air strata has been reached, is easily estimated from pressure gradient data.

Surface features and constantly changing meteorological conditions within the surface boundary layer greatly compound the problem of estimating spore movement. Although the resulting mixing actions are very important and necessary from the standpoint of reducing radiational heat loads and temperature and humidity gradients, they make it extremely difficult to measure accurately the coefficient of mass exchange.

The mass exchange coefficient, which is proportional to eddy diffusivity, gives the rate at which suspended matter is transported by eddies in turbulent motion. The magnitude of the exchange coefficient varies greatly in both space and time but normally increases with height. Although surface irregularities make it quite difficult to measure the mass exchange coefficient near natural surfaces, the coefficient may be estimated from smoke dispersion techniques, energy balance procedures, or the aerodynamic approach.

Precise vertical profile measurements of temperature, humidity, and wind are required to estimate exchange coefficients with any degree of accuracy. This necessitates use of the most accurate and sensitive instrument in well-designed field studies. The dimensions of experimental plots and surface uniformity of the crop under investigation are very important in determining proper placement of sensors. This must include consideration of both the distance to upwind discontinuities and inhomogeneous surface features which may act as sources and sinks of heat and water vapor. The distance to such areas, referred to as fetch, dictates the height to which micrometeorological parameters may be measured. This is especially important in studies where treatments applied to relatively small plots affect the microclimate of each plot. For example, significant errors will result if measurements are taken over a relatively small, irrigated plot adjacent to a large, nonirrigated area. Measured values taken over these small plots are greatly influenced by the condition of the air prior to reaching the study plot. Changing from a warm to cool and a dry to moist surface requires a given fetch if the profiles are to be representative. Measurements taken high above the canopy are not representative of the surface unless the fetch is large. The smaller the fetch, the lower the height to which we are restricted to obtain representative measurements. Near the surface the gradients are greater and more easily measured. The recommended ratio of maximum sensor height to fetch is not a universal figure, but is roughly between 1/20 and 1/100. A more precise estimate of this recommended distance would require a detailed analysis of characteristic data.

VI. MEASURING IRRADIANCE

Solar radiation is the primary source of energy available to the earth's surface for evaporating moisture, heating the soil, air, and plant material, and for metabolic processes. Partitioning radiation measurements for a given geographical location has provided useful energy balance equations for estimating quantities such as evapotranspiration, sensible heat flux, and temperature differences between crop foliage and the ambient air. Previous work in the area of forecasting and simulating plant disease has centered primarily on temperature, relative humidity, rainfall, and dew. A majority of the investigations have discounted the role of the quantity and quality of solar energy. Although past studies reveal that "light" has a definite effect on pathogens and on diseases they cause, the utility of the data is questionable because many of the investigators lacked the required instrumentation to characterize the treatments accurately.

A. Photometry

Photometry and radiometry are the two primary measures of illumination. Although each has its proper place in science, they are frequently utilized in the improper area of investigation. We must recognize that these are two basically different measurements which cannot be simply equated with each other.

Photometry measures the intensity of visible electromagnetic radiation, i.e., visible light as seen by the standard human eye. The response of the human eye is not constant over the entire visible range, 400 to 700 nm, but resembles a normal distribution response which peaks at ca. 550 nm. Based on this response, photometric measurements are only appropriate when evaluating a light source from the standpoint of its effect on human vision and is therefore a very important measurement for the illuminating engineer. It is unfortunate that photometric measurements have been incorrectly used in areas not specifically pertinent to human vision, e.g., plant growth and disease development.

Photometers are specifically designed to measure the intensity of visible radiation. They generally consist of a sensor (e.g., a phototube, photocell, or photomultiplier tube) and a filter which adjusts the system response to conform to the standard luminous curve. The measuring device commonly provides a direct readout in units of footcandles, candles, footlamberts, etc.

B. Radiometry

The radiometric measurement of energy is not restricted to the visible portion of the electromagnetic spectrum. Therefore it is useful for measurement in the ultraviolet and infrared regions as well as in the visible spectrum.

The basic unit of radiant flux, i.e., the total radiant power emitted from a source, is expressed in watts. The radiometric measurement of energy is useful because a unit watt of radiation in the ultraviolet region is equal to a unit watt in the visible region. It is totally independent of eye response and wavelength. In contrast, the photometric system would require approximately ten units of blue light to equal one unit of green light.

Instruments which detect all wavelengths of energy or selected broad wavelength bands may be broadly classified as: (a) radiometers–sensors which respond to the total flux of energy of all wavelengths received on a single surface, (b) pyrheliometers–sensors which measure only direct solar radiation, (c) pyronometers–sensors which measure the combined intensity of incoming direct solar radiation and diffuse sky radiation, and (d) net radiometers–sensors which respond to the difference in energy of all wavelengths received on two surfaces.

Sensors utilized to obtain radiant flux measurements are commonly composed of a specially treated flat surface that absorbs incident radiation, a thermopile, and hemispherical filters. In principle, the absorbed energy creates a temperature difference between the sensing element and a reference normally at ambient air temperature. The resulting thermoelectric output is nearly proportional to the radiant flux density. The sensor will respond to both short- and long-wave radiation if a bare element, or one covered with thin polyethylene hemispheres, is exposed to the energy source. The addition of special hemispheric filters, however, allows the investigator to limit the sensor response to specific wave bands. Instruments of these general configurations are available for measuring total, net, direct short-wave, and sky radiation. Depending upon the type of sensor, the instruments commonly have a near linear response with sensitivities of 5 to 10 mV/cal/cm^2/min, response times ranging from ca. 10 sec to 2 min, and accuracies of 2 to 10%.

Another form of radiometer uses photocells or photomultiplier tubes as sensing detectors. The sensitivity of these instruments, however, varies with wavelength. Therefore absolute measurements of radiant flux require sensor calibration as a function of wavelength.

Recent advances in radiometry have led to the development of portable spectroradiometers. These precision instruments utilize a monochrometer to focus the wavelength of interest onto a detector. Therefore the measurements are not restricted to available cutoff or narrow bandpass filters. The use of fiberoptic probes and telescopic accessories allows this relatively large measuring system to be used in many host–pathogen environments. Calibrated to measure the absolute intensity per unit wavelength over a broad spectrum, the systems have accuracies of ca. 15% in the ultraviolet and 10% in the visible and near-infrared regions. Instruments of this type greatly aid in characterizing the radiant environment to which plants and disease-causing organisms are subjected.

VII. ACQUIRING METEOROLOGICAL DATA

Instrumentation technology has resulted in the development of commercial devices which respond to the aforementioned meteorological parameters and produce electrical outputs related to the magnitude of the measured value. Electrically operated sensors are also available for the measurement of associated variables, e.g., rainfall, soil moisture, leaf wetness, and chemical composition of the atmosphere. Signals from the sensors may be recorded either manually from indicating instruments or electronically via strip chart recorders or automated data acquisition systems. It is, however, no longer practical to observe, record, and analyze large volumes of data manually. To obtain the full potential from available instrumentation, we must integrate sensor elements, data monitoring units, and computer capabilities. This approach not only expedites the monitoring process, but facilitates rapid data reduction and analysis and decision making. In addition, an integrated network eliminates human error in transcribing data and minimizes data processing errors.

Automatic data acquisition systems encourage the accumulation of unnecessary data. Investigators often find they have collected much data that is either irrelevant or impossible to interpret. In designing a system to meet specific objecitves, one must carefully answer such questions as: (a) what parameters are to be measured, (b) what sensors are to be employed for maintenance-free operation, (c) where sensors are to be positioned to obtain representative measurements, (d) what degree of error is acceptable, (e) what the required sampling time is, and (f) how the data are to be processed to obtain the most pertinent information.

Centralized systems that utilize digital telemetry techniques are well suited for monitoring systems that involve a large number of sensors. These systems should be programmable, real time units operated by modular subprograms which may be readily altered at an operator console. The systems should have the capability to store and reduce data for given time periods and store the raw and summarized data on peripheral equipment for later analysis.

These centralized systems allow frequent check of the data to determine if the system, sensors, etc., are functioning properly. A major advantage of this type of unit is the ability to observe on a real time basis the predicted effect of environmental conditions on disease development in crops. These on-line systems are really tools which help us forecast, simulate, and understand disease development. They are invaluable in verifying forecasting systems and computer simulators of plant disease and they allow operator adjustment and decision making during the epidemic. The testing procedures, however, are not always accomplished by the computer alone. They frequently require interaction from professionals to make unprogrammed decisions in response to unexpected events.

The major problem in collecting concurrent meteorological and biological data is the potentially large volume of information that must be processed. Automatic data processing of raw data from measurement devices is practical from the standpoint of time, cost, and accuracy. Experience has shown, however, that it is not feasible to monitor and/or control all experiments via the computer. Short-term studies and those in which instrumentation design is rapidly changing require more time and expense to initiate than the results justify. Studies of this scope require the aid of a technician to transpose the data and computer process it in batch mode rather than via the on-line system approach. Additional information must be manually entered, even if most of the meteorological data are acquired via a direct on-line approach. If on-line remote entry to the computer via a communication system is not available, there is little advantage to having the information in a computerized system. The remote entry system allows access to the data and provides a means to include associated biological information. If we do not, however, carefully design our data monitoring and retrieval network, more work may be required to enter the data into the computer system than is required to record, analyze, and report the data manually.

At the present time the computer can only improve the efficiency of the researcher on long-term investigations where little design change takes place during the study. Advances in minicomputers, microprocessors, and coupling devices that interface sensors to remote entry systems of large computers have greatly aided in monitoring data from smaller scale investigations. The rate at which new and more advanced systems are being placed on the market makes it difficult to predict the advantageous characteristics future data systems will possess. I do not, however, foresee in the immediate future justification on a cost/return basis for the total replacement of manually documenting data.

The centralized system type of installation in conjunction with digital computers and mathematical models will improve the science that underlies the art of disease forecasting and identify areas which need additional investigation. In addition to data acquisition, conversion, and analysis, computerized systems may be utilized to maintain a data base of meteorological conditions and disease information and to provide interactive process control for the environmental studies, e.g., growth chamber and greenhouse climates. Although digital computers are very useful for data acquisition and data analysis, they are very slow and costly when utilized for the display of output from complex mathematical models and time simulated events. These latter tasks may more efficiently be accomplished via the hybrid computer.

Future development of sensors more suitable for continuous monitoring of all relevant variables, and knowledge of the biological process, will provide algorithms applicable to natural ecosystems that may be completely controlled by computer programmed decisions. Computers will not, however, replace all laboratory and field men. There is no system that can set up its own analysis nor is there any likelihood that a system will be developed in the immediate future that will assure continuous uninterrupted operation. The computer cannot make all the corrective actions. This requires professionals with expertise in instrumentation and biology.

Suggested References

The National Bureau of Standards (Superintendent of Documents, U.S. Government Printing Office, Washington, D.C. 20402) is a prime source for information related to measurement and instrumentation. Additional references for individuals desiring to explore this area include:

Anonymous. (1961). Supplement on Instruments and Apparatus. Part 3. Temperature measurement. ASME Power Test Codes. PTC 19.3–1961. The Am. Soc. of Mechanical Engineers. 118 pp.

Anonymous. (1976). Measurement Fundamentals–Basic Instrumentation Lecture Notes and Study Guide (R. L. Moore, ed) Instrument Soc. of Am., Pittsburgh, Pa. 303 pp.

Hofmann, G. (1965). Hints on Measurement Techniques used in Microclimatologic and Micrometeorologic Investigations. In "The Climate Near the Ground" (by R. Geiger). Harvard University Press. Cambridge. pp 520–542.

Middleton, W. E. K., and Spilhaus, A. F. (1964). Meteorological Instruments. University of Toronto Press. 286 pp.

Monteith, J. L. (1972). Survey of Instruments for Micro-meterology. Blackwell Scientific Publications, Oxford and London. 263 pp.

Monteith, J. L. (1975). Vegetation and the Atmosphere. Vol. 1, Principles. Academic Press, New York. 278 pp.

Munn, R. E. (1966). Descriptive Micrometeorology. Academic Press, New York. 245 pp.

Platt, R. B. and Griffiths, J. F. (1964). Environmental Measurement and Interpretation. Reinhold Publishing Corp. New York. 235 pp.

Chapter 6

Pathometry: The Measurement of Plant Disease

JAMES G. HORSFALL AND ELLIS B. COWLING

Plant pathologists of the world earn a reasonably good living with a little butter on their bread occasionally because the society that pays them is afraid of plant disease. Even city people worry about black spot on their roses and fire in their tulips. Society shuns moldy food, scabby apples, and spoiled potatoes. It sees or hears of dramatic diseases like the potato famine, maize blight, and rice blast. It has watched blister rust kill white pines, blight kill chestnuts, Dutch elm disease destroy elms, and blue mold sweep across Europe killing cigarette tobacco as it went.

119

It is the less dramatic diseases that concern us here. Although it can see the dying elms along the streets, society in general is only vaguely aware of how much damage it suffers from the depredations of other plant pathogens. That is because we have not told society about the losses, and that is because we do not know, and that is because we have not researched it very well.

Lyman (1918) lamented 60 years ago: "How can we expect practical men to be properly impressed with our work and to vote large sums of money for its support . . . when we have only vague guesses to give them and when we don't take the trouble to make careful estimates." We can measure diseases somewhat better than we could 60 years ago, but we have very far to go.

I. WHY MEASURE DISEASE?

Large (1953) introduced an intriguing term for disease measurement —pathometry—and we have appropriated it for our chapter title. As Pennypacker says in Chapter 5 of this volume, we must measure if we are to understand, and so we must measure disease if we are to understand it. For most of the life of our profession the measurement of plant disease and the losses from it have been in the realm of "abracadabra." We have been content to "guess" how serious the loss was. We used our guesses as a charm to ward off our doubts.

The chief reason for our poor showing in measuring loss was that we could not measure the independent variable—the disease. Only when we can measure disease well can be demonstrate how much loss it causes.

LeClerg (1964) lists some reasons for measuring loss. He really means measuring disease because this is the horse that must go before the cart. By adding a few reasons of our own, we submit the following list: (1) researchers and extension people can decide priorities; (2) administrators will have powerful data to present to legislators; (3) industry will have the data to decide on priorities in research and sales; (4) the crop reporting services will be better able to forecast crop production; (5) environmental protection agencies can produce better cost/benefit ratios to use in deciding what compounds and how much to approve; (6) research agencies can decide (i) whether their cherished varieties are doing well; (ii) whether they are losing effectiveness; (iii) whether a new fungicide or nematicide is performing up to expectations; or (iv) whether it is losing ground to resistant strains of the pathogens and how fast; and finally (7) whether the farmer can establish for his farm a better estimate of his economic threshold. With such information he can calculate how much loss he must sustain before he can afford the cost of control.

II. SCOPE OF THE CHAPTER

Our job in this chapter is to discuss the measurement of disease on whole living plants. This is where our ignorance lies. Once we can measure disease, we are home free on losses. In this chapter, we will devote only a brief section to the mathematics of relating disease amount to loss because Main (1977) has already done this so well in Chapter 4, Volume I, of this treatise. Also, we shall not discuss the measurement of the more biochemical results of disease such as respiration, photosynthesis, or water movement. These are more truly in the province of the authors of the chapters on disease components which follow in Volume III of this treatise.

III. OUR IGNORANCE IS PROFOUND

As we have said, our ignorance of plant disease intensity and plant disease loss is profound because we have researched extensively most aspects of plant disease except intensity and loss.

The relative paucity of research is dramatically documented in Volume 4 of the *Annual Review of Phytopathology*. In that volume Large (1966) took only 77 references to review our ignorance of it. We must have known twice as much about the biochemistry of pectic enzymes in diseased tissue at that time because there were 162 references on this subject in the same volume. Similarly, the 1968 list of American plant pathologists revealed only 67 members who were interested in plant disease losses, as compared to 383 unduplicated names in the two physiology categories. That is a ratio of 1 to 6.

According to Chester (1950), W. D. Moore once wrote, "It is too bad that so many have contributed so little to this very important subject." A few like Lyman, Neil Stevens, and Chester in the United States; Beaumont, W. C. Moore, and Grainger in Great Britain; James in Canada, and Chiarappa of the Food and Agriculture Organization have "shaken the tambourine and rattled the beggar's tin cup" for research support. By and large, the results have by no means matched the need, however, and most of our ignorance remains.

IV. WHY IS OUR IGNORANCE SO PROFOUND?

Why are we more ignorant of the overall damage from disease than we are of the internal damage to biochemical processes? There are different explanations for each generation and each country. We suggest the following probabilities, at least for the United States, where we are most familiar and where about one-third of the world's plant pathologists work. (1) Many plant pathologists think that pathometry is not very scientific. They think it is not likely to produce a Nobel prize or even a distinguished professorship. We have professors of disease physiology and even professors of ornamental disease control but no professors of pathometry. (2) The "elite" journals are unlikely to accept articles on the subject. (3) Most departments have not given promotion and tenure credit for such "primitive research." (4) No Federal granting agencies of the 1950's and 1960's gave a grant for such research. (5) Most plant pathologists think (i) that the methods are untrustworthy and (ii) that visual methods, the most effective so far, are "subjective" and therefore beneath their dignity as "objective thinkers." (6) The United States has too much food anyway. Farmers' prices have been depressed for most of 60 years. To measure loss and reduce it would simply add to the surplus. (7) It is hot, dusty work and consumes much "valuable" time that could be used more "profitably" operating an electrophoresis gadget.

V. THE WORM IS TURNING

Considerable evidence suggests, however, that the worm is turning. For example, James (1974), writing on losses 8 years after Large (1966), needed 106 references not 77. This is a measure of progress. It is instructive to examine the sociology of the matter. Suppose we compare three situations (1) in the United States that has had a surfeit of food for about 60 years, (2) Great Britain on an island that must import much of her food through sometimes dangerous waters, and (3) the Food and Agriculture Organization that must deal with the "Hungry Nations" where food is short.

A. The Story in the United States

War mercilessly shows the stress zones in a nation. During World War I the great wheat rust epidemic of 1916 cost the United States 200 million bushels of wheat and the Canadians 100 million bushels. The following year was the "year of the wheatless days." A plant disease *could* make America hungry!

In the middle of the year of the wheatless days, the United States Department of Agriculture hurried to establish a Plant Disease Survey with G. R. Lyman in charge. He promptly encountered among his colleagues their emphasis on the pathogen and a deemphasis on the damage it causes. The very next year he felt impelled to write his now famous lament quoted above (Lyman, 1918).

When the war was over, the United States was saddled with a large overproduction capacity for food. A severe agricultural depression set in and lasted until the next war. During that interval interest in disease damage decreased drastically.

Neil Stevens struggled hard to sustain interest in the matter, but he, too, was discouraged and left. The survey stumbled along during the great depression, rose up briefly during World War II, and died in 1945 when, as Paul Miller, long associated with the survey, wrote later to Horsfall that it "was discontinued owing to lack of support and little interest on the part of some Bureau officials" (of the United States Department of Agriculture).

K. Starr Chester tried hard to stir up interest in 1950 and later, as did E. L. LeClerg and W. D. McClellan who organized a symposium at the annual meeting of the American Phytopathological Society in 1963. Ten years later, Milton Schroth and his committee organized several symposia at the Second International Congress of Plant Pathology in Minnesota.

There is other evidence of heightened interest. The journal *Phytopathology* has recently begun to print on each cover page the list in that issue of research papers on disease detection and losses. Also it has begun to include the key word *loss* in the index. Up to that time, the editors clearly thought that it was too unimportant to list.

We even found some data to quantify the trend in the United States. Using McCallan's (1969) technique we determined the percentage of members of the American Phytopathological Society who listed themselves in the periodic membership lists as interested in disease assessment—in 1963, 0.6%; in 1968, 2.5%; and in 1974, 5.0%. The numbers look small but at least they are rising.

B. The Story in Great Britain

World War I scared the British more than it scared the Americans. They were living on a small island and submarines sank the ships bringing food from abroad. After 100 years of eating Australian wheat and Argentine beef, they had forgotten until then the warning of their countryman, the Reverend Mr. Malthus that the population would out-run its food supply.

They, too, set up a plant disease survey. When the war was over they tended to neglect it somewhat, but not as much as people on the other side of the Atlantic. With Hitler coming to power in the 1930's they created a Committee of the British Mycological Society to examine methods of assessing disease intensity and loss in their crops (Anony-mous, 1933).

W. C. Moore seems to have been the cement that glued the British effort together. He was a continuing member of the committee and he helped to develop a grading system which was published in 1943. In 1949, as retiring president of the Association of Applied Biologists, he devoted his presidential address to the matter. In 1950 he began publi-cation from his laboratory of a journal, *Plant Pathology*, devoted ex-tensively to research in the field. He brought the distinguished researcher, E. C. Large, into his laboratory to continue his work on assessing disease.

The British have a greater urge to have data on disease losses than the Americans and they have pushed the research much further.

C. The Story in the FAO

The Food and Agriculture Organization of the United Nations, with its concern for the "Hungry Nations," shows an enthusiastic interest in assessing disease losses. They publish a Plant Protection Bulletin that deals *inter alia* with such matters. Chiarappa seems to be a driving force at FAO. In 1967, the FAO convened a "Symposium on Crop Losses" to emphasize the need for the development and use of experimental methods to measure crop losses quantitatively—in short, to deal with Lyman's complaint as stated above, a half century later. Chiarappa edited the report (1971) and later wrote an excellent analysis of the need for and methods to produce more reliable disease loss estimates (Chiarappa, 1972). The FAO issued a supplement in 1973 and convened another symposium in 1975.

We must conclude from this brief digression into sociology that con-cern with disease losses is in direct proportion to the need for food.

VI. THE PARAMETERS TO BE MEASURED

Disease offers three parameters for measurement: incidence, intensity (often called severity), and yield which, of course, is the reverse of loss. Incidence can be defined as the number of plant units infected, such as whole plants, leaves, fruits, tubers, twigs, etc. Intensity can be defined as the area or volume of plant tissue that is diseased. Yield is the amount of crop harvested by the farmer. Loss is the diminution of the crop.

A. Incidence

Incidence is the most popular parameter of disease to measure because it is the easiest and quickest. One can count accurately and reproducibly the number of smutted heads of wheat, scabbed apples, wilted elms, or virus-infected sugar beets. One can express the numbers as percentages and transform them if one likes into logs, probits, or logits. Incidence is often used to measure the spread of disease through a field, county, or country. Incidence may be equated with intensity and loss where a single lesion is fatal or nearly so, as in crown gall, neck blast on rice, plum pocket, brown rot of peach, head smut of cereals, and some wilt diseases.

The relation of incidence to intensity and loss is only tenuous at best, however, with self-limiting lesions such as leaf spots, apple scab, and foliage rusts.

B. Intensity

If we are to understand disease, we must surely know how intense it is. How sick is the plant? In humans a disease that produces 3 or 4°C of fever is much more intense than one that produces only 1°C. A single leaf is enough to establish incidence, but it is not as intense or as destructive as many spots that cause the leaf to fall. One can measure intensity by counting individual lesions, but the time consumed is much greater than with a visual method which will be described shortly. It is so high that we think it is not worth the effort. Croxall *et al.* (1952) agree.

Intensity is closely related to loss. Once conversion factors are established, loss can be measured by intensity and measured more easily than yield.

C. Loss

Loss from plant disease is what plant pathology is all about. For the first 50 years of our science, we used yield as a measure of loss from disease. It was practical and objective. We assumed that loss is governed by disease, and therefore that yield is governed by loss. The difficulty is that not all yield is governed by loss and not all loss is governed by disease. The weather is heavily involved as well. Despite the huge importance of loss to our profession, we need in our research a better measure of disease than yield. Yield is too gross a measure.

VII. MEASURING INTENSITY BY VISUAL OBSERVATION

A. The Pioneering Cobb Scale

N. A. Cobb was an inventive genius who is too little understood by plant pathologists. Born in America, he went to Australia toward the end of the last century where he devised the first scale to measure disease intensity (1892). This is the famous Cobb scale for cereal leaf rusts. He drew sketches of infected leaves showing diagrammatically five degrees of rust ranging from 1 to 50% of the leaf area covered by pustules. By comparing the sketches with real leaves, he could derive a measure of the intensity of the rust.

The idea of the Cobb scale has been extended to many other diseases. Dixon and Doodson (1971) of Great Britain have published many scales and keys for measuring disease on many crops.

Tehon (1927) and Tehon and Stout (1930) have published diagrammatic sketches of leaf-spotting diseases on cereals and fruits. Similarly, Ullstrup *et al.* (1945) have published sketches of whole maize plants with various amounts of leaf spotting by *Helminthosporium turcicum*. Many investigators, such as Croxall *et al.* (1952), are using visual estimates instead as will be discussed in the next section. Croxall *et al.* state that the use of diagrams is too slow.

B. The McKinney Index

Circa 1920 L. R. Jones and his colleagues and students at Wisconsin (Jones, 1921) began to look into the effects of weather on plant disease and devised the famous Wisconsin soil temperature tanks. They had to have a workable measure of disease intensity if they were to quantify the effects of the weather. They immediately decided that yield was not a useful parameter.

They were forced to use a visual estimate of disease. Initially, they expressed their data as percentage of infected plants. This was incidence, but as McKinney (1923), one of Jones' students, pointed out, this parameter was not enough. To my knowledge McKinney was the first to use a visual, quantitative estimate of disease, which he called a "numerical rating." His system is as follows: No disease = 0.00 rating, very slight = 0.75, slight = 1.00, moderate = 2.00, and abundant = 3.00. His method has been copied freely and rediscovered many times. Horsfall

(1930) used it for a scale to measure the intensity of disease on clovers and grasses. He also used the McKinney system in a statewide survey of these diseases.

C. The Human Eye is a Photocell

Since there were no real photocells in McKinney's day, he did not think of it as such but he was in fact using his eye as a photocell. The human eye is a highly portable, rapid, and astonishingly precise photocell. Let him who doubts its accuracy try to win a prize from a weight estimator at a carnival.

Pathologists use the human photocell for incidence studies to count diseased plants or diseased spots and also to use it for intensity studies to measure areas or volumes of infection. Many pathologists would define the former as "objective" and the latter as "subjective." We hold that both are objective, provided the brain behind the eye does not know which treatment it is reading.

D. A Study of the Precision of the Method

Having found that McKinney's grades are reasonably good, Horsfall and Heuberger (1942) used them extensively in their fungicide trials for measuring the intensity of the attack of *Alternaria solani* on their experimental tomatoes. At the end of one year, when infestation was particularly severe, they used the following three parameters to determine the amount of disease on all of the individual plants on four replicate field plots of 12 fungicidal treatments plus a check. The parameters were: (1) incidence as a percentage of leaves killed and as percentage of stem end rot, (2) intensity slightly modified as McKinney numbers, and (3) loss at the end of the season as yield of green fruits and yield of green vines less fruit. The data, taken from Horsfall and Heuberger (1942), appear in Fig. 1. They were astonished at how well the McKinney so-called subjective index matched the so-called objective counts of stem end rot and dead leaves. The extrapolated curves intersect the abscissa at the zero point.

Curve B in Fig. 1 is worth further consideration. In a sense, it can be considered as a comparison of incidence (dead leaves) and intensity (McKinney index). The comparison is not absolutely clear-cut, however, because dead leaves are a part of the McKinney index. When the intensity was essentially 100%, the incidence registered as dead leaves was only 65%. The data emphasize that incidence does not give as accurate a picture as disease intensity. It does not register correctly the number of

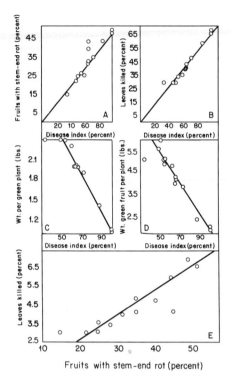

Fig. 1. A graphic comparison of methods of measuring magnitude of infection by *Alternaria solani* on tomatoes (from Horsfall and Heuberger, 1942) (A) Relationship between infection index and percentage of fruits affected by stem end rot. (B) Relationship between infection index and percentage of leaves killed. (C) Relationship between infection index and green weight of plants less fruit. (D) Relationship between infection index and weight of green fruits at the end of the season. (E) Relationship between percentage of fruits affected by stem end rot and percentage of leaves killed.

successes scored by the fungus. The so-called subjective method seems to do a better job than the so-called objective one.

Figure 1 shows how disease was related to the green weight of the green fruit at frost time and the weight of the green plant minus the fruit. The agreement was excellent. Loss was linear with disease grade. We conclude that the human eye is in fact a useful photocell for measuring disease intensity and, therefore, loss.

There are data to examine the criticism that the grading system is subjective. Marsh *et al.* (1937) had their fungicide trials examined by six observers and reported that "a satisfactory and close agreement was obtained between the different observers." Horsfall and Heuberger

(1942) also examined the agreement between observers. They asked three people to read their plots, two who had used the system before and one who had not. The variation was less than 1%.

It is clear that the human eye is a good photocell, provided the identity of the treatments is hidden from the observer.

E. The Eye Reads in Logarithms

After further consideration of the McKinney system, Horsfall and Barratt (1945) decided that the system could be improved at the upper and lower ends. For example, if all plants were to fall into McKinney's group 1, the mean would show more disease than really was there. This weakness in the system did not cause any significant difficulty with the test depicted in Fig. 1, however, because disease was fairly intense that year.

About this time they discovered by chance the Weber–Fechner law which says that visual acuity is proportional to the logarithm of the intensity of the stimulus. Music also can be heard in terms of logarithms. The string for middle C on the piano vibrates twice as fast as the string for the C one octave below or one-half as fast for the C one octave above.

The Weber–Fechner law provided the rationale for a new system of grading in which the grades were adapted to the readout capability of the human photocell. They were placed logarithmically, not arithmetically. At the same time, Horsfall and Barratt (1945) came to realize that the eye actually reads diseased tissue below 50% and healthy tissue above 50%. Therefore, they set the grades to go both ways from 50% with a ratio of two. In the downward direction the grades were set at 50 to 25%, 25 to 12%, 12 to 6%, 6 to 3%, 3 to 0%, and, of course, 0%. In the upward direction the grades were set at 50 to 75%, 75 to 88%, 88 to 94%, 94 to 97%, 97 to 100%, and, of course, 100%. Some have called this the HB system.

One can create a special graph paper for this by cutting two pieces of semilog paper at the 5 level, upending one, and gluing the two together at the 5 level. A graph plotted on such paper appears in Fig. 2 (taken from Horsfall, 1945). The graph paper looks very much like arithmetic probability paper; Chester (1950) concluded that it was.

A stranger to the Weber–Fechner law has difficulty at first in believing that he reads 3 to 6% with the same accuracy as 25 to 50%, but a test in some field plots will soon verify this. When he discovers this, he is apprehensive that he cannot do better between 25 and 50%. This is really not as serious a problem as it appears to be. By examining a large number of plants, the mean amount of disease turns out to be accurate.

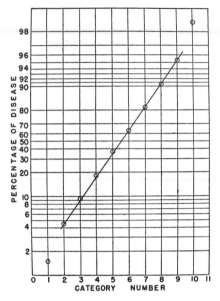

Fig. 2. Calibration curve for converting graded disease readings into percentage (after Horsfall, 1945). Reprinted by permission of the copyright owner.

From the graph in Fig. 1 he can convert mean grade to percentage of disease intensity or its reciprocal percentage disease control.

The grades based on the Weber–Fechner law have been widely used. In 1968 Redman *et al.* (1969) of Eli Lilly and Co. (Elanco) prepared a table for converting field readings to percentage disease or percentage disease control. They devised the table for use in their fungicide testing but they now use a computer program instead of the table to translate field readings on plots to percentage disease control.

Brown wrote us recently that he has distributed approximately 225 copies of the conversion tables and that the United States Environmental Protection Agency suggests that plant disease intensity be rated by the logarithmic system.

F. Logarithmic Aspects of Other Systems

Knowing the logarithmic aspect of visual acuity, one wonders if earlier grades may not have had a logarithmic component without being so recognized. Chester (1950) investigated this. According to him Gassner (1915) arranged his cereal rust grades in a logarithmic progression. Rusakov's (1921) scale of rustiness of cereals is essentially logarithmic as is

TABLE I

Comparing Scales of Disease Intensity

Source	Scale									
Moore (1949)	0.1	1	5	10	25	50	75	—	95	—
Horsfall and Barratt (1945)	—	3	6	12	25	50	75	88	94	97

Melchers' and Parker's (1922) modification of the Cobb scale. Reichert *et al.* (1944) used a log scale for grape diseases in Israel.

It seems probable that the eye of all disease observers reads in logs whether they accept this premise or not. Upon reexamination and armed with the additional knowledge that the observer reads diseased tissue below 50 percent and healthy tissue above 50 percent, we can see the astonishing logarithmic features of the well-known scale of the British Mycological Society (Moore, 1949). This is compared with the HB grades in Table I.

Similarly, Marsh *et al.* (1937) used grades of late blight of potato as follows: 0 = no blight, 2 = scarce lesions, 4 = occasional blight lesions, 6 = frequent blight lesions, 8 = abundant blight lesions, some defoliation, 10 = complete defoliation. The intensity of disease was probably seen in logs and the numbers, therefore, should have been 1, 2, 4, and 8, with 0 at the bottom and 100 at the top of the scale.

Both Chester (1950) and James (1974) agree that the logarithmic scale for reading disease is "satisfactory." James, however, says that the log scale is "appropriate" only when the relationship between disease and yield is logarithmic and not when it is arithmetic. James appears to forget that the readout from the human photocell is logarithmic and wholly independent of the disease–yield relationship. Since the eye reads in logs, the grades must be in logs regardless of the relationship between disease and loss.

Some investigators have worried about the width of the classes in the middle ranges and have attempted to fill these with intermediate grades. However plausible this seems, it ignores the characteristic of the readout from the human photocell. Therefore, it is probably a case of "love's labor lost."

G. Usefulness of Intensity Measures in Research

Having developed a method of quantifying disease, it can be used for many phases of research. For example, Barratt and Richards (1944) and Barratt (1945) were able to quantify differences in varietal susceptibility in tomatoes to *Alternaria solani* by reading disease at frequent intervals during the season. They plotted their data as arithmetic time and probit disease. From this they calculated what would now be called ET_{50}, the effective time for 50% of disease to develop. This varied directly with resistance. They also could calculate the slope of the curve. This becomes van der Plank's r factor or rate of disease development.

Recently, van der Swet *et al.* (1970) have used the log scale of disease for breeding pears for resistance to fire blight. Prabhu and Prakash (1973) have used it for studying weather factors for leaf blight of wheat in India. Brown of the Eli Lilly Company uses it extensively in fungicide trials.

H. Intensity of Disease in Crops—Surveys

Thus far we have been discussing measuring devices for the intensity of disease on the individual leaf, fruit, or whole plant. In a volume devoted to disease in populations, we must extend the measures to populations of plants in a field, state, or country.

One simply applies the yardstick for an individual plant to all the plants in a sample of the population. Here, however, we encounter problems of sampling. How many fields? How many samples in a field? Is it sufficient to sample along roadsides or must one walk through the fields?

Having determined from surveys the intensity of disease on the ground, then one can turn to aerial surveys as pioneered by Colwell (1956). To the aerial people, the survey data provide "ground truth." Once this is keyed to the aerial photographs, one can then survey huge acreages without walking. The technique was used by Bauer *et al.* (1971) in their well-known "corn blight watch" following the corn blight epidemic of 1970 in the United States.

VIII. RELATING INTENSITY TO LOSS

Once intensity in the crop has been established, then one has to relate this to loss. This has to be done experimentally, of course, and then applied to the field data. Since Main (1977) has discussed these techniques in some detail, only a brief summary need be given here.

A popular device is the critical-point model in which regression analysis is used to relate disease as the independent variable and percentage loss in yield as the dependent variable. The Horsfall–Heuberger (1942) example of this has been discussed above.

Another is the multiple-point model using the disease progress curve as was done by Barratt and Richards (1944). Here, multiple regression analysis is used to relate disease to loss. Watson *et al.* (1946) developed a good example of this. Van der Plank (1963) proposed a modification of the multiple-point model in which he used the area under the curve to measure loss.

Calpouzos *et al.* (1976), using wheat rust, progressed still further and created a response-surface model—a three-dimensional graph showing the effect of date of onset of disease and the rate of development of the epidemic.

James (1974) called attention to the methods of relating disease to loss by using various procedures for keeping disease under control in some plots and making yield comparisons with plots where the disease was allowed to run rampant. Romig and Calpouzos (1970) varied the number and timing of fungicidal sprays to induce various types of epidemics of wheat stem rust for study.

One can also use isogenic lines with different susceptibilities. Kassanis and Varma (1967) studied losses in viruses by using virus-free clones produced from virus-free meristems.

IX. WHERE NEXT?

It is clear that our ignorance of the relation of disease intensity to loss is receding rather rapidly.

Since intensity can be measured effectively by the human photocell, and since intensity can be related directly to loss, we are now in a position to compile a handbook of diseases and crops where we already know the relationship. Admittedly there would be some errors in using such a list for policy-making, but it would be vastly better than the present lists compiled in a guessing game. Horsfall once suggested to the *Plant Disease Reporter* that it publish such a handbook.

Preece (1971) started us off in the right direction. He named four outstanding examples in the United Kingdom: yellows of sugar beet (Hull, 1953), potato blight (Large, 1958), powdery mildew of cereals (Large and Doling, 1962), and barley leaf blotch (James, 1969). There are additional examples in the Chiarappa-edited manual (1971) of the Food and Agriculture Organization.

If research is to continue and if surveys are to be made, someone will have to do it. Who will? Horsfall (1973) suggested in the disease-loss symposium at the Second International Congress of Plant Pathology that "for a few years now we [i.e., the research agencies and the universities] give triple credit toward promotion for the disease appraisers." Cowling makes much the same point in his concluding remarks in Chapter 17 of this volume. We hope pathometry will come of age in the last quarter of this twentieth century.

References

Anonymous. (1933). Symposium and discussion on the measurement of disease intensity. *Trans. Br. Mycol. Soc.* 18, 174–186.

Barratt, R. W. (1945). Intraseasonal advance of disease to evaluate fungicides or genetical differences. *Phytopathology* 35, 654 (abstr.).

Barratt,R.W.,and Richards,M. F.,(1944).*Alternaria* blight versus the genus, *Lycopersicon*. *N.H., Agric. Exp. Stn., Bull.* 82, 1–25.

Bauer, M., Nelson, D., Johannsen, C., and Davis, S., eds. (1973). "The 1971 Corn Blight Watch Experiment," Final Rep., Vol. III. Natl. Aeron. Space Admin. Earth Resources Program, Houston, Texas.

Calpouzos, L., Roelfs, A. P., Madson, M. E., Martin, F. B., Welsh, J. R., and Wilcoxson, R. D. (1976). A new model to measure yield losses caused by stem rust in spring wheat. *Minn., Agric. Exp. Stn., Tech Bull.* 307, 1–83.

Chester, K. S. (1950). Plant disease losses: their appraisal and interpretation. *Plant Dis. Rep., Supply.* 193, 189–362.

Chiarappa, L., ed. (1971). "Crop Loss Assessment Methods." FAO, Rome.

Chiarappa, L., Chiang, H. C., and Smith, R. F. (1972). Plant pests and diseases: assessments of crop losses. *Science* 176, 769–773.

Cobb, N. A. (1892). Contribution to our economic knowledge of the Australian rusts (Uredineae). *Agric. Gaz. N.S. Wales* 3, 60–68.

Colwell, R. N. (1956). Determining the prevalence of certain cereal crop diseases by means of aerial photography. *Hilgardia* 26, 223–286.

Croxall, H. E., Gwynne, D. C., and Jenkins, J. E. E. (1952). The rapid assessment of apple scab on leaves. *Plant Pathol.* 1, 39–41.

Dixon, G. R., and Doodson, J. K. (1971). Assessment keys for some diseases of vegetable, fodder, and forage crops. *J. Natl. Inst. Agric. Bot. (G.B.)* 23, 299–307.

Gassner, G. (1915). Die Getreideroste und ihr Auftreten im subtropischen ostlichen Sudamerika. *Zentralbl. Bakteriol. Parasitenkd., Infektionskr., Abt. 2* 44, 305–381.

Horsfall, J. G. (1930). A study of meadow-crop diseases in New York. *Cornell Univ. N.Y., Agric. Exp. Stn., Mem.* 130, 1–139.

Horsfall, J. G. (1945). "Fungicides and Their Action." Chronica Botanica, Waltham, Massachusetts.

Horsfall, J. G. (1973). Does the scope of the research match the scope of the need? *Abstr. 0020 Pap., Int. Congr. Plant Pathol., 2nd, 1973.*

Horsfall, J. G., and Barratt, R. W. (1945) An improved grading system for measuring plant disease. *Phytopathology* 35, 655 (abstr.).

Horsfall, J. G., and Heuberger, J. W. (1942). Measuring magnitude of a defoliation disease of tomatoes. *Phytopathology* 32, 226–232.

Hull, R. (1953). Assessments of losses in sugar beet due to virus yellows in Great Britain, 1942–52. *Plant Pathol.* 2, 39–43.

James, W. C. (1969). A survey of foliar diseases of spring barley in England and Wales in 1967. *Ann. Appl. Biol.* 63, 253–263.

James, W. C. (1974). Assessment of plant diseases and losses. *Annu. Rev. Phytopathol.* 12, 27–48.

Jones, L. R. (1921). Experimental work on the relation of soil temperature to disease in plants. *Trans. Wis. Acad. Sci., Arts Lett.* **20**, 433–459.

Kassanis, B., and Varma, A. (1967). The production of virus-free clones of some British potato varieties. *Ann. Appl. Biol.* **59**, 447–450.

Large, E. C. (1953). Some recent developments in fungus disease survey work in England and Wales. *Ann. Appl. Biol.* **40**, 594–599.

Large, E. C. (1958). Losses caused by potato blight in England and Wales. *Plant Pathol.* **7**, 39–48.

Large, E. C. (1966). Measuring plant disease. *Annu. Rev. Phytopathol.* **4**, 9–28.

Large, E. C., and Doling, D. A. (1962). The measurement of cereal mildew and its effect on yield. *Plant Pathol.* **11**, 47–57.

LeClerg, E. L. (1964). Crop losses due to plant diseases in the United States. *Phytopathology* **54**, 1309–1313.

Lyman, G. R. (1918). The relation of phytopathologists to plant disease survey work. *Phytopathology* **8**, 219–228.

McCallan, S. E. A. (1969). A perspective on plant pathology. *Annu. Rev. Phytopathol.* **7**, 1–12.

McKinney, H. H. (1923). Influence of soil temperature and moisture on infection of wheat seedlings by *Helminthosporium sativum*. *J. Agric. Res.* **26**, 195–218.

Main, C. E. (1977). Crop destruction—the raison d'être of plant pathology. *In* "Plant Disease: An Advanced Treatise" (J. G. Horsfall and E. B. Cowling, eds.), Vol. 1, pp. 55–78. Academic Press, New York.

Marsh, R. W., Martin, H., and Munson, R. G. (1937). Studies upon the copper fungicides. III. The distribution of fungicidal properties among certain copper compounds. *Ann. Appl. Biol.* **24**, 853–866.

Melchers, L. E., and Parker, J. H. (1922). Rust resistance in winter wheat varieties. *U.S., Dep. Agric., Bull.* **1046**, 1–32.

Moore, W. C. (1943). The measurement of plant diseases in the field. *Trans. Br. Mycol. Soc.* **26**, 28–35.

Moore, W. C. (1949). The significance of plant diseases in Great Britain. *Ann. Appl. Biol.* **36**, 295–306.

Prabhu, A. S., and Prakash, V. (1973). The relation of temperature and leaf wetness to the development of leaf blight of wheat. *Plant Dis. Rep.* **57**, 1000–1004.

Preece. T. F. (1971). Disease assessment. *In* "Diseases of Crop Plants" (J. H. Western, ed.), pp. 8–20. Wiley, New York.

Redman, C. E., King, E. P., and Brown, I. F., Jr. (1969). "Tables for Conversion of Barratt-Horsfall Rating Scores to Estimated Mean Percentages," Mimeo, unnumbered. Elanco Products Co., Indianapolis, Indiana.

Reichart, I., Minz, G., Palti, J., and Hochberg, N. (1944). Trials for the control of grape vine diseases. *Rehovoth Agric. Res. Sta. Bul.* **5**,1–12.

Romig, R. W., and Calpouzos, L. (1970). The relationship between stem rust and loss in yield of spring wheat. *Phytopathology* **60**, 1801–1805.

Rusakov, L. F. (1927). A combination scale for estimating the development of rusts. *Bolezni Rast.* **16**, 179–185 (in Russian) (cited by Chester, 1950).

Tehon, L. R. (1927). Epidemic diseases of grain crops in Illinois. 1922 to 1926. The measure of their prevalence and destructiveness and an interpretation of weather relations based on wheat leaf rust data. *Ill. Nat. Hist. Surv., Bull.* **17**, 1–96.

Tehon, L. R., and Stout, G. L. (1930). Epidemic diseases of fruit trees in Illinois, 1922–1928. *Ill. Nat. Hist. Surv., Bull.* **18**, 415–502.

Ullstrup, A. J., Elliott, C., and Hoppe, P. E. (1945). "Report of the Subcommittee on

Methods of Reporting Corn Disease Ratings," Mimeo, unnumbered Publ. U.S. Dep. Agric. Div. Cereal Crops Dis., Washington, D.C.

van der Plank, J. E. (1963). "Plant Diseases: Epidemics and Control." Academic Press, New York.

van der Zwet, T., Oitto, W. A., and Brooks, H. J. (1970). Scoring system for rating the severity of fire blight in pear. *Plant Dis. Rep.* **54**, 835–839.

Watson, M. A., Watson, D. J., and Hull, R. (1946). Factors affecting the loss of yield of sugar beet caused by beet yellows virus. I. Rate and date of infection; date of sowing and harvesting. *J. Agric. Sci.* **36**, 151–156.

Chapter 7

Inoculum Potential

RALPH BAKER

I. THE CONCEPT

Inoculum potential has had good press. Almost all compendiums, texts, and tomes treat it. If the literature is any indication, scholars use the term in various senses. Yet its properties are such that it can be treated with great precision and detail. This anomaly should not disturb us. Wald (1958) assured us that "all great ideas come in pairs, the one the negation of the other, and both containing elements of truth." Undoubtedly inoculum potential is a great idea.

Horsfall (1932) initiated the idea by equating it with inoculum density. Garrett (1956) considered that there was more to potential than the number of propagules threatening the host; he felt activity was influenced profoundly by environment. In the graceful style characteristic of his nationality, he defined inoculum potential as the energy available for colonization of a substrate (host) at the surface of the substrate to be colonized. Energy! Some viewed this word as a challenge (Baker, 1965), while others viewed it as a treacherous phrase—energy in plant pathology was a vain substitute for sense. Dimond and Horsfall (1960) entered the fray. After all, the practical measurement of inoculum po-

137

tential is host response. In this sense, it is "the result of the action of the environment, the vigor of the pathogen to establish an infection, the susceptibility of the host and the amount of inoculum present." This could be a definition of disease. Baker (1965, p. 415) provides a diagram relating all of these definitions.

Credimus ut cognoscamus (St. Augustine). To expand developing theory, I believe Garrett's definition to be most useful.

The subject of inoculum potential could occupy all the volumes of this treatise—small wonder every reviewer takes a different tack (Dimond and Horsfall, 1960, 1965; Garrett, 1960; Baker, 1965; Wilhelm, 1966). Perhaps it is now opportune to develop quantitative aspects. These were at the heart of the great idea in the first place.

First, Eq. (1) puts inoculum potential in perspective (Baker, 1971a).

$$\text{Disease} = \text{inoculum potential} \times \text{disease potential} \qquad (1)$$

For quantitative analysis, inoculum potential is inoculum density (made up of propagules having a certain amount of virulence and vigor) as modified by capacity (or environmental) factors. Disease potential is the susceptibility of the host over the period of its life cycle as influenced by disease proneness.

In this chapter, we will first treat events leading to the presence of propagules in the infection court. Then the impact of this inoculum on the host will be discussed. The approach will be to quantify these events when possible, although such dismemberment leads to disjointure. If the treatment appears disjointed, it is because these ventures into quantitative analysis are just beginning. Models yet unproved can be synthesized but "nothing is achieved until thoroughly attempted" (Sidney).

II. THE INOCULUM

A. The Quantities Produced

Heterotropic plant pathogens typically increase at the expense of a host, although commensal activity in rhizosphere or phyllosphere and saprophytic growth occurs in some situations.

A single cotton plant may contain more than 25×10^4 microsclerotia of *Verticillium* (Evans *et al.*, 1966). These are incorporated into soil when plant debris enters the ground. Cultivation distributes the propagules in soil where they may germinate and sporulate or produce secondary microsclerotia several times in the vicinity of organic matter or the rhizosphere of nonhost plants (Emmatty and Green, 1969; Farley *et al.*, 1971; Lacy and Horner, 1966; Menzies and Griebel, 1967) and still have enough inoculum potential to induce disease. Such examples of the qualitative aspects of the inoculum production can be found repeatedly in the literature for both foliage and root diseases. Critical quantitative analyses of the etiology of host–pathogen relationships are found less frequently.

Populations of pathogens are measured in two ways: (1) directly, by assaying the actual or relative inoculum level and (2) indirectly, by measuring the host colonization or the damage that inoculum can produce on a crop.

Soil microbiologists have developed cunning methods for direct or indirect means of determining inoculum levels in soils (Tsao, 1970). Various types of spore traps can be used for airborne inoculum (e.g., Hirst et al., 1967). There is a tendency among foliage disease pathologists, however, to use the more indirect method of assaying populations of lesions. Airborne inoculum is capricious; quantities are not necessarily related to development of disease since capacity factors play such an important role in epidemics (e.g., Kerr and Shanmuganathan, 1966).

Infection in the host eventually produces the inoculum with which we should be concerned in this chapter. Equations describing, analyzing, and/or predicting propagule formation and, correspondingly, inoculum potential as it changes with time have been summarized by Kranz (1974). Within the substrate of the host, production of inoculum might well take the form of Blackman's (1919) compound interest law of growth. In the logarithmic form, it is:

$$x = \ln x_0 + rt \qquad (2)$$

where x_0 is the amount of inoculum at time $t = 0$ and r the relative rate of sporulation. The compound interest, as expressed in Eq. (2), however, increases to infinity. This nonbiological characteristic is eliminated by the logit equation (Yule, 1925) which provides a mathematical description of the typical S-shaped growth curve. In effect, this is the reasoning used by van der Plank (1963) in his analysis of compound interest epidemics but could have wider application in analyses of populations of soil-borne organisms over long time periods and sequences of crops.

Scientists can view the soil-borne pathogens from a secure viewpoint. They are fixed in the three-dimensional enclosure of soil. In cultivated crops, propagules are distributed randomly (Nash and Snyder, 1962), making mathematical definitions of positions in space and time possible. Position and distance from the below-ground infection court predictably influence disease development.

Distance relationships, as a function of inoculum density, were calculated by Baker and McClintock (1965) and corrected by McCoy and Powelson (1974). In the latter paper (in Fig. 2B), distance between propagules decreases prodigiously as the number rises to 3000 and less rapidly to 10,000 propagules/g of soil. Distance relationships between pathogen and host infection court also assume similar proportions. These figures turn out to be near the maximum for many of those found for field populations of many pathogens attacking fixed infection courts in soil (e.g., Nash and Snyder, 1962). When constrained by space relationships, are these pathogens so endowed with reason that they evolve to a population that expends little energy in unnecessary multiplication that would not bring them much closer (on average) to their hosts?

There are other pathogens that compensate by employing different mechanisms. For example, the propagules of *Rhizoctonia solani* are low in naturally infested soils, one to nine propagules/g dried soil (Ko and Hora, 1971). This mercurial fungus is unique in its remote sensing ability. It responds quickly to infection sites when located within the rhizosphere (Baker and Martinson, 1965). Apparently it also is synergistic; it induces proportionately more disease than the sum of its parts. With these attributes, who needs a high population?

Again, some hosts virtually "disseminate themselves" to soil pathogens because their infection courts (root tips) invade inoculum distributed in the soil substrate. Penetration only occurs in the root tips. In such a situation a few propagules can encounter a great amount of host tissue. Thus Ashworth *et al.* (1972) reported 3.5 microsclerotia of *Verticillium*/g of soil could induce 100% disease in cotton.

These observations and correlations, if valid, may eventually become an ecological principle: The inoculum density of a given soil-borne pathogen usually is no greater than that needed for reasonably efficient positioning in three-dimensional space to cover the available infection courts.

Space relationships predictably influence propagules that can swim to their infection courts. Incidence and severity of *Phytophthora fragariae* infection in strawberry plants are inversely related to the distance of seedlings from an inoculum source (Maas, 1976).

B. Types of Inoculum

Not all propagules are equal. Obviously, genetic capacities differ among races. Ecological, morphological, and physiological characteristics of various pathogens influence inoculum potential, especially as it relates to survival.

Richly endowed propagules may have more inoculum potential than poor ones. Phillips (1965) produced conidia of *Fusarium roseum* on media of high and low nutrient content. At relatively low inoculum densities, the "fat" conidia induced greater damage on carnations than did undernourished ones at equal inoculum densities. The size of the propagules also may be correlated with inoculum potential. Propagule size of *R. solani* has effects on hyphal growth, infectivity, and virulence to bean seedlings (Henis and Ben-Yephet, 1970). Even if viable, not all propagules are capable of or are able to induce infections.

C. Survival

Once inoculum is produced, it is subjected to the vicissitudes of the environment. In this phase the inoculum density eventually encountered by the host is determined by the survival characteristics of the pathogen.

The death rate of units of inoculum per unit of time, designated r by Papavizas (1965), is the logical descriptive measurement of survival. Three equations and transformations have been proposed for calculation of r. The half-life equation (Yarwood and Sylvester, 1959) predicts the time required for a 50% death of inoculum. Dimond and Horsfall (1965) suggest the initial lag phase (resulting when propagules are made up of more than one cell) may cause erroneous results when this equation is used and prefer the semilogarithmic transformation of survival rate. In this, the lag phase is compensated for and adequately conforms to the tendency of inoculum to die in soil at a logarithmic rate. The log-probit transformation (Bliss, 1935) has also been proposed (Baker, 1971b) for characterizing survival data. Tendency of propagules to die should follow a normal distribution and time (the "toxin") should operate logarithmically.

The last two transformations have been applied to survival studies of *Rhizoctonia* in soil by Benson and Baker (1974c). The nontransformed data often showed an initial increase 2–4 days after inoculum was added. This is apparently related to the well-known capacity for this fungus to grow through soil. Upon further incubation of infested soil, however, the population declined rapidly except under extremely favorable conditions for survival. This phase was noted in other studies (e.g., Ellerbrock and Lorbeer, 1977; Guy and Baker, 1977), using different pathogens. This is explained by the assumptions in the log-probit transformation, i.e., normally distributed components of the population having various abilities to survive. To illustrate this, Benson and Baker (1974c) added three different levels of propagules of *R. solani* to soil and determined r. The higher the amount of inoculum added initially, the higher the ultimate inoculum density maintained by the fungus. After rapid death of a large proportion of the inoculum added to soil, the population remained relatively stable, indicating persistent and relatively long-term survival of residual propagules.

Transformations of data add increments of knowledge about the characteristics of the system. Straight lines generated from sigmoid-shaped survival curves can readily be interpolated to provide information on position of the curve; usually this is the point in time required for 50% of the initial inoculum to die. Extrapolations may predict the time when all inoculum is extinguished, and slope values indicate the rate at which this occurs. Thus, the importance of various environmental parameters to the rate of survival can be compared by objective mathematical analysis.

Benson and Baker (1974c) found good agreement in data analysis using either the semilogarithmic or log–probit transformations. Thus they are complementary and the assumptions upon which they are based appear valid.

III. INOCULUM AND INFECTIONS

A. The Inoculum Density–Disease Curve

The most comprehensive treatment of the inoculum density–disease curve (ID–D) to date has been given by van der Plank (1975) in the first two chapters of his book on the "Principles of Plant Infection." The idealized form of the curve is presented in Fig. 1, although all of its parts may not be manifested in every host–pathogen system.

The characteristics of the initial portion of the ID–D curve (Fig. 1A) did not receive critical attention until van der Plank formulated his "law of the origin." Using numerous examples from the literature, van der Plank demonstrated that Gäumann's (1946) idea of a numerical threshold of infection (i.e., a minimum number of spores greater than 1 is necessary to establish disease) is invalid. He summarizes his treatment of the subject as an experimental law: "When disease is plotted against inoculum, both on arithmetic scales, the curve starts at the origin."

This must be true because at low inoculum densities propagules act independently (Garrett, 1960). In a perfect situation, each propagule produces a single infection. The slope of an ID–D curve would have a value of 1. Nature provides no examples of such efficiency, but, suppose 25% produced successful infections. The slope of the nontransformed ID–D curve would differ but the origin would not. A graphic illustration of this has been published (Baker, 1968, p. 281).

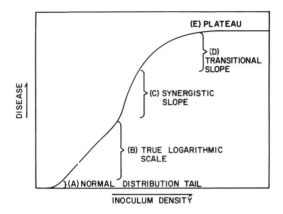

Fig. 1. Idealized elements of the nontransformed inoculum density–disease curve (modified from Baker, 1971a).

Even so, the curve in Fig. 1 gently moves upward to the right of the origin. This is predicted from the normal distribution of host susceptibilities. A calculation of the normal curve gives exponential "tails" at either end. These result from the mathematical assumption that a normal curve reflects the interaction of a number of different inputs and many factors. Certainly disease systems are among the most complex interrelationships known and this mathematical assumption should be valid. Is this normal distribution tail actually seen in experiments? Examples of the data points conforming to this idea can be seen in the papers of Glynne (1925), Martinson (1963), Richardson and Munnecke (1964), Baker (1970), Benson and Baker (1974b), Boelema (1976), and Hanounik et al. (1977). The limited study of this portion of the curve, however, precludes unqualified confirmation at this time.

Add more inoculum and infections increase in proportion producing a logarithmic increase (Fig. 1B). Transformations involving semilogarithmic, log–probit, or log–log conversions are most appropriate for this phase (as treated later).

As inoculum level increases even more, there may be opportunity for propagules to help each other, to pool energy, and synergism may result (Fig. 1C). Synergism of this type was termed facultative by van der Plank (1975), as contrasted with obligate synergism which involves the dubious concept of threshold of infection.

There are at least two methods for detecting facultative synergism in an inoculum density–disease curve. A plot of points relating infections to low amounts of inoculum typically originates at the origin and progresses proportionally in a linear fashion. There may be occasions, however, when further increases of inoculum result in more than a proportional increase in the number of infections. A regression line drawn through these points falls to the right of the origin on the x axis. The classic example of this phenomenon often cited by van der Plank (1963, 1975) is the study of inoculum potential relationships of stem rust of wheat by Petersen (1959). The other method for detecting synergism is to transform the ID–D curve so that it can be compared to a model. If the slope of the transformed curve is greater than predicted, synergism is a possible explanation (Baker, 1971a; Peto, 1953).

Facultative synergism may occur infrequently in plant diseases. The only soil-borne pathogen so far clearly demonstrating this phenomenon is *R. solani* (Baker, 1971a; Benson and Baker, 1974b). This capability may be related to the high potential of the pathogen for competitive saprophytism coupled with its operation in the relatively large volume of soil included in the rhizosphere (Baker and Martinson, 1965). A rhizosphere, as contrasted with a rhizoplane influence, incites proportionately more propagules to activity. Nutrients in exudates from the infection court would be expected to increase inoculum potential even further, providing the opportunity to pool energy. Indeed, Benson and Baker (1974a) observed that *R. solani* forms a "sweater" of hypha around a radish seed. This is indicative of the potential for growth of the fungus in such an infection court.

As hosts and/or host substrates become limiting and multiple infections increase, propagules begin to compete for sites on the infection court, and a transitional slope (Fig. 1D) is reached on the ID–D curve. This eventually ends in a plateau (Fig. 1E) when all hosts are dead or there is no increase in host response. While the theory explaining this portion of the curve is straightforward, plots of points in this area have profound influences on transformations and the mathematical relationships involved in modeling as explained later.

There are reports that the curve may reach a maximum and then diminish with the addition of more inoculum. Such instances may all be special cases. When Davison and Vaughan (1964) applied large numbers of uredospores of *Uromyces phaseoli* to bean leaves, the number of pustules diminished below that induced by lower inoculum densities. The same thing was reported by Domsch (1953) for *Erysiphe graminis* on barley. These can be explained as antagonistic interactions between inoculum units. Rao and Rao (1966) produced similar curves for disease development in cotton in inoculum potential studies with *Fusarium oxysporum* f. sp. *vasinfectum*. Here inoculum was added in an oatmeal–sand culture to soil and, at higher inoculum densities, the added organic material induced biological control. The same phenomenon was reported by Khan (1972). These results indicate the importance of using "natural" inoculum for experimentation.

B. Transformations and Modeling

Transformations may be used as tools not only for straightening ID–D curves but for interpolation and extrapolation. They provide simpler solutions to mathematical problems and provide clues to mechanisms. Their most important functions may be in allowing a better understanding of the biological phenomena involved in a given system providing theory expansions. Transformations, however, are based on assumptions and Kranz (1974) presents a very timely plea for not applying them blindly to any experimental situation: "check suitability first by verification of underlying distribution."

The models and transformations used for the relationship are reviewed elsewhere (Baker, 1971a; Bald, 1970; Dimond and Horsfall, 1965). Briefly, three transformations are suggested.

The semilogarithmic transformation is based on Gregory's (1948) multiple-infection correction. As inoculum density increases, individual infection sites may be invaded by a pathogen many times, but would be recorded only once as diseased. Typically $\ln 1/(1-y)$ [which is van der Plank's (1963) version of the multiple infection transformation] is plotted on the ordinate and the inoculum densities distributed on the abscissa. Plots of points may result in a straight line because infections are more proportional to inoculum density than visual observations of disease per unit. At high inoculum densities, however, inoculum units can compete for single infection courts and corrections may not be adequate.

Logarithmic-probability or probit transformation relates disease (per unit) in probits in a linear fashion to the logarithm of inoculum density. Dimond and Horsfall (1965) consider this a realistic transformation since it is based on normal distribution of susceptibilities of the host and response to a toxin (inoculum, in this case) is logarithmic rather than arithmetic.

The log–log transformation (Baker *et al.*, 1967) used for pathogens distributed in the three-dimensional space of soil, is based on ecological principles. Usually propagules are dormant in soil and germinate in response to the presence of substrate—in the case of a host, exudations from the infection court. Models can be fashioned relating inoculum density to the number of infections expected in combinations of fixed or motile inoculum and fixed or moving infection courts. For inoculum distributed about a fixed infection court, the slope value of a trans-

formed log–log should be 1.0 if successful propagules operate in the
volume of a rhizosphere and 0.67 if they operate only on the rhizoplane.
In systems where inoculum is invaded by a moving infection court (e.g.,
a moving root tip), the slope of the transformed inoculum density–
disease curve should also be 0.67. In practice, disease readings are corrected for multiple infections using $\ln 1/(1 - y)$ where y is disease per
unit. Tests of these models using data reported in the literature suggested
corroboration of predicted slope values (Baker, 1971a). Since then, the
attest of experimentation has continued. Mitchell (1975) demonstrated
a slope of near 1.0 for *Pythium* infection of rye, as predicted (Baker,
1971a), and, in a system involving *R. solani* damping-off of radish, slope
values conformed to the model (Benson and Baker, 1974a,b). Again
slopes were not significantly different from the predicted 0.67 values for
Fusarium wilt of pea (Guy and Baker, 1977) and *Cylindrocladium* black-
rot of peanut (Hanounik *et al.*, 1977).

Not all biological processes are linear even when transformations are
utilized. A growing body of evidence indicates, however, that models involved with the log–log transformation are valid as long as points are
used from the appropriate section of the inoculum density–disease curve.
There may be apparent differences in slope values from those predicted
for three reasons: (1) Obviously facultative synergism may give points
reflecting proportionally greater energy per unit of inoculum than those
at lower population levels. In regression analysis this increases the slope
values above those predicted in the log–log transformation (Baker,
1971a) and above 2.0 in the log–probit analysis (Peto, 1953). (2) If
points are taken from either the normal distribution tail, the transitional
slope, or plateau in the log–log transformation, slope values from regressions are different from those in the logarithmic phase. For example,
if all points are plotted from the data contained in Table I in the paper
by Hanounik *et al.* (1977), slope values of inoculum density–disease
curves are very low. Using points that conform to linearity in the semilogarithmic transformation (this indicates the successful operation of
the multiple infection correction) brings the slopes very close to predicted values for *Cylindrocladium* black-rot of peanut, as shown in their
Fig. 1. (3) Dimond and Horsfall (1965) suggested that slopes of inoculum density–disease curves give information on the mechanism of disease action. Analysis of appropriate data in the literature, however, in
dicated no changes in slope values due to environment, fungitoxins, or
different host susceptibilities (Baker, 1971a). Benson and Baker (1974b),
however, found transformed inoculum density–disease curves for *Rhizoctonia* damping-off of radish to be parallel at 22°, 26°, and 30°C but
significantly different at 15° and 20°C. Careful analysis of the component

parts of the host–pathogen–environmental interaction suggested that nonlinear values of disease potential at the lower temperatures were responsible for the slight changes in slope values. This phenomenon may not be common in other disease systems as considerable synergism exists in *Rhizoctonia* damping-off. This complicates the basic geometry involved.

C. Environmental and Host Influences

Given the log–log phase of an ID–D curve, what may alter its properties? Capacity factors operate to incite propagules to greater or less activity, shifting the position of the curve to the left or right respectively. The position is conventionally defined by the effective dosage of inoculum giving 50% disease incidence (ED_{50}). Competitive saprophytic ability can be correlated with inoculum potential of a primitive parasite like *R. solani* Martinson (1963). Thus, Benson and Baker (1974b) considered the effects of soil pH, moisture, and temperature on growth of this fungus in soil and selected the last factor for intensive study of its effect on ID–D curves. Slope values were parallel at 22°, 26°, and 30°C. The lowest ED_{50} value was at 30°C, indicating that this was the temperature most conducive to disease development. Similarly, a fungistatic chemical shifted ID–D curves for the same disease system to the right with no changes in slope (Benson and Baker, 1974a).

Another factor influencing the position of the curve is the relative resistance or susceptibility of the host. All transformed ID–D curves (log–log phase) comparing hosts varying in resistance have so far been parallel (Baker, 1971a; Hanounik *et al.*, 1977) as predicted by models (Baker, *et al.*, 1967).

It is of interest to know how the curve behaves with hosts having increased resistance. Can the addition of hordes of propagules finally break down all defenses of the host? This does not appear to be the case in the systems so far examined. For instance, Hanounik *et al.* (1977), using the disease system (Cylindrocladium black rot of peanut), generated ID–D curves for the Florigiant (susceptible) and Spancross (resistant) varieties of peanuts. In the nontransformed plot of points, the transitional slope and plateau begin at inoculum densities above 20 propagules/10 g soil for both varieties, but much less disease was observed at all inoculum densities in the resistant Spancross. This can best be explained in terms of normal distribution of host susceptibilities. In either case, the most susceptible were affected at lower inoculum densities, but the proportion affected was much less in the resistant variety. On log–log and log–probit transformations of data, this situation is reflected in a

greater intercept value on the ordinate for the susceptible variety Florigiant.

D. Position of Inoculum in the Infection Court

For pathogens inducing disease on above-ground portions of plants,
the approach for modeling of inoculum–infection interactions on plane
surfaces, such as stem or leaves, is quite straightforward. Approaches
to this were summarized by Kranz (1974). When host and inoculum are
distributed in the three-dimensional substrate of soil, however, some
interesting interrelationships occur that are related to time and space.

Inoculum in soil is usually dormant and becomes active only when
suitable substrates incite germination of propagules. Consider the plight
of propagules that must await the presence of a fixed infection court
like a subsurface hypocotyl or seed. Only those units in the vicinity of
such an infection court have an opportunity to germinate, penetrate,
or infect. For those that attack moving infection courts, however, the
opportunities are more numerous. Examples are the *Fusarium* (e.g.,
Nyvall and Haglund, 1972) and *Verticillium* wilt pathogens (Garber
and Houston, 1966). Root tips moving through soil must activate a
larger proportion of the inoculum in soil than fixed infection courts. The
resulting multiple infections on a single plant increase the number of
infected xylem elements, which increase the subsequent development
of symptoms that characterize this disease system (e.g., Bugbee and
Sappenfield, 1968). The model suggests (Baker *et al.*, 1967) that increasing inoculum densities in this system yield slopes of 0.67 for log–
log transformed ID–D curves. Slopes near this value are reported for
this and related systems (Baker, 1971a; Guy and Baker, 1977). It follows that root density should have profound effects: the more roots, the
greater proportion that encounter inoculum, and the more infections.
All increments of additional infection courts, however, should not increase the slope values of an ID–D curve (log–log), but only move the
position of the curve to the left. A little inoculum can go a long way.
Given reasonable efficiencies for propagules that infect and relatively
uniform susceptible hosts, the model predicts that even very low numbers of propagules should induce numerous infections in roots and the
plants should uniformly display symptoms, provided the environment is
not limiting.

There is controversy over this prediction. Symptom expression in
vascular wilt diseases is influenced profoundly by environment. Pathologists working with vegetatively propagated plants are especially aware
of the hosts' ability to "carry" vascular wilt pathogens internally without

manifesting symptoms until the environment is conducive (e.g., Baker and Phillips, 1962). This infers profound influences of capacity factors especially concerned with disease potential. DeVay *et al.* (1974) reported that the relationship between the concentration of microsclerotia of *V. dahliae* in field soils and percentage of diseased plants was not significantly different from zero; thus the inoculum level is probably not the limiting factor in disease development. Reducing proportions of inoculum, by removal of cotton stalks, leaves, and upper roots predictably would have little effect on incidence of *Verticillium* wilt (Garber and Carter, 1970). In contrast, Ashworth *et al.* (1972) correlated the prevalence of *Verticillium* wilt in cotton fields with the concentration of microsclerotia in the soil. The latest data published on this by Ashworth *et al.* (1976) are plotted in Fig. 2. They show wide deviation from the predicted slope values (0.67) of log–log ID–D curves. If environmental influences are relatively constant between fields or have little effect on symptom expression, then slope values should be near those predicted, as was found (Baker, 1971a) for similar field assays of inoculum and disease incidence for *Sclerotium rolfsii* in sugar beets (Leach and Davey, 1938). Given relatively constant host susceptibility, the only thing that can make slope values in Fig. 2 deviate so widely from those predicted would be the operation of capacity factors. Thus, prediction of the suitability of fields for growing sugar beets, with tolerable economic losses due to rot, can be predicted from sclerotium population, whereas predictions will be less accurate for *Verticillium* wilt.

The time–space relationships involved as a moving root tip invades

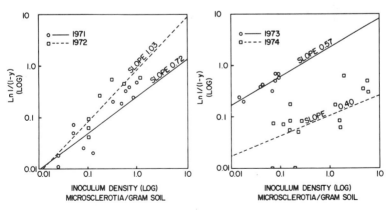

Fig. 2. Log–log transformations of the data collected in field sampling of microsclerotia in soil and incidence of *Verticillium* wilt of cotton by Ashworth *et al.* (1976). The multiple infection correction ln $1/(1-y)$ is applied to the point where it corrects adequately for disease incidence values, in this case below 50%.

soil that contains inoculum, should present possibilities for modeling. Griffin (1969) provided us with a start toward this goal with an elegant collection of critical data. The peanut roots he observed grew at the average rate of 0.56 mm/hr. Up to 73.4% of the chlamydospores of *F. oxysporum* germinated 13 mm from the root tip. This indicates good germination about 23 hr after first encounter with host. The period may be less since this was a 3-day test and most growth occurred during the last 1.5 days. Germination of the spores in Griffin's system was relatively high. This may not be true for others. For instance, Whalley and Taylor (1976) found germination of chlamydospores on pea roots to be no more than 30%.

For the pathogen, there is a race to penetrate before the susceptible infection court passes by and tissue becomes resistant. How much time does the pathogen have to germinate, penetrate, and infect? What are the environmental influences that affect this interaction and what quantitative analyses can be applied in modeling the phenomena? Answers to the problems of "engineering" encountered by the pathogen should make fascinating study.

The models (Baker *et al.*, 1967) predict only one possibility for reasonably large slope value differences in the ID–D curve within a single host-pathogen system. This could occur when the rhizosphere influence under one set of conditions (yielding a slope value of the log–log ID–D curve of 1.0) is reduced to a rhizoplane (slope value 0.67). Rouse (1976) demonstrated this phenomenon using *Rhizoctonia* damping-off of radish.

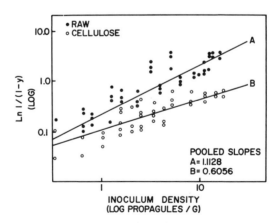

Fig. 3.   Pooled log–log slope values of ID–D curves for *R. solani* preemergence damping-off of radish in raw soils as compared to soils amended with cellulose. Slope values are not significantly different from those predicted for raw soil (1.0) and cellulose-amended soil (0.67) (from Rouse, 1976).

In raw soil, slope values of the ID–D curve (log–log transformation) were near 1 (Fig. 3). In cellulose-amended soil values were not significantly different from 0.67. The amendment, having a wide C/N ratio, affords biological control by stimulating soil microflora to immobilize the soil nitrogen (Baker, 1968) essential for germination and penetration by the pathogen. The only available nitrogen in such a system is that produced in exudates from the seed—these are soon immobilized by competition. Activity of the pathogen is thus limited to the rhizoplane.

IV. THE EQUATION FOR INOCULUM POTENTIAL

Perhaps it is appropriate to define inoculum potential for soil-borne pathogens in mathematical terms though exact numbers cannot be applied to all elements of the equation.

The question first arises as to the units of inoculum potential. To define inoculum potential in terms of energy is a guileful spell unsuited to present technology, so we shall fall back on the traditional units of infections.

Let s equal the number of infections, x the inoculum density, m the slope of the inoculum density–infection curve, and k a parameter (Baker et al., 1967):

$$s = k(x)^m \tag{3}$$

If the only parameter is inoculum density, for soil-borne pathogens the value of m would be 1 or 2/3 on a log–log basis, depending on the host–pathogen relationship involved, although the value could be modified by synergism. The parameter (k) would define the position of the curve (log–log) and would be a function of virulence (v), nutritional status of the propagules (n), and environmental influences on germination and penetration efficiency (f). Thus, on a nontransformed basis, efficiencies of propagules would influence the slope of the inoculum density–infection curve and Eq. (4) would result. Combining these factors and making the relationships linear for soil-borne pathogens, Eq.

$$k = (vnf)^m \tag{4}$$

(5) is derived. This is a mathematical description of inoculum potential

$$\log s = m(\log x + \log v + \log n + \log f) \tag{5}$$

(sensu Garrett, 1956) in Eq. (1).

To obtain the amount of disease per unit (y), the multiple infection correction can be used:

$$s = \ln \frac{1}{1 - y} \tag{6}$$

Solving for y solely as a function of inoculum potential:

$$y = 1 - e^{-s} \tag{7}$$

or

$$y = 1 - e^{-(xvnf)^m} \tag{8}$$

It bears repeating: this does not describe the transitional or plateau portions of the ID–D curve. Here, any value of inoculum potential induces disease in per unit measurements approaching 1.

V. APPLICATIONS

The data derived from studies involving inoculum potential can be an integral and important part of systems analysis leading to the simulation of epidemics. Many plant pathologists are not proponents of mathematical expressions of disease development. Even so, there is agreement that assembly and analysis of the fundamental parts of a system lead to a better understanding of the whole. Perhaps the most important benefit lies in an understanding of which factors are important and which are insignificant.

Nowhere is this more evident than in studies of comparative epidemiology. Major comparisons revolve around the significance or insignificance of the various factors contributing to an epidemic. For example, in comparing root and foliage diseases, many pathogens of the latter are relatively independent of nutrients during germination and are strongly affected by prevailing wide fluctuations in environment. In contrast to root pathogens, they have a relatively short generation time during the development of epidemics. Compound interest transformations may be applied in a single season but theoretical predictions over long periods are complex. Conversely, soil-borne pathogens survive as propagules capable of being defined in time and place in the three-dimensional space of soil. Profoundly influenced by nutrients and the antagonism found in soil, they experience less dramatic changes in the relatively stable subterranean environment during a growing season. Generation time is relatively long, and simple interest or log–probit transformations may apply in a single season. For those types of root diseases (discussed earlier) in which inoculum density is the most important factor in disease development, reasonably simple correlations can be made between the number of propagules present in the field and a prediction of losses.

Integrated pest management has become a new "catch phrase" for the 1970's. If appreciable advances are to be made in this area, it is obvious that the various components proposed for integration be quantified. Further, comparisons between control measures should have a quantitative as well as qualitative basis. Inoculum potential relationships afford a splendid foundation for such analyses.

For example, Rouse (1976) determined the amount of biological control of *Rhizoctonia* damping-off achieved with chitin (Henis *et al.*, 1967). In contrast to ID–D (transformed log–log) curves in which the cellulose amendment reduces the slope values (Fig. 3), the chitin amendment merely shifted the position of the curve to the right; slopes between raw soil controls and chitin-amended treatments were not significantly different. A comparison of the efficiency of the two compounds in achieving biological control is possible using the following equation:

$$\Delta A = \int_{x = I_0}^{x = I} (b_1 - b_1')x + (b_0 - b_0')dx \qquad (9)$$

where ΔA equals the change in the area under the ID–D curve due to control, b_0 and b_1 are the regression coefficients for the raw soil curve, b_0' and b_1' are the regression coefficients for the second treatment curve, and I and I_0 are the inoculum density limits for integration (arbitrarily chosen as the ED_{10} and ED_{90} values of the controls). Using this approach the overall efficiency of biological control can be compared for the two systems (Table I). In this case, cellulose was more efficient than chitin.

Comprehensive analyses of the behavior of inoculum in relation to crop sequence, environmental incidence, population changes, and disease,

TABLE I

Index of Control Achieved by Chitin and Cellulose Amendments for *Rhizoctonia* Damping-Off of Radish from ID–D Transformed Log–Log Curves [a]

Experiment number	Amendment	
	Cellulose	Chitin
1	0.32380	0.28594
2	0.63858	0.40312
3	0.87114	—
Average	0.61117	0.34453

[a] From Rouse, 1976.

from the standpoint of the theory developed above, are just beginning to be of interest to plant pathologists. For example, Papavizas *et al.* (1975) and Herr (1976) attempted to follow populations of *R. solani* in the field and relate this to survival characteristics and disease. Katan *et al.* (1976) followed the number of propagules following repeated incorporation of plant tissues into infested soil. Such detailed approaches based on a solid theoretical background should be of considerable value in our understanding of epidemiology.

Finally, the most worthwhile application to scientists is that inoculum potential studies provide an opportunity for an analysis of how nature works. Using developing theory, we can ask intelligent questions of our experiments and obtain replies giving insight into mechanisms. This is what science is all about.

References

Ashworth, L. J., Jr., McCutcheon, O. D., and George, A. G. (1972). *Verticillium albo-atrum:* The quantitative relationship between inoculum density and infection of cotton. *Phytopathology* **62**, 901–903.

Ashworth, L. J., Jr., Huisman, O. C., Grogan, R. G., and Harper, D. M. (1976). Copper-induced fungistasis of microsclerotia of *Verticillium albo-atrum* and its influence on infection of cotton in the field. *Phytopathology* **66**, 970–977.

Baker, R. (1965). The dynamics of inoculum. *In* "Ecology of Soil-Borne Plant Pathogens" (K. F. Baker and W. C. Snyder, ed.), pp. 395–419. Univ. of California Press, Berkeley.

Baker, R. (1968). Mechanisms of biological control of soil-borne pathogens. *Annu. Rev. Phytopathol.* **6**, 263–294.

Baker, R. (1970). Use of population studies in research on plant pathogens. *In* "Root Diseases and Soil-Borne Pathogens" (T. A. Toussoun, R. V. Bega, and P. E. Nelson, eds.), pp. 11–15. Univ. of Calif. Press, Berkeley and Los Angeles.

Baker, R. (1971a). Analyses involving inoculum density of soil-borne plant pathogens in epidemiology. *Phytopathology* **61**, 1280–1292.

Baker, R. (1971b). Simulators in epidemiology—can a simulator of root disease be built? *Adv. Study Inst.: Epidemiol. Plant Dis., Wageningen, Neth.* **1**, 1–2.

Baker, R., and McClintock, D. L. (1965). Populations of pathogens in soil. *Phytopathology* **55**, 495.

Baker, R., and Martinson, C. A. (1965). Epidemiology of diseases caused by *Rhizoctonia solani*. *In* "Rhizoctonia solani: Biology and Pathology" (J. R. Parmeter, ed.), pp. 72–188. Univ. of California Press, Berkeley and Los Angeles.

Baker, R., and Phillips, D. J. (1962). Obtaining pathogen-free stock by shoot tip culture. *Phytopathology* **52**, 1242–1244.

Baker, R., Maurer, C. L., and Maurer, R. A. (1967). Ecology of plant pathogens in soil. VIII. Mathematical models and inoculum density. *Phytopathology* **57**, 662–666.

Bald, J. G. (1970). Measurement of host reactions to soil-borne pathogens. *In* "Root Diseases and Soil-Borne Pathogens" (T. A. Toussoun, R. V. Bega, and P. E. Nelson, eds.), pp. 37–41. Univ. of California Press, Berkeley.

Benson, D. M., and Baker, R. (1974a). Epidemiology of *Rhizoctonia solani* preemergence damping-off of radish: Influence of pentachloronitrobenzene. *Phytopathology* **64**, 38–40.

Benson, D. M., and Baker, R. (1974b). Epidemiology of *Rhizoctonia solani* preemergence damping-off of radish: Inoculum potential and disease potential interaction. *Phytopathology* **64**, 957–962.

Benson, D. M., and Baker, R. (1974c). Epidemiology of *Rhizoctonia solani* preemergence damping-off of radish survival. *Phytopatholgy* **64**, 1163–1168.

Blackman, V. H. (1919). The compound interest law and plant growth. *Ann. Bot. (London)* **33**, 353–360.

Bliss, C. I. (1935). The calculation of the dosage-mortality curve. *Ann. Appl. Biol.* **22**, 134–167.

Boelema, B. H. (1976). Infectivity titrations with *Corynebacterium michiganense* in tomato seedlings. I. The effect of the age of the seedlings at the time of inoculation. *Phytophylactica* **8**, 41–46.

Bugbee, W. M., and Sappenfield, W. P. (1968). Varietal reaction of cotton after stem or root inoculations with *Fusarium oxysporum* f. sp. *vasinfectum. Phytopathology* **58**, 212–214.

Davison, A. D., and Vaughan, E. K. (1964). Effect of uredospore concentration on determination of races of *Uromyces phaseoli* var. *phaseoli. Phytopathology* **54**, 336–338.

DeVay, J. E., Forrester, L. L., Garber, R. H., and Butterfield, E. J. (1974). Characteristics and concentration of propagules of *Verticillium dahliae* in air-dried field soils in relation to prevalence of *Verticillium* wilt of cotton. *Phytopathology* **64**, 22–29.

Dimond, A. E., and Horsfall, J. G. (1960). Inoculum and the diseased population. *In* "Plant Pathology: An Advanced Treatise" (J. G. Horsfall and A. E. Dimond, ed.), Vol. 3, pp. 1–22. Academic Press, New York.

Dimond, A. E., and Horsfall, J. G. (1965). The theory of inoculum. *In* "Ecology of Soil-Borne Pathogens" (K. F. Baker and W. C. Snyder, ed.), pp. 404–415. Univ. of California Press, Berkeley.

Domsch, K. H. (1953). Über den Einflusz photo periodischer Behandlung auf die Befallsintensität beim Gerstenmehltan. *Arch. Mikrobiol.* **19**, 287–318.

Ellerbrock, L. A., and Lorbeer, J. W. (1977). Survival of sclerotia and conidia of *Botrytis squamosa. Phytopathology* **67**,219–225.

Emmatty, D. A., and Green, R. J., Jr. (1969). Fungistasis and the behavior of the microsclerotia of *Verticillium albo-atrum* in soil. *Phytopathology* **59**, 1590–1595.

Evans, G., Snyder, W. C., and Wilhelm, S. (1966). Inoculum increase of the *Verticillium* wilt fungus in cotton. *Phytopathology* **56**, 590–594.

Farley, J. D., Wilhelm, S., and Snyder, W. C. (1971). Repeated germination and sporulation of microsclerotia of *Verticillium albo-atrum* in soil. *Phytopathology* **61**, 260–264.

Garber, R. H., and Carter, L. (1970). Relationship of crop residue management to *Verticillium* wilt of cotton. *Proc. Beltwide Cotton Prod. Res. Conf., 1970* p. 25.

Garber, R. H., and Houston, B. R. (1966). Penetration and development of *Verticillium albo-atrum* in the cotton plant. *Phytopathology* **56**, 1121–1126.

Garrett, S. D. (1956). "Biology of Root-infecting Fungi." Cambridge Univ. Press, London and New York.

Garrett, S. D. (1960). Inoculum potential. *In* "Plant Pathology: An Advanced Treatise" (J. G. Horsfall and A. E. Dimond, eds.), Vol. 3, pp. 23–56. Academic Press, New York.

Gäumann, E. (1946). "Pflanzliche Infektionslehre." Birkhaeuser, Basel.

Glynne, M. D. (1925). Infection experiments with wart disease of potatoes *Synchytrum endobioticum* (Schilb.) Perc. *Ann. Appl. Biol.* 12, 34–60.

Gregory, P. H. (1948). The multiple-infection transformation. *Ann. Appl. Biol.* 35, 412–417.

Griffin, G. J. (1969). *Fusarium oxysporum* and *Aspergillus flavus* spore germination in the rhizosphere of peanut. *Phytopathology* 59, 1214–1218.

Guy, S. O., and Baker, R. (1977). Inoculum potential in relation to biological control of *Fusarium* wilt of peas. *Phytopathology* 67, 72–78.

Hanounik, S. B., Pirie, W. R., and Osborne, W. W. (1977). Influence of soil chemical treatment and host genotype on the inoculum density-disease relationships of *Cylindrocladium* black-rot of peanut. *Plant Dis. Rep.* 61, 431–435.

Henis, Y., and Ben-Yephet, Y. (1970). Effect of propagule size of *Rhizoctonia solani* on saprophytic growth, infectivity, and virulence on bean seedlings. *Phytopathology* 60, 1351–1356.

Henis, Y., Sneh, B., and Katan, J. (1967). Effect of organic amendments on *Rhizoctonia* and accompanying microflora in soil. *Can. J. Microbiol.* 13, 648–650.

Herr, L. J. (1976). In field survival of *Rhizoctonia solani* in soil in diseased sugarbeets. *Can. J. Microbiol.* 22, 983–988.

Hirst, J. M., Stedman, O. J., and Hogg, W. H. (1967). Long distance spore transport, methods of measurement, vertical spore profiles, and the detection of immigrant spores. *J. Gen. Microbiol.* 48, 329–355.

Horsfall, J. G. (1932). Dusting tomato seed with copper sulphate monohydrate for combating damping-off. *N.Y., Agric. Exp. Stn., Geneva, Tech. Bull.* 198, 1–34.

Katan, J., Sofer, S., and Lisker, N. (1976). Effect of incorporating plant tissues into soil on the increased incidence of root diseases and inoculum build-up. *Am. Phytopathol. Soc. Proc.* 3, 286.

Kerr, A., and Shanmuganathan, N. (1966). Epidemiology of tea blister blight (*Exobasidium vexans*). I. Sporulation. *Trans. Br. Mycol. Soc.* 49, 139–145.

Khan, I. D. (1972). Effect of peak vigor and inoculum potential on competitive pathogenic ability and interaction of four root rot pathogens in soil. *Z. Pflanzenkr. Pflanzenschutz* 79, 714–728.

Ko, W., and Hora, F. K. (1971). A selective medium for the quantitative determination of *Rhizoctonia solani* in soil. *Phytopathology* 61, 707–710.

Kranz, J. (1974). The role and scope of mathematical analysis and modeling in epidemiology. *In* "Epidemics of Plant Diseases: Mathematical Analysis and Modeling" (J. Kranz, ed.), pp. 7–54. Springer-Verlag, Berlin and New York.

Lacy, M. L., and Horner, C. E. (1966). Behavior of *Verticillium dahliae* in the rhizosphere and on roots of plants susceptible, resistant, and immune to wilt. *Phytopathology* 56, 427–430.

Leach, L. D., and Davey, A. E. (1938). Determining the sclerotial population of *Sclerotium rolfsii* by soil analyses and predicting losses of sugar beets on the basis of these analyses. *J. Agric. Res.* 56, 619–632.

Maas, J. L. (1976). Comparison of *Phytophthora fragariae* inoculum sources for infection of strawberry plants. *Plant Dis. Rep.* 60, 219–221.

McCoy, M. L., and Powelson, R. L. (1974). A model for determining spacial distribution of soil-borne propagules. *Phytopathology* 64, 145–147.

Martinson, C. A. (1963). Inoculum potential relationships of *Rhizoctonia solani* measured with soil microbiological sampling tubes. *Phytopathology* 53, 634–638.

Menzies, J. D., and Griebel, G. E. (1967). Survival and saprophytic growth of *Verticillium dahliae* in uncropped soil. *Phytopathology* 57, 703–709.

Mitchell, D. J. (1975). Density of *Pythium myriotylum* oospores in soil in relation to infection of rye. *Phytopathology* **65**, 570–575.

Nash, S. M., and Snyder, W. C. (1962). Quantitative estimations by plate counts of propagules of the bean root rot *Fusarium* in field soils. *Phytopathology* **52**, 567–572.

Nyvall, R. F., and Haglund, W. A. (1972). Sites of infection of *Fusarium oxysporum* f. *pisi* race 5 on peas. *Phytopathology* **62**, 1419–1424.

Papavizas, G. C. (1965). Summary and synthesis of papers on antagonism and inoculum potential. *In* "Ecology of Soil-Borne Plant Pathogens" (K. F. Baker and W. C. Snyder, eds.), pp. 436–439. Univ. of California Press, Berkeley.

Papavizas, G. C., Adams, P. B., Lumsden, J. A., Lewis, R. L., Dow, R. L., Ayers, W. A., and Kantzes, J. G. (1975). Ecology and epidemiology of *Rhizoctonia solani* in field soil. *Phytopathology* **65**, 871–877.

Petersen, L. J. (1959). Relations between inoculum density and infection of wheat by uredospores of *Puccinia graminis* var. *tritici*. *Phytopathology* **49**, 607–614.

Peto, S. (1953). A dose response equation for the invasion of microorganisms. *Biometrics* **9**, 320–335.

Phillips, D. J. (1965). Ecology of plant pathogens in soil. IV. Pathogenicity of macroconidia of *Fusarium roseum* f. sp. *cerealis* produced on media of high and low nutrient content. *Phytopathology* **55**, 328–329.

Rao, M. V., and Rao, A. S. (1966). The influence of inoculum potential of *Fusarium oxysporum* f. *vasinfectum* on its development in cotton roots. *Phytopathol. Z.* **56**, 393–397.

Richardson, L. T., and Munnecke, D. E. (1964). Effective fungicide dosage in relation to inoculum concentration in soil. *Can. J. Bot.* **42**, 301–306.

Rouse, D. I. (1976). Model for biological control: *Rhizoctonia solani*. M.Sc. Thesis, Colorado State University, Fort Collins.

Tsao, P. H. (1970). Selective media for isolation of pathogenic fungi. *Annu. Rev. Phytopathol.* **6**, 157–186.

Van der Plank, J. E. (1963). "Plant Diseases: Epidemics and Control." Academic Press, New York.

Van der Plank, J. E. (1975). "Principles of Plant Infection." Academic Press, New York.

Wald, G. (1958). Innovation in biology. *Sci. Am.* **199**, 100–113.

Whalley, W. M., and Taylor, G. S. (1976). Germination of chlamydospores of physiologic races of *Fusarium oxysporum* f. *pisi* in soil adjacent to susceptible and resistant pea cultivars. *Trans. Br. Mycol. Soc.* **66**, 7–12.

Wilhelm, S. (1966). Chemical treatments and inoculum potential of soil. *Annu. Rev. Phytopathol.* **4**, 53–78.

Yarwood, C. E., and Sylvester, E. S. (1959). The half-life concept of longevity of plant pathogens. *Plant Dis. Rep.* **43**, 125–128.

Yule, G. V. (1925). The growth of populations and the factors which control it. *J. Roy. Stat. Soc.* **88**, 1–58.

Chapter 8

Dispersal in Time and Space: Aerial Pathogens

DONALD E. AYLOR

I. INTRODUCTION

Dispersal of the pathogen is pivotal to epidemiology. Without it there can be no epidemic. Yet dispersal remains the most elusive of those elements required to understand and simulate the dynamic course of a disease. This elusiveness resides in part in the physical details of the removal and deposition processes and in the stochastic nature of wind and weather, but mostly in the lack of cohesiveness of an integrative model.

The aerial dispersal, or spread through space, of fungal spores involves three highly interdependent events: liberation, transport, and deposition—

159

or, if you please, takeoff, flight, and landing. Most would agree that turbulence in the air enhances dispersal of airborne spores, but the details are still mostly unspecified. It is my aim here to show (1) how spores are liberated, (2) how turbulence aids their transport in the atmosphere, and (3) how they are deposited on the leaves in the canopies. I then propose to interrelate these factors in the overall process of dispersal. I shall limit my discussion to passive liberation of dry-spore fungi which encompass many pathogens of agricultural importance. I shall also limit my analysis to downwind distances of less than 40 m from the source of inoculum.

Dispersal can be represented mathematically by the following functional equations clearly illustrating the interdependency of liberation Q, diffusional transport T, and deposition D:

$$C(\bar{x}, t) = Q(t)\, T(\bar{x}, t) \tag{1}$$
$$D(\bar{x}, t) = R(\bar{x}, t)\, C(\bar{x}, t) \tag{2}$$

These equations state that D ($\#/m^2$ sec), the number of spores deposited per unit leaf area and per unit time at some vector distance $\bar{x}$ from a source of spores at time t, is given once the local air concentration C (spores/m^3) and the rate of deposition R (m/sec) are known. The concentration, on the other hand, is expressed in terms of the source strength Q, i.e., the number of spores released each instant by the source and the diffusional characteristics of the atmosphere T (sec/m^3).

II. HOW SPORES ARE LIBERATED

While there exists a direct relationship in Eq. (1) between the number of spores in the air (C) and the number liberated (Q), attempts to quantify Q simply by measuring C are inadequate because of our imperfect knowledge of T, the transport. Uncertainty about turbulent transport dictates that we study Q by direct observation of conidia attached to their conidiophores rather than by inference from the observed airborne concentration of spores. This methodology of direct observation is the one followed here.

A. Force Required to Liberate Conidia

To evaluate direct and indirect (e.g., flapping of leaves) effects of air movement on the liberation of conidia we need to know the force that must act on the spore to effect its removal, since mechanisms can readily be reduced to forces. Removal forces have been measured for *Helminthosporium maydis* (Aylor, 1975a) and for *Erysiphe graminis* (Bainbridge and Legg, 1976). These two fungi illustrate the wide range of force required for spore removal. The removal of a mature *H. maydis* conidium requires a force equivalent to about 2000 times its own weight, but mature conidia of *E. graminis* can be removed by only a small fraction of this force.

Spores are not removed continuously with increasing wind speed, but rather most are removed only above a minimum speed characteristic for a species, as was suggested by Stepanov (1935, as cited by Gregory, 1973). This critical speed, for *H. maydis* (Waggoner, 1973), has recently been equated to a critical force (Aylor, 1975a). Certainly size and, seemingly also, strength of attachment exhibit considerable variations within a population of conidia. Interestingly, however, the force required for removal can be remarkably uniform, as shown in Fig. 1. The open circles and solid curve in Fig. 1 represent the observed removal of *H. maydis* conidia as a function of the force applied: the force on a conidium has been normalized by that due to its own weight yielding relative force, a dimensionless quantity. Here the force was applied by a centrifuge and wind currents were eliminated. When the variability in spore size and in the radial distance of individual spores from the axis of rotation is accounted for, removal closely approximates the ideal (shown in Fig. 1 by the dashed line), i.e., no removal for actual force less than about 0.018 dynes with complete removal occurring at this force. This indicates that the strength of the attachment of the spore to its sporophore is essentially independent of individual size.

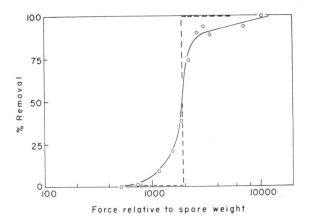

Force relative to spore weight

Fig. 1. The removal of *H. maydis* conidia from their conidiophores by centrifugal force. The open circles (O) are a result of the force being normalized by the weight of the spore. The dashed line, an idealization, is approximated when the variation in size among individual spores is accounted for in determining the actual force in dynes.

B. Strength of Attachment and Ultrastructure

Further understanding of the removal process comes from examining the mechanical strength and structure of the conidium–conidiophore attachment. Detailed knowledge of the mechanical connection coupled with histochemical studies of the connecting material could ultimately suggest strategies for limiting the flight of spores.

Micrographs of the attachment site for *H. maydis* (Aist *et al.*, 1976) reveal that, for this fungus, attachment of mature conidia to conidiophores is by a narrow isthmus of cell wall material only about 0.4 μm thick. Using the known structure, a mechanical model was devised which, together with the measured force of removal, yielded for the strength of the isthmus material an ultimate tensile stress approximately equal to that for wood fiber. It is also interesting that prolonged loading by the applied force of the centrifuge did not significantly increase spore removal compared to cases when the load was applied only briefly. By analogy with engineering materials it appears that this material is more nearly crystalline than amorphous. It remains for biochemists to further elucidate the composition of this material.

C. Wind Required for Removal

Spores are not removed solely at one wind speed but are, of course, removed over a range of speeds. The narrowness of this range for *H. maydis* suggested by Fig. 1 encourages us, however, to model their removal by assuming no removal below some threshold speed and complete removal at or above this speed.

Aerodynamics allows us to equate removal force with an air speed requisite for spore liberation (Aylor, 1975a). For *H. maydis* this speed is about 5 m/sec. Removal requires that air of this speed impinge *directly* on the conidium. Wind near a surface is slowed by viscosity; typically, then, spores reside well within the consequent "boundary layer" near the leaf. The usual steady state analysis of this boundary layer (Schlichting, 1960) indicates that with the exception of leaf edges, an average wind of fully 25 m/sec is required to ensure the requisite 5 m/sec at spore height. Spores of *H. maydis* are, nevertheless, liberated when the average wind in the field is only moderate (Aylor and Lukens, 1974). This seeming paradox can be explained in terms of the transient growth of the viscous boundary layer initiated by a brief but energetic gust of wind (Aylor and Parlange, 1975).

The growth of the boundary layer is illustrated in Fig. 2A. When air above a leaf is started suddenly from rest, the velocity near the surface

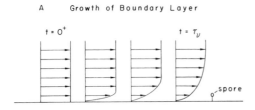

A   Growth of Boundary Layer

$t = 0^+$ $t = \tau_\nu$

spore

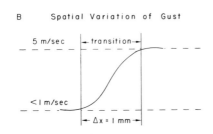

B   Spatial Variation of Gust

5 m/sec   ← transition →

<1 m/sec

← Δx = 1 mm →

Fig. 2. Characteristics of air flow important to spore removal. Transient growth of the viscous boundary layer (A). The length of the arrow represents the relative magnitude of air speed. Spatial variation of a gust of wind which can sweep spores from a sizable area before the boundary layer has time to develop (B).

at the instant of starting $t = 0^+$ is everywhere equal to the bulk air velocity. Immediately thereafter, viscosity begins to slow the air adjacent to the surface and the boundary layer begins to evolve. At some relatively long time $t = \tau\nu$, a dynamic equilibrium exists between the force due to viscosity and inertia; then the boundary layer is fully grown and steady. A schematic of a spore is also shown in Fig. 2A for comparison of scale. By this means, 5 m/sec can be imposed on the spore, if for only a very brief time, by a 5 m/sec gust. Calculations by the formulae of Schlichting (1960) show that the air speed at a spore height of about 150 μm maintains 95% of its free stream value for only 10^{-3} to 10^{-4} sec. This time, while brief, is sufficiently long for a gust to sweep spores from a sizable area in advance of boundary layer growth.

The gust structure which affects this removal over a large area is illustrated in Fig. 2B. The solid line represents the spatial variation of wind speed near the surface due to a sudden gust or puff. Such a transition from slow to fast was observed to occur within a distance of about 1 mm (Aylor and Parlange, 1975). At a given point in space the time for the passage of the transition τ_t is approximately $\tau_t = \Delta x / UPF$ where Δx is the width of the transition and UPF is a mnemonic symbol for the speed of the puff. For the present case τ_t is about 2×10^{-4} sec, which indicates that the transition can be swept over the leaf surface faster than the boundary layer is growing.

Clearly, not all spores, e.g., those of *E. graminis,* require a relative air speed as high as 5 m/sec but some may require more. In any case the principle described here should be universally useful in understanding most cases of passive liberation by wind. While others have appreciated that turbulence enhances spore removal (e.g., Rich and Waggoner, 1962; Eversmeyer *et al.,* 1973; Grace and Collins, 1976), the details of the mechanism have not been clear heretofore.

When the average wind is slight the occurrence of energetic gusts of the type discussed above is rare (Aylor and Parlange, 1975). Even less is known about their frequency of occurrence throughout the space of a plant canopy. Recent advances in anemometry hold promise for uncovering the details of turbulence within canopies (Shaw *et al.,* 1974). Within a corn canopy, speeds some 15 times greater than the local average have been recorded (R. H. Shaw, personal communication, 1977), but so far little is known about the frequency of occurrence or the structure of these gusts.

D. Other Mechanisms of Liberation

Other means of dry liberation often discussed are leaf flapping, abrasion due to rubbing of contiguous leaves, and the puff phenomena associated with the splash of a raindrop impacting on a surface (Hirst and Stedman, 1963). It seems that the phenomenon underlying Hirst and Stedman's splash and puff mechanism is essentially that of the unsteady boundary layer already explained in Section II,C. However, a quantitative analysis should be most interesting.

Abrasion of a lesion by another leaf can readily dislodge spores, but only a small portion of the surface of a leaf is actually abraded by other foliage in a canopy. Reflecting on the complexities of leaf morphology and habit of corn, for example, it is not hard to see that this is true. Nevertheless, precise quantification of rubbing remains difficult and is expected to vary considerably among crops. When corn leaves were restrained from touching, the removal of spores did not differ from those unrestrained when the leaf area index (LAI) was 3.2 or 3.9. The removal was slightly but not significantly greater when LAI was up to 5.2 (Aylor and Lukens, 1974).

Leaf flapping, or the oscillatory movement of leaves caused by wind, can apparently impart sufficient acceleration or force to remove weakly held conidia, such as the mature ones of *E. graminis* (Bainbridge and Legg, 1976). The force due to an accelerating leaf is small, however. It usually amounts to a force of only about 0.15 times the weight of a conidium. Knowing the removal force explained in Section II,A, one can readily evaluate this mechanism for any fungus. Certainly leaf flapping is not significant for *H. maydis*.

E. Reentrainment of Deposited Spores

Once spores have been deposited on plants they may be reentrained by wind. This removal would reduce estimates of trap dose and of deposition rate and would also alter the specification of Q, the source strength of spores. The physics described in Section II,C also pertains here. An important difference is that the spore now rests on the leaf and is held by adhesion rather than by a structural connection. Of course, the stickiness of spores can vary greatly. The force of adhesion of ragweed pollen and *Lycopodium* spores to corn leaves and pine needles has been measured and their removal from corn plants in the field has been observed (Aylor, 1976a). Conclusions from that study are that removal becomes more difficult with time exposed to weather (of course, once infection pegs enter the host, removal is nil) and removal is extremely variable from place to place on a leaf. To underscore the important but still unspecified connection between this spatial variability, our discussion of removal of conidia from conidiophores, and our goal of predicting source strength Q, some representative data for the removal by wind of pollen from replicate 0.5 cm² areas on corn leaves on plants near the edge of a 0.1 ha field are shown in Fig. 3. While average removal did increase with wind speed and turbulence, the range of percentage removal was great.

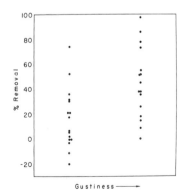

Fig. 3. Removal of ragweed pollen (*Ambrosia elatior*) from corn (*Zea mays L.*) leaves by wind. Increased gustiness increases average removal but its variability over the surface of the leaf is great.

F. The Source Strength Q

The above discussion, while aiding our understanding of passive liberation, poignantly underscores our present inability to specify Q completely from first principles. It is unmistakably clear, however, that Q is not a constant simply to be plugged into diffusion formulae as can be done for material emitted from a factory chimney, but rather $Q = f [u(t), t]$, i.e., Q depends explicitly on instantaneous wind speed u as well as time. The importance of this relationship to transport will be amplified somewhat in Section III and some consequences will be illustrated in Section V.

III. HOW SPORES ARE TRANSPORTED THROUGH THE ATMOSPHERE

The Gaussian Plume and Puff Models

A unique formulation for turbulent diffusion in the atmosphere still does not exist. Instead several alternative formulations have been applied to special problems (Gifford, 1968). For the present purpose of demonstrating the interrelationship of liberation, transport, and deposition in epidemiology we shall discuss the Gaussian description of atmospheric diffusion. The Gaussian formulation is suggested by analogy with Fickian diffusion and also because material carried by the random fluidic movements of the atmosphere, when averaged over a long enough time, does indeed tend toward a normal distribution in space. The distribution of spores through space is described by variances in the three spatial dimensions, and it is assumed that these variances are uncorrelated.

An extremely important difference between Fickian diffusion and atmospheric turbulence exists, however. The standard deviations which gauge diffusion in the atmosphere are not properties of the fluid as in molecular diffusion, but rather are properties of the flow itself. As such they depend on the speed of the wind, on the aerodynamic roughness of the terrain, and crucially on the thermal stability of the atmosphere. Importantly, these standard deviations grow as the time that the material has been exposed to atmospheric motion increases and, since distance from the source is related to travel time by wind speed, they grow with distance from the source. Curves showing how these standard deviations increase with distance and atmospheric stability over level grassland are given by Gifford (1968).

The diffusional transport $T'(x, y, z, t)$ (sec/m^3) of spores continuously released into a steady wind of average speed U from a point source at height H above the ground can be written (Pasquill, 1962)

$$T'(x, y, z, t) = (2\pi\sigma_y\sigma_z U)^{-1} \exp\left(\frac{-y^2}{2\sigma_y^2}\right) \exp\left[-\frac{(H - z)^2}{2\sigma_z^2}\right] \tag{3}$$

where σ_y and σ_x, the diffusion coefficients in the crosswind y and vertical z directions, are functions of downwind distance x and atmospheric stability. The reciprocal of wind speed, i.e., $(U)^{-1}$ can be interpreted as a dilution factor since particles that are sequentially emitted from a point every fractional second are separated by the wind through a distance proportional to U. So far, Eq. (3) does not account for deposition. This problem will be dealt with in Section V.

The formalism underlying Eq. (3) assumes a stationary homogeneous region of turbulence (Gifford, 1968), but we wish to apply the theory near the ground where these idealizations are not met. Fortunately, experiments performed near the ground provide some help by supplying σ's which incorporate the effects of inhomogeneous turbulence (Gifford, 1968). These σ's are restricted in their use in part because the experiments were performed over level grassland but also because the averaging time was always about 10 min. Turbulence consists of random fluid movements (eddies) having a tremendous range of lengths; thus the opportunity for eddies of all sizes to influence the dispersion of spores increases with time of exposure and σ increases with averaging time (Slade, 1968). Furthermore, Eq. (3) assumes that the wind is steady on the average with respect to both direction and speed, requiring additional care in its application (Luna and Church, 1974).

According to our discussion of removal (Section II) an ideal source, which continuously and uniformly spews forth material in the manner of a factory chimney, oversimplifies our situation for fungi. Furthermore, wind and atmospheric stability can vary with time and space and this is not accounted for by Eq. (3). Therefore, we turn our attention away from the continuous plume and consider the transport of a puff of material instantaneously released into the atmosphere. In this case the diffusional transport $T''(x, y, z, t)(\text{m}^{-3})$ of the advected puff is given by Turner (1970):

$$T''(x, y, z, t) = [(2\pi)^{3/2} S_x S_y S_z]^{-1} \exp\left[\frac{-(x - Ut)^2}{2 S_x^2}\right] \exp\left(\frac{-y^2}{2 S_y^2}\right) \exp\left[\frac{-(H - z)^2}{2 S_z^2}\right] \tag{4}$$

where the S's are now functions of time rather than of distance and the growth rate of the S's is fundamentally different from that of the σ's (Gifford, 1957; Turner, 1970). Moreover, unlike Eq. (3) where diffusion in the downwind direction x was neglected, Eq. (4) depends explicitly on diffusion in the x direction and requires a knowledge of S_x. It should also be noted that the reciprocal of U does not appear in Eq. (4) as it did in Eq. (3); the main direct effect of U is to transport the center of puff to distance x in time t. Further approximations and assumptions required to evaluate Eq. (4) for a specific example will be deferred until Section V.

Our goal is to determine the density of inoculum that accumulates on new host material over some period of interest, say a few hours or a day. This requires that we add the effects of many puffs. In Section V we will adapt this cumulative puff method to interrelate liberation, transport, and deposition. This account of diffusional transport has been necessarily brief. The reader can gain much by consulting the following works on spore transport: Gregory (1973), Schrödter (1960), Waggoner (1962), and Waggoner (1965).

IV. HOW SPORES ARE DEPOSITED

The basic mechanisms governing deposition have been well understood for some time (Gregory, 1973; Chamberlain, 1967; Davies, 1966). Excluding submicron-sized particles, for which Brownian diffusion is important, sedimentation due to gravity and impaction, governed by the inertia of the particle, are the most important processes regulating particle deposition to plant surfaces. Still, a most important task remains for aerobiologists, i.e., to determine collection efficiencies for different spores by various plant parts both as individual elements and as constituents of a plant canopy.

A. Particle Parameters

The most important physical characteristic of the spore, as regards deposition, is its measured rate of settling in still air, i.e., its terminal velocity v_s. From v_s, sedimentation can be evaluated directly, and with the help of v_s, impaction efficiencies can also be determined. Terminal velocities for most common fungal spores range from somewhat less than 0.1 to nearly 3.0 cm/sec. Gregory (1973) gives some measured values of v_s and some prescriptions for estimating v_s for spores of various shapes and sizes.

Impaction is governed by the inertia of the spore, i.e., its mass and initial velocity and the aerodynamic flow field around the leaf or target; the aerodynamic flow field depends on the size and shape of target and the air speed past the target. Impaction efficiencies E_i (May and Clifford, 1967; Gregory, 1973) are expressed in terms of a parameter P:

$$P = \frac{(u_i v_s)}{(g\,L)}$$

where g is the acceleration of gravity, L is a characteristic dimension of the target, u_i is the initial speed of the projectile before it enters the fluid flow field influenced by the target (usually the wind speed), and v_s is terminal velocity as before. Small values of P resulting from a small particle, a large target, or slow wind mean that the particle is likely to remain in the air stream and be diverted around the target. Consequently, impaction is inefficient. Large P, on the other hand, due to a massive particle, a small target, or fast wind results in high E_i. May and Clifford (1967) have presented experimental and theoretical curves of E_i versus P for cylinders, spheres, ribbons, and discs that can be used to estimate E_i for various spores on plant parts of diverse shapes.

B. Net Collection Efficiency of Plant Surfaces

Not all spores settling or impacting on a plant will necessarily remain there to germinate. As discussed in Section II,E, some may be reentrained. Fewer spores are captured by dry or nonsticky surfaces than by wet or sticky ones. Chamberlain (1966, 1967) has attributed this to rebound of the particles from the target during impaction.

A first principles analysis of bounce-off is extremely difficult. This is because the mechanics involve an impulsive interaction, and further, this interaction is between materials of unknown elastic properties. One has only to visualize a spiny ragweed pollen colliding with a furrowed leaf or stem (Aylor, 1976a) to appreciate the magnitude of the mechanical problem. In the case of blow-off (cf. Section II,E), a difficult physiochemical problem, i.e., specifying the adhesion of particles to foliage, was supplanted by measurement. If they could be devised, high-speed micromotion pictures showing rebound might aid the quantification of rebounding particles.

In any case Chamberlain (1966, 1967) has shown that bounce-off can reduce the collection efficiencies by impaction significantly below the theoretically expected values and below experimental efficiencies that are obtained when the target surface is either wet or otherwise sticky. Therefore, considerable work remains to be done in order to fully quantify impaction efficiencies. Sedimentation, on the other hand, is a gentle process and bounce-off should rarely occur. Consequently, sedimentation efficiencies are at present much better understood than impaction efficiencies.

C. Deposition in Real Canopies

The testing of mechanisms of deposition in wind tunnels is indispensable but final evaluations must still be made in real crops that experience the much greater turbulence of the atmosphere. Canopies reduce the wind speed but increase turbulence (Shaw *et al.*, 1974). They also present the spore with a labyrinth of target elements.

Deposition of ragweed pollen from a steady point source to leaves in a crop of corn has recently been observed (Aylor, 1975b). The observations of pollen were made by direct counts of deposited pollen in acrylic plastic replicas of leaves. Experiments were conducted on different days when the average wind speed at twice the crop height was 2.0 and 3.8 m/sec. The unique leaf habit and leaf shape of maize and the method of counting pollen on small areas from various places on the leaf offered a means of evaluating the relative importance of sedimentation and impaction in deposition. Figure 4 shows the percentage of the pollen grains deposited per unit of a corn leaf area (A) when it was nearly horizontal, (B) when it was inclined from 30° to 60° from the horizontal, and (C) when it was nearly vertical. Deposition was greatest on more nearly horizontal than on more nearly vertical leaf sections.

Physical considerations show that impaction efficiency should be greatest for narrow objects that are oriented perpendicularly to the wind (cf. Section IV,A). Corn leaves vary considerably in width along their length, being wide for the more horizontal sections and narrower for the more nearly vertical sections. Even though impaction efficiency is greatest on the narrower vertical leaf sections, the least pollen was caught there.

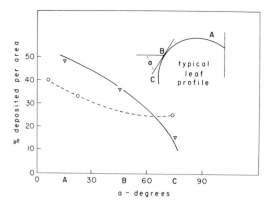

Fig. 4. The number of pollen grains deposited on sections of corn leaf characterized by angle classes A, B, and C compared to the total number collected for all angle classes when the average wind was 2.0 m/sec (▽) or 3.8 m/sec (O). The angle (a) specifies the inclination of the tangent to the leaf surface from the horizontal.

The results as depicted in Fig. 4 indicate that, for moderate winds in a corn crop, sedimentation and not impaction is the main mechanism for depositing spore-sized particles from a steady source. Values of the impaction parameter P also indicate that impaction is nil in these winds for particles 20 μm in size and for leaves as wide as corn. Suzuki (1975) found that conidia of the rice blast disease *Pyricularia oryzae* also are deposited more heavily on horizontal than on inclined leaf surfaces and, like Fig. 4, the effect is more pronounced for slower winds.

The distribution of pollen in Fig. 4 for the 2.0 m/sec wind (inverted triangle) follows closely that expected for sedimentation on inclined surfaces, i.e., deposition is reduced according to the cosine of angle a. For the faster wind, pollen is more evenly distributed (open circle). Since the impaction efficiency for the average wind inside the canopy is expected to be nil, this tendency for a more uniform distribution could be due to either greater impaction, caused by enhanced turbulence, or to reentrainment or redistribution of pollen by wind. Very few grains were found on the underside of leaves; the few found, however, were near the top of the canopy where the wind was strongest.

Finally, deposition velocities v_g computed from measured air concentrations of pollen grains, measured deposition per unit leaf area, and observed wind speed (Chamberlain, 1966), were essentially equal to v_s for all positions in the canopy during 2.0 m/sec wind and also for the lower leaves during fast wind. On upper leaves, however, when the wind was 3.8 m/sec, v_g/v_s was about 3 with a standard deviation of 1. Chamberlain (1966) found that v_g/v_s ranged from 1 to 2 for deposition of *Lycopodium* on grassland for essentially the same range of friction velocity.

Deposition per unit leaf area and per unit time D is expressed in terms of a deposition rate R and concentration C (Section I). R can be specified separately for sedimentation by R_s and for impaction by R_i; in a crop R is usually a combination of both. The deposition rate by sedimentation is given simply by $R_s = v_s$. For impaction on vertically oriented surfaces

$$R_i(z) = bE_i(P)u_i(z)$$

where E_i is impaction efficiency, b is an adhesion efficiency which must be experimentally determined *ad hoc*, and u_i is the initial speed of the particle, as in Section IV,A. R_s is independent of z but R_i varies with z both because of its explicit dependence on u_i and because of its implicit dependence on u_i through P and b. Of course, both R_s and R_i can be diminished by reentrainment, but the complexity of Fig. 3 suggests that this will not be specified easily.

D. Depletion of the Spore Cloud or Plume

As discussed in Section III, deposition reduces the concentration C beyond that due to turbulence alone. There are two fundamental ways of expressing this depletion of airborne particles. When the depletion occurs throughout space, as in a canopy, this loss is specified in the governing equation itself (Gregory, 1945; Slade, 1968).

On a larger scale, a canopy or ground surface might more properly be expressed as a boundary plane for the atmosphere above, depletion being considered to occur along this plane. In this case the formulations given by Horst (1976) or Overcamp (1976) for correcting the concentration given by the Gaussian plume must be used. In Section V the only examples given are for spore transport when depletion occurs uniformly throughout space and not at the ground plane. Thus the present discussion is restricted to distances relatively close to the source of inoculum, i.e., before an appreciable portion of the cloud or plume intercepts the ground. This limited region allows us to focus on the interaction between turbulence and dispersal.

V. INTEGRATION

In this section I shall combine liberation, transport, diffusion, and deposition to estimate the number of spores deposited per unit area on foliage away from a point source of infection in a crop. Deposition patterns, derived from the usual Gaussian plume [Eq. (3)] and having a uniformly continuous source of spores, are contrasted with the patterns expected when spores are removed infrequently but by fast gusts, as described in Section II. For this comparison a constant average wind direction is assumed and attention is focused along the downwind x axis, i.e., along the plume centerline.

A. Deposition Velocity and Depletion of the Spore Cloud

If the source is located at midcanopy, then within the first 30 or 40 m for neutral atmospheric conditions, the bulk of the diffusing material travels within the foliage and the depletion due to deposition is assumed to occur uniformly throughout space. This allows deposition to be computed relatively simply. For great distances, however, most of the spores still in the air travel above the crop and deposition occurs at the boundary of the cloud contacting the crop rather than uniformly from the entire cloud. Calculation of deposition for this case is considerably more complex (Horst, 1976) and will not be considered here.

In the present discussion the number of continuously released spores still airborne at distance x, $Q(x)$, varies with distance according to (Gregory, 1945; Slade, 1968)

$$Q(x) = Q(O) \exp\left[-\int_o^x v_g' \int_{-\infty}^x T'(x, y, H) \, dy \, dx\right] \tag{5}$$

where v_g', defined below, is related to the deposition velocity defined in Section IV and $Q(O)$ is the number of spores released by the source. Combining Eqs. (3) and (5) leads to

$$Q(x)/Q(O) = \exp\left[-\frac{v_g' x^{0.125}}{0.125(\pi)^{1/2} UC_z}\right] \tag{6}$$

where C_z is a coefficient dependent upon the intensity of turbulence, and where v_g', the deposition rate reckoned on a unit land area basis, is related to v_g, the deposition rate reckoned on a unit leaf area basis, by

$$v_g' = \int_0^{\pi/2} v_g \, \mathrm{LAI}(\theta) \, d\theta$$

where LAI is leaf area per land area. Depending on leaf width and particle speed u_i, v_g may also depend on u_i. Table I contains values of $v_g(u_i)$ for a ribbonlike corn leaf and a spore with v_s of 1 to 2 cm/sec. The values were determined using curves of E_i versus P (cf. Section IV,A) given by May and Clifford (1967) and assuming a leaf width of 5 cm. For speeds u_i below 550 cm/sec, deposition by impaction is essentially nil for particles this size and leaves this wide. As explained earlier, v_g to a narrower leaf (e.g., rice) for these same conditions is expected to be much greater.

B. The Gaussian Plume

The transport T' along the center line of a Gaussian plume, i.e., $y = O$, $z = H$ [cf. Eq. (3)], given by

$$T'(x, O, H) = (2\pi\sigma_y\sigma_z U)^{-1} \tag{7}$$

is plotted in Fig. 5A for ·Class "D" stability (Gifford, 1968) and for average wind speeds of 0.5, 1.0, and 3.0 m/sec. According to Table I, for a corn leaf $v_g = v_s$ for these speeds. Therefore, combining Eqs. (2), (6), and (7), where $R(x, t)$ from Eq. (2) is now just the constant v_s, and multiplying by the total time of release gives $\overline{D}'(x)$, the total number of spores deposited per unit horizontal leaf area with downwind distance. The results of this calculation, shown in Fig. 5B, reflect the greater deposition at lower wind speeds dictated by the dependence of Eq. (6) on U^{-1}.

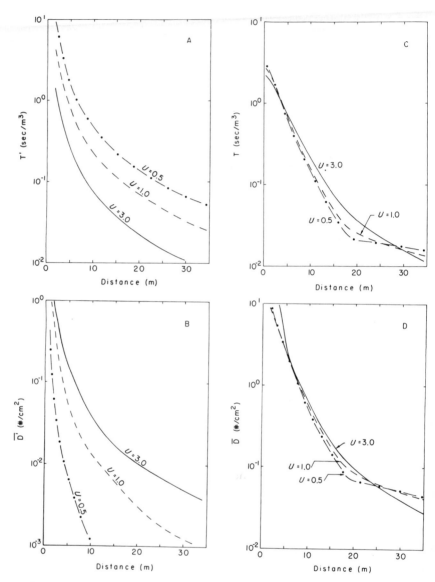

Fig. 5. Comparison of transport and deposition computed by using the plume model [Eq. (7)] with a steady continuous removal of spores (A and B) and by using the advected puff model [Eq. (8)] with infrequent spore removal by gusts (C and D). Calculations for (A) and (B) are based on Class "D" atmospheric stability (Gifford, 1968). For (B) and (D) it is assumed that a total of 1 million spores are released in each case. The parameter values used to obtain the results in (C) and (D) are: UTHLD = 5 m/sec and U = 3.0 m/sec, UPRIM = 7.22 m/sec; and U = 1.0 m/sec, UPRIM = 5.84 m/sec; and U = 0.5 m/sec, UPRIM = 5.44 m/sec.

TABLE I

Deposition Velocity v_g to a Ribbonlike Corn
Leaf 5 cm Wide Due to Impaction for
Various Particle Speeds u_t

u_t (cm/sec)	v_g (cm/sec)
700	35
650	16
600	9
580	6
550	nil

C. The Advected Gaussian Puff

The transport T (sec/m³) along the center line of a passing puff is obtained by integrating Eq. (4) with $y = O$, $z = H$ over the time of passage for the puff, i.e., $\Delta t'(x)$ at x, or

$$T(x, O, H) = \int_{\Delta t'(x)} [(2\pi)^{3/2} S_x S_y S_z]^{-1} \exp\left\{-\frac{[x - UPF(t)t]^2}{2 S_x^2}\right\} dt \qquad (8)$$

where the standard deviations, the S's are functions of time, and UPF, the speed of the puff carrying the spores, is also a function of time. The transport T is plotted in Fig. 5C for later comparison with T' for the plume.

Defining the crosswind integrated transport by $\overline{T}$

$$\overline{T}(x, H) = \int_{\Delta t'(x)} \int_{-\infty}^{\infty} [(2\pi)^{3/2} S_x S_y S_z]^{-1} \exp\left\{-\frac{[x - UPF(t)t]^2}{2 S_x^2}\right\} \exp\left(-\frac{y^2}{2 S_y^2}\right) dy \, dt \qquad (9)$$

the deposition depletes the spores in the air according to

$$Q(x)/Q(O) = \exp\left(-\int_0^x v_g' \{UPF [x'(t)]\} \overline{T}(x', H) \, dx'\right) \qquad (10)$$

The total number of spores per unit leaf deposited along the x axis is obtained by combining Eqs. (2), (8), and (10) giving

$$\overline{D}(x) = v_g \{UPF [x(t)]\} Q(x) T(x, O, H) \qquad (11)$$

where v_g is functional of x.

To carry out these evaluations for Figs. 5C and 5D, the S's in Eqs. (8) and (9) had to be specified as functions of time. According to our model, spores are liberated periodically by turbulence only when the air speed at spore height due to a puff reaches UPRIM which equals or exceeds

some threshold speed UTHLD. For *H. maydis* UTHLD is about 5 m/sec.
(Section II,C). The speed of the puff UPF carrying the liberated spores
is, of course, initially UPRIM and decays exponentially to the average
speed U with a time constant TL that is related to the Lagrangian time
scale of turbulence (Gifford, 1968). During this initial period while UPF
exceeds U the S's are all assumed equal and to increase by relative diffusion (Csanady, 1973; Aylor, 1976b), i.e.,

$$S = S_o + e^{1/2} t^{3/2} \tag{12}$$

where S_o is the initial standard deviation of the puff of spores taken to
be equal to the width of a corn leaf, and e is the rate of eddy–energy
transfer. Once UPF decays to U, the puff is convected at the constant
speed U and S_x is taken equal to S_y (Lumley and Panofsky, 1964):

$$S_x = S_y = c_x(Ut)^{0.875} + c_1$$
$$S_z = c_z(Ut)^{0.875} + c_2$$

where c_1 and c_2 are constants used to match S_x and S_z at the time
TPRIM $= -$TL ln(U/UPRIM). The integrations of Eqs. (8) and (9)
are carried out numerically realizing that, at a given distance x, the
maximum contribution to the integral occurs at a time when the instantaneous position of the puff center located at XCONV(t) is at x, i.e.,
$\int_o^t \mathrm{UPF}(t)dt = x$. Applying the values of Table I to Eqs. (10) and (11)
and recalling the definition of v_g' the results for $\overline{D}(x)$ shown in Fig. 5D
are obtained. The number deposited per unit area of leaf depends on
the location on its surface (e.g., see Fig. 4). Within the first few meters
from the source in the fastest wind, i.e., 3 m/sec, deposit is calculated
for the nearly vertical leaf sections. Otherwise deposit is calculated for
the nearly horizontal leaf sections.

D. Comparison of Results

Several interesting contrasts are apparent in Fig. 5. First it should be
stated that the advected puff model with UTHLD $= 0$ gives results almost identical to the plume model (Fig. 5A). This is a good check on
the method of evaluating Eq. (8) since, in principle, the integration of
Eq. (4) over time can be shown to be equivalent to Eq. (3) within a
close approximation (Sutton, 1953). However, when UTHLD $= 5$, the
transport T for the puff (Fig. 5C) is strikingly different from T' for the
case of the plume (Fig. 5A).

Since the initial size of the puff is taken to be the leaf width for all
speeds, and since in all cases the puff is convected quite rapidly over
the first few meters, the transport for these wind speeds is similar. This
contrasts with the sizable differences that are apparent near the source

according to the plume model (Fig. 5A). Perhaps more interesting is the apparent reversal in T compared to T' with average wind speed U. The values of T (Fig. 5C) result because the material moves considerably more rapidly than U for the initial period, i.e., t is less than TPRIM. Since the puff has less time to diffuse, its growth is less [Gifford, 1968; see also Eq. (12)] and consequently it is more concentrated than if UTHLD had been 0.

Another striking contrast is made between Figs. 5B and 5D for $\overline{D}'$ is $\overline{D}$, i.e., the total number of spores deposited per unit leaf area. Both $\overline{D}'$ and $\overline{D}$ are based on a total release of 1 million spores. Since deposition from the plume, in our example, is due only to sedimentation, $\overline{D}'$ is greater for the smaller U very near the source (not shown on graph). This reduces concentration greatly. Consequently $\overline{D}'$ is less for the smaller U farther from the source even though T' from Fig. 5A indicates an opposite trend. $\overline{D}$ for the puff, on the other hand, is greater for the larger U near the source but less farther away. This is because in the present model the spores are traveling faster than U for the first few meters and net deposition by impaction is appreciable near the source. Bounce-off and reentrainment have been neglected in these calculations.

Some interesting conclusions can be drawn from the model about deposition gradients near the source for conidia that have different strengths of attachment to the conidiophores. Very weakly held conidia, like those of mature *E. graminis*, should disperse according to Fig. 5B, while those more strongly held, such as *H. maydis*, should disperse according to Fig. 5D. Weakly held conidia not relying on fast gusts for removal can disperse during slight and reasonably steady wind. Under these conditions most will be deposited quite near the source. In faster winds, and for a crop and a spore such that impaction efficiency E_i is small, fewer conidia will be deposited nearby and more will be deposited far away (Fig. 5B). Strongly held conidia, on the other hand, may disperse in quite a different way; they may be more heavily deposited nearby in faster than in slower average winds and be more heavily deposited at a distance in slower than in faster average winds.

VI. CHALLENGES

Much about the aerial dispersal of spores has been left unsaid, and this is necessarily so. Many questions remain open, challenging not just the meteorologist but the physicist, the aerodynamicist, and the plant pathologist as well.

The present model illustrates the importance of the strength of conidium–conidiophore attachment and its relation to the critical wind speed required for spore removal. Plant pathologists can measure the force required to detach various kinds of spores. Determining the ultrastructure of the attachment and the chemical composition of the material comprising this attachment offers a difficult but important challenge. This knowledge could underlie a logical strategy to control the dispersal of certain spores. Models of the type developed here can help predict how possible alterations in the strength of the attachment might modify dispersion.

We must improve our understanding of aerodynamic flow around plant leaves in the uncontrolled turbulent flow of natural wind. While I have considered here the important, unsteady feature of boundary layer flow around a leaf, the underlying basis is still two-dimensional. The air flow around a leaf is extremely complex and usually it is three-dimensional. The region of the leaf where the shear stress on a spore is greatest is ill defined and moves from point to point on the leaf. The progress on three-dimensional boundary layers now being made by fluid dynamicists should be of help here when applied to ventilation in crops.

The net trapping efficiency of spores by various plant elements remains open, particularly under the conditions of impaction when particles might rebound. The physics underlying particle rebound involves the impulsive interaction of inelastic materials, and present knowledge makes a complete solution almost impossible. While efforts to uncover the physics is being made, it seems that knowledge of net collection efficiency during impaction must continue to depend on *ad hoc* experiments.

The greatest challenges in modeling the aerial dispersal of spores remain for the micrometeorologist. The structure of the turbulent flow within canopies, which has long been studied, is now beginning to yield somewhat to recently developed anemometers. Even so, the present aerobiological model asks for a different analysis than is usual. Most problems of interest to the micrometeorologist, e.g., heat and mass exchange within canopies, depend largely on average wind rather than on infrequent peak gusts as spore entrainment does. Not only do we require knowledge about the frequency of occurrence and spatial variation of these gusts but, importantly, we must know their trajectories. In the present model I have assumed that these gusts follow, on the average, the direction of the mean wind and that their speed decays to that of the mean in a time related to the Lagrangian time scale of turbulence. These assumptions remain to be proven.

Considerations of dispersal have been limited here to the first few meters where it is assumed that the bulk of the diffusing spores are moving through the canopy. To follow the diffusing cloud out of the canopy and reasonably far from the source requires further formulation. Allowance for changes in wind direction and vertical position of the inoculum source within the canopy remain to be properly integrated into dispersal models.

Finally, dispersal gradients do not necessarily reflect infection gradients (Gregory, 1973). To obtain infection gradients requires further integration of the life cycle of the fungus and of the environment. The objective of aerobiological models is to supply the simplest algorithms that nevertheless contain the fundamental physics of dispersal. Their use, however, goes beyond that of a submodel for a disease simulator. Fully developed aerobiological models will supply, in their own right, a logical basis for the strategies needed to control epidemics.

References

Aist, J. R., Aylor, D. E., and Parlange, J.-Y. (1976). Ultrastructure and mechanics of the conidium-conidiophore attachment of *Helminthosporium maydis*. *Phytopathology* **66**, 1050–1055.

Ayler, D. E. (1975a). Force required to detach conidia of *Helminthosporium maydis*. *Plant Physiol.* **55**, 99–101.

Aylor, D. E. (1975b). Deposition of particles in a plant canopy. *J. Appl. Meteorol.* **14**, 52–57.

Aylor, D. E. (1976a). Resuspension of particles from plant surfaces by wind. *In* "The Symposium on the Atmosphere-Surface Exchange of Particles and Gases" (G. Sehmel, ed.), U.S. Energy Research and Development Admin. Doc. No. CONF-740921, pp. 791–812. ERDA, Oak Ridge, Tennessee.

Aylor, D. E. (1976b). Estimating peak concentrations of pheromone in the forest. *In* "Perspectives in Forest Entomology" (J. F. Anderson and H. K. Kaya, eds.), pp. 177–188. Academic Press, New York.

Aylor, D. E., and Lukens, R. J. (1974). Liberation of *Helminthosporium maydis* spores by wind in the field. *Phytopathology* **64**, 1136–1138.

Aylor, D. E., and Parlange, J.-Y. (1975). Ventilation required to entrain small particles from leaves. *Plant Physiol.* **56**, 97–99.

Bainbridge, A., and Legg, B. J. (1976). Release of barley-mildew conidia from shaken leaves. *Trans. Br. Mycol. Soc.* **66**, 495–498.

Chamberlain, A. C. (1966). Transport of Lycopodium spores and other small particles to rough surfaces. *Proc. R. Soc. London, Ser. A* **296**, 45–70.

Chamberlain, A. C. (1967). Deposition of particles to natural surfaces. *Symp. Soc. Gen. Microbiol.* **17**, 138–164.

Csanady, G. T. (1973). "Turbulent Diffusion in the Environment." Reidel Publ., Dordrecht, Netherlands.

Davies, C. N. (1966). Deposition from moving aerosols. *In* "Aerosol Science" (C. N. Davies, ed.), pp. 393–446. Academic Press, New York.

Eversmeyer, M. G., Kramer, C. L., and Burleigh, J. R. (1973). Vertical spore concentrations of three wheat pathogens above a wheat field. *Phytopathology* **63**, 211–218.

Gifford, F. A., Jr. (1957). Relative diffusion of smoke puffs. *J. Meteorol.* **14**, 410–414.

Gifford, F. A., Jr. (1968). An outline of theories of diffusion in the lower layers of the atmosphere. *In* "Meteorology and Atomic Energy" (D. H. Slade, ed.), pp. 65–116. U.S. At. Energy Comm., Oak Ridge, Tennessee.

Grace, J., and Collins, M. A. (1976). Spore liberation from leaves by wind. *In* "Microbiology of Aerial Plant Surfaces" (C. H. Dickinson and T. F. Preece, eds.), pp. 185–198. Academic Press, New York.

Gregory, P. H. (1945). The dispersion of airborne spores. *Trans. Br. Mycol. Soc.* **28**, 26–72.

Gregory, P. H. (1973). "The Microbiology of the Atmosphere." Wiley, New York.

Hirst, J. M., and Stedman, O. J. (1963). Dry liberation of fungus spores by raindrops. *J. Gen. Microbiol.* **33**, 335–344.

Horst, T. W. (1976). A surface depletion model for deposition from a Gaussian plume. *In* "The Symposium on the Atmosphere-Surface Exchange of Particles and Gases" (G. Sehmel, ed.), U.S. Energy Research and Development Admin. Doc. No. CONF–740921, pp. 423–433. ERDA, Oak Ridge, Tennessee.

Lumley, J. L., and Panofsky, H. A. (1964). "The Structure of Atmospheric Turbulence." Wiley (Interscience), New York.

Luna, R. E., and Church, H. W. (1974). Estimation of long-term concentrations using a universal wind speed distribution. *J. Appl. Meteorol.* **13**, 910–916.

May, K. R., and Clifford, R. (1967). The impaction of aerosol particles on cylinders, spheres, ribbons and discs. *Ann. Occup. Hyg.* **10**, 83–95.

Overcamp, T. J. (1976). A general Gaussian diffusion-deposition model for elevated point sources. *J. Appl. Meteorol.* **15**, 1167–1171.

Pasquill, F. (1962). "Atmospheric Diffusion." Van Nostrand-Reinhold, Princeton, New Jersey.

Rich, S., and Waggoner, P. E. (1962). Atmospheric concentration of *Cladosporium* spores. *Science* **137**, 962–965.

Schlichting, H. (1960). "Boundary Layer Theory." McGraw-Hill, New York.

Schrodter, H. (1960). Dispersal by air and water—the flight and landing. *In* "Plant Pathology: An Advanced Treatise" (J. G. Horsfall and A. E. Dimond, eds.), Vol. 3, pp. 169–227. Academic Press, New York.

Shaw, R. H., Silversides, R. H., and Thurtell, G. W. (1974). Some observations of turbulence and turbulent transport within and above plant canopies. *Boundary-Layer Meteorol.* **5**, 429–449.

Slade, D. H. (1968). "Meteorolgy and Atomic Energy." U.S. At. Energy Comm., Oak Ridge, Tennessee.

Sutton, O. G. (1953). "Micrometeorology." McGraw-Hill, New York.

Suzuki, H. (1975). Meteorological factors in the epidemiology of rice blast. *Annu. Rev. Phytopathol.* **13**, 239–256.

Turner, D. B. (1970). "Workbook of Atmospheric Dispersion Estimates." Environ. Protect. Agency, Research Triangle Park, North Carolina.

Waggoner, P. E. (1962). Weather, space, time, and chance of infection. *Phytopathology* **52**, 1100–1108.

Waggoner, P. E. (1965). Microclimate and disease. *Annu. Rev. Phytopathol.* **3**, 103–126.

Waggoner, P. E. (1973). The removal of *Helminthosporium maydis* spores by wind. *Phytopathology* **63**, 1252–1255.

Chapter 9

Dispersal in Time and Space: Soil Pathogens

H. R. WALLACE

I. INTRODUCTION

In this chapter the thesis is proposed that the combination of motility and ability to orientate increases the chance of survival of soil pathogens faced with microenvironmental fluctuations. In this respect they are no different from those organisms above soil level which must disperse to avoid extinction in a changing environment. Like other organisms that lack the ability to disperse and that live in a transient environment, soil pathogens can survive extended periods in the absence of a host plant and at other environmental extremes until conditions are conducive to infection. Soil pathogens have characteristics other than quiescence that increase the chance of their infecting the host and maintaining their infectivity. Passive dispersal of soil pathogens by animals, plants, wind, and water occurs frequently, intensively, and over several miles. The cosmo-

181

politan distribution of some soil pathogens can be attributed to their recent dispersal through human activities.

The frequent use of terms such as survival, natural selection, and adaptation may arouse some misgivings because hypotheses in which they are involved are untestable and may even have an air of "make-believe" about them. However, I have used them freely in this chapter because evolutionary theory is a convenient framework on which to build a discussion of dispersal and useful questions often emerge that lead to experimentation. Furthermore, for the sake of clarity I have tried to indicate the environmental constraints that appear to have endowed an adaptation with survival value (Wallace, 1973).

II. THE ECOLOGICAL SIGNIFICANCE OF DISPERSAL

The places in which plants and animals live are continually changing. If an area of bare land were to be observed for many years, two features would be apparent. First, different species of plants would invade, colonize, decline, become extinct, and be replaced by other species in a more or less distinct succession. Second, the distribution of any one plant species would change during the time it colonized the area. Such changes reflect the problem of survival for plant species because the more frequently a species is suppressed and becomes extinct during succession, the greater the need for dispersal and a high rate of seed production. Without these characteristics the species would be doomed to universal extinction.

Furthermore, it follows that animals and parasites associated with these plants must also disperse to survive. However, there are some organisms whose dispersiveness is greatly restricted because they live in the soil. Some have an aerial phase when dispersal occurs, e.g., soil fungi which occasionally form fruiting bodies from which spores are released. But for those organisms like nematodes, bacteria, and soil fungi which are wholly confined to the soil, dispersal over large distances is a chancy and fortuitous business. In addition, when conditions become unsuitable for growth and reproduction, they have one avenue of survival left, they become quiescent until conditions revert to a more favorable state. Soil pathogens, the subject of this essay, fall into this category. They can be loosely defined as those pathogens in which at least part of their life cycle is subterranean. Airborne passive dispersal and quiescence will be discussed later in this chapter but there is a further important aspect of dispersal of soil pathogens which seems to have received inadequate attention—dispersal at the microenvironmental level.

Many soil pathogens have a motile stage. The dispersive power of the flagellate bacteria, fungal zoospores, fungal mycelia, and the nematodes may have little survival value when the host plant disappears following harvesting, succession, or fire. But it is valid to ask whether there are not any other environmental factors in the soil, the degree and range of whose fluctuations pose similar problems in survival for which the motility of the soil pathogen confers an advantage. Perhaps there are changes at the microenvironmental level which are matched by the dispersive ability of the soil pathogens. To explore this idea it will be necessary to consider the properties of the microenvironments of the soils, particularly their patchy distribution and transient nature.

III. THE ENVIRONMENT OF THE SOIL PATHOGEN

Soil consists essentially of aggregates or crumbs with pore spaces between and within them. The size, frequency, and shape of the aggregates determine the size and tortuosity of the pore spaces. Thus, intercrumb spaces are wider than intracrumb spaces and as the distribution of water in soil is largely determined by the pore dimensions, these differences have a profound influence on the microenvironment. For example, a saturated soil releases its water in stages as matric potential decreases during drainage. The larger pores empty first, followed successively by pores of decreasing diameter. Thus, at a particular matric potential, e.g., that equivalent to field capacity, most intercrumb spaces are empty while the spaces inside crumbs are filled with water. Consequently, the intercrumb spaces may contain air, unlike the spaces within crumbs. As diffusion of oxygen and CO_2 in water is so much slower than in air, the diffusion of oxygen into and CO_2 out of water-filled crumbs will be much slower than through the air-filled intercrumb spaces. Thus adequate soil aeration for soil pathogens is dependent on diffusion at the microstructural level. Many water-saturated crumbs greater than about 5 mm diameter have no oxygen at their center which explains the widespread occurrence of localized pockets of anaerobic conditions in most soils. Consequently, aerobic and anaerobic soil microorganisms live in close proximity to each other. Furthermore, a microorganism may find itself in an aerobic environment at one moment and an anaerobic one a few hours later. The dimensions of the pores may also determine which organisms can occupy them; most nematodes, for example, are confined to channels 30–100 mμ diameter whereas bacteria, because they are so much smaller, can live within much smaller spaces. Clearly, gradients of oxygen and CO_2, as well as water content, occur over distances on the order of 1 mm or less.

Soil organic matter also has a profound influence on microorganisms

in soil, particularly those, including some pathogens, which are saprophytic. Soil organic matter is distributed at discrete sites within the soil
and determines to a great extent which parts of the available pore space
are occupied. Thus, Gray and colleagues (1968) state that the total
available surface of soil area colonized by bacteria is on the order of
0.04% and Griffin (1972) asserts that such bacteria usually occur as discrete colonies of one or more species. Furthermore the variation in numbers of fungal propagules within a clod of soil (Dobbs and Hinson, 1960)
probably reflects variation in the distribution of environmental components like nutrient concentration and pore size (Griffin, 1972).

For those soil pathogens which are obligate plant parasites, distribution
and abundance are dictated by the presence of roots, but even roots are
anisotropic as shown by the aggregation of microorganisms at discrete
sites in the rhizoplane. If, as Bowen and Rovira (1976) indicate, exudates
from roots have a very limited range in their influence on microorganisms
such that only those in the rhizosphere are affected, then it seems likely
that chemical gradients from organic matter will have an equally limited
range on the same order of magnitude as the effective range over which
gradients of oxygen, CO_2, and water content operate.

Contributing to this heterogeneity is the fact that soil conditions are
continually changing at any one site. The temperature of the soil profile,
for example, has diurnal and seasonal cycles. Rainfall, evaporation, and
drainage control the amount of water in the soil at various depths; fluctuations are more sporadic than temperature but seasonal cycles are
apparent. Thus, steady states are seldom achieved; in fact, it is likely that
fluctuations in soil processes with a short time period are a frequent
occurrence, especially in the surface layers of the soil. Such rapid changes
are caused by rainstorms or changes in radiation and temperature; they
decay with depth until a more or less steady state is reached (Zaslavsky,
1969). As most soil pathogens reach their highest numbers in the top 20
cm of the soil, they will be exposed to these fluctuations.

To study the ecology of soil pathogens it is necessary to sample the
soid, to assess the number of organisms in the samples, and to measure
environmental components associated with these organisms. The abundance of organisms can then be correlated with levels of environmental
components and their distribution explained. But, if the soil is so intensely
anisotropic, how can samples of sufficiently small size and of sufficient
number be taken to establish such relationships? With nematodes, at least,
the problem of assessing numbers in a population in the field is extremely
laborious (Proctor and Marks, 1974; Smith and Wallace, 1976); with
bacteria the task is even more formidable. Griffin (1972) has stated the
problem succinctly and I only mention it here to stress the point that in

considering the environment of pathogens and their dispersal in soil we are dealing with a medium which presents greater technical problems than those in other ecological fields.

The general picture of the soil is one of extreme anisotropy as well as of continual and rapid change. The question that arises now is whether soil pathogens are mobile enough to escape from sites whose environments are becoming or have become unfavorable. To attempt an answer we must first consider their mobility and ability to disperse.

IV. THE MOBILITY OF PATHOGENS IN SOIL

A. Active Motility

To move actively through the soil a pathogen must develop sufficient force to overcome the resistances associated with surface tension, adsorption, and physical obstruction. Active propulsion may be achieved through flagellar movement, as in bacteria and fungal zoospores, or by muscular activity along the whole body, as in nematodes. Motility may also be achieved by growth, as shown by the fungal mycelium. Fungi, bacteria, and nematodes are so different in the ways in which they move that it is necessary to consider each in turn to assess how quickly and how far they can spread through the soil pore spaces.

In terms of motility and dispersal, soil pathogens and nonpathogens are probably similar, hence, where data are lacking on pathogens I have included nonpathogens in the discussion.

1. Bacteria

Bacteria require continuous water pathways to move. They cannot bridge air-filled pores in the soil. Furthermore, the pore necks must be sufficiently wide because bacteria, like other soil microorganisms, with the exception of some nematodes, cannot displace soil aggregates; they are confined to the existing geometry of the pore spaces. Thus, Griffin and Quail (1968) calculated that *Pseudomonas aeruginosa*, a rod-shaped bacterium measuring about 0.5 to 3.0 μm, is impeded by pore necks of 1 to 1.5 μm diameter. This size of pore neck requires a matric potential of -3 to -3.2 bars to drain the pore, and if the matric potential exceeds this value, continuity of water pathways is broken so that movement is inhibited. Their experiments did not contradict this hypothesis and results indicated that at high matric potentials, i.e., wetter than field capacity, this bacterium moved up to 2 cm in 24 hr. Hamdi (1971) also stressed the necessity for continuous water pathways for movement of *Rhizobium* which traveled up to 3 cm in 5 days in soil at matric potentials greater than −150 cm water. These values are of the same order as those published by Thornton and Gangulee (1926) with nodule bacteria (2.5 cm in 24 hr). A word of caution is expressed by Wong and Griffin (1976) who suggest that studies of bacterial movement in artificial media may give excessive values for speed of movement. They attribute this to the lack of a component with a high surface charge density on to which bacteria become adsorbed. Such colloidal adsorption occurs in natural soils where bacterial speeds of 0.5 cm in 2 days were recorded compared with 2.6 cm in 2 days in artificial soils. The order of speed for bacteria in soil is therefore about 0.5 to 1.0 cm/day.

2. Fungal Zoospores

Although it is likely that zoospores, like bacteria, require high matric potentials for movement in soil, quantitative data are lacking. Lacey (1967) states that zoospores of *Phytophthora infestans* traveled about 1.3 cm in 2 weeks in wet soils. In contrast Hickman and Ho (1966) found that zoospores of *Pythium aphanidermatum* moved at speeds up to 14.4 cm/hr over short periods of chemotactically directed movement; however, these observations were made in artificial aquatic conditions, so they have little relevance to dispersal in soil.

3. Fungal Mycelium

Some idea of the range and rate of spread of fungal mycelium through soil can be obtained from the following examples. *Phytophthora cinnamomi* moved at least 4.5 m uphill in 22 months (Zentmyer and Richards, 1952). Runner hyphae of *Gaeumannomyces graminis* grew along adjoining host roots within a cereal crop for a radius of 1.5 m in a growing season (Shipton, 1972). Mycelial growth of the same fungus from infected wheat plants was about 1.5 m in a season (Wehrle and Ogilvie, 1956). The rate of spread of *Rhizoctonia solani* in nonsteam-sterilized soil in glasshouse flats was estimated by Dimock (1941) to be 25 cm/month. In natural unsterilized soil the growth of *R. solani* amounted to 21.2 cm in 23 days (Blair, 1943). For the same fungus, Shurtleff (1953) estimated the rate of mycelial growth at 2.5 cm/day. Finally, Redfern (1973) states that rhizomorphs of *Armillaria mellea* grew through field soil at about 1 m/year. For these few examples of fungi of quite different habit it is evident that their rate of mycelial growth is of the order of 0.5 cm/day.

Some fungi have a low dispersive ability in soil. Trujillo and Snyder (1963), for example, suggest that *Fusarium oxysporum* f. *cubense* is dispersed chiefly by wind, rainfall, and flooding which transport chlamydospores from infected root tissues. Burke (1965) found that *Fusarium solani* f. *phaseoli* failed to spread through soil from infected to noninfected plants 1.3 to 5 cm apart. Similar views have been expressed for *Sclerotium cepivorum* by Scott (1956a,b).

The growth of fungi in soil has been described by Griffin (1972) in his book where he makes the important point that, unlike those of bacteria, fungal mycelium can spread along the walls of drained pores or even across pores. Consequently, fungi can spread through soils at matric potentials that are too dry for bacterial and zoospore movement, and, as we will see later, for nematodes too.

4. Nematodes

Like bacteria, nematodes move most quickly in soil whose pores have just been drained of water. The dimensions of plant nematodes are such that the size of soil pores through which most of them move most quickly is in the range 30 to 100 μm. Such pores drain at matric potentials of -0.1 and -0.03 bar, respectively (Wallace, 1968). Data on the rate at which nematodes disperse through soil come from controlled experiments under artificial conditions and from observations in the field. Endo (1959) found that *Pratylenchus zeae* moved about 0.1 cm/day in sandy loam in containers and Rode (1962), using somewhat similar experimental methods, assessed the speed of *Heterodera rostochiensis* at about 0.3 cm/day. Tarjan (1971) found that *Radopholus similis, Pratylenchus coffeae,* and *Tylenchulus semipenetrans* move at about 0.4, 0.2, and 0.1 cm/day, respectively, although speeds varied according to soil type. The average speed of *Pratylenchus penetrans* in three Ontario soils under optimum conditions was about 0.3 cm/day (Townshend and Webber, 1971). *Radopholus similis* moved at 0.7 cm/day (Feldmesser *et al.,* 1960), *Ditylenchus dipsaci* at 0.5 to 1.0 cm/day (Webster and Greet, 1967), and *Tylenchulus semipenetrans,* 0.1 cm/day (Baines, 1974). Speeds in soil of 0.1 to 0.5 cm/day appear to be reasonable approximations for nematodes.

5. Conclusion

So far, the motility of pathogens in soil has been described in terms of rate of spread, i.e., the distance a population of pathogens has dispersed in a given time. However, such movements fail to indicate whether individuals in the population move at a continuous steady speed and whether they follow a tortuous route. Rate of dispersal is thus the end effect of events occurring over an extended period (days or weeks). As might be expected much higher speeds have been recorded over short periods of time. Speeds up to 14.4 cm/hr for zoospores of *Pythium aphanidermatum* (Hickman and Ho, 1966) and 20 cm/hr for the nematode *Heterodera schachtii* (Wallace, 1958) are examples. Thus, if an approximation had to be given, rates of spread would be on the order of 0.5 cm/day or about 0.2 mm/hr, whereas short-term speeds of individuals might be up to 100 times this value. Fungal mycelium has a much slower short-term rate of growth than the short-term speeds of zoospores, bacteria, and nematodes. For example, the germ tube of *Pythium ultimum* reached 0.72 cm/day (Griffin, 1972) which is little more than the long-term rate of spread. However, because it can often spread through empty pores in soil, mycelium is probably less sporadic in its dispersal than other pathogens, i.e., the relatively slow growth of mycelium is somewhat offset by the longer time available for growth.

The picture that emerges, albeit dimly, from these observations is that soil pathogens, apart from fungal mycelium, probably disperse in a series of short "bursts" whose frequency and length are determined by environmental conditions. For bacteria, zoospores, and nematodes matric potentials corresponding to field capacity may provide the most suitable conditions for dispersal because there are continuous water pathways at such levels of water content. The short-term speeds and range of dispersal of soil pathogens are probably sufficiently great to enable them to vacate sites which are becoming unfavorable and to reach other more favorable sites several millimeters away, even where environmental changes occur over a period of about 24 hr. Furthermore, the ability to orientate in gradients of oxygen, CO_2, temperature, and soil water would enhance the chance of finding more favorable microenvironments. Nematodes have this ability but whether fungi and bacteria have it also is unknown.

B. Passive Dispersal by Water in the Soil

The ability of a pore to transmit water is proportional to the fourth power of its radius, so when a soil begins to drain, its hydraulic conductivity falls rapidly because the first pores to drain are the largest and most highly conductive. It follows from this that (1) soil pathogens in the larger pore spaces are more likely to be passively transported through the soil than those within soil crumbs; (2) as potential energy gradients in soil are largely vertical, most passive dispersal will be downward; (3) values for hydraulic conductivity in soil usually vary between 2.5 and 0.025 cm/hr. Thus, bacteria, zoospores, and nematodes which are most subject to passive dispersal in soil have short-term speeds which are greater. Consequently, when a soil is draining after rainfall or irrigation there is little reason to think that motile soil pathogens will be flushed through the soil. It also seems likely that few nonmotile spores or bacteria would be transported down the soil profile because they would be adsorbed on to the surface of particles. Similarly nonmotile nematodes would be trapped across pore necks.

Experimental evidence does not contradict these views. Zan (1962) measured the downward movement of zoospores of *Phytophthora infestans* in columns of sand and soil and concluded that it was only by using sand or amounts of water in excess of normal rainfall that downward movement occurred. Measurements of the downward movement of nematodes in sand columns indicated that excessively high flow rates of water or large pores were required to obtain downward transport. At flow rates approaching that occurring in nature, passive dispersal was very small (Wallace, 1959). There is little evidence that even nonmotile spores are carried far by percolating water. Hepple (1960), for example, concluded that the rate of infiltration of water through soil is seldom high enough to move fungal spores. Measurements of the movement and retention of two strains of the bacterium *Klebsiella aerogenes* in saturated soil columns indicate that soil type, pH, and bacterial size are important factors influencing downward movement (Bitton *et al.*, 1974). The hydrophobic nature of the spore surface of actinomycetes and the low percolation rate in natural soils account for their low rate of passive transport (Ruddick and Williams, 1972). Thus, soil pathogens are probably not dispersed more than a few centimeters by percolating water and even then it is mostly downward.

C. Passive Dispersal by Other Soil Organisms

Organisms in the soil influence each other in numerous ways including the dispersal of soil pathogens by other organisms usually referred to as vectors. Vectors are usually larger and more mobile than the organisms they carry. Thus earthworms transport soil fungi (Hutchinson and Kamel, 1956), saprozoic nematodes transport plant pathogenic bacteria (Chantanao and Jensen, 1969) and fungi (Jensen, 1967; Jensen and Siemer, 1971), plant parasitic nematodes (Martelli, 1975) and fungi (Hewitt and Grogan, 1967; Smith *et al.*, 1969) act as vectors of viruses, and so on. There are numerous reports of associations of this kind. Some are fortuitous. Others such as the plant nematode–virus associations are obligatory and specific. There is no doubt that many more associations will be described in the near future. For the obligatory association, the vector is the most important factor contributing to the dispersal of the pathogen, so much so that to describe the dispersal of the vector is to describe the dispersal of the organism it carries. The dispersal of vectors will not be considered in this chapter. Nonobligatory chance transport by other soil organisms may be a very important aspect of the dispersal of soil pathogens. However, what little evidence there is to support this view is largely anecdotal so until experimental data are forthcoming it may be unwise to explore this topic further.

V. THE MAINTENANCE OF INFECTIVITY

In spite of their ability to disperse between microsites and so escape the hazards of a changing microenvironment, it is inevitable that sooner or later soil pathogens will be subjected to conditions that not only inhibit dispersal but threaten survival itself. Absence of a host plant is probably the chief hazard.

Four constraints or "problems" facing the pathogen are discussed here: (1) environmental conditions that threaten survival, e.g., desiccation, anaerobic conditions, absence of a host; (2) uninhibited germination or hatch that would be disadvantageous if conditions were unsuitable for dispersal; (3) absence of a host to infect following dispersal, which would threaten survival; and (4) consumption of energy reserves during dispersal either by active movement or growth. The last problem is that the pathogen might successfully reach the host plant but lose so much energy that it is unable to infect the host after arrival.

Let us now consider some of the characteristics evolved by soil pathogens which appear to counter these exigencies.

A. Survival in a Hostile Environment

Those viruses which infect plant roots are dispersed by nematode or fungal vectors. Thus the problem of survival in soil hardly arises. So, although viruses like tobacco necrosis virus are released into the soil from the host plant root and can be recovered from soil by its fungal vector, *Olpidium brassicae,* for at least 11 weeks after it is no longer detectable by mechanical assay, such particles are probably only a small proportion of those in the disintegrating host tissues (Smith *et al.,* 1969).

Although few plant pathogenic bacteria appear to be successful soil inhabitants, some do have a soil phase. However, the conditions under which they can survive and possibly reproduce in soil are little understood. Perhaps their chief characteristic is their ability to become quiescent for long periods without the formation of any specialized resistant structures.

Plant nematodes, like bacteria, do not have any resting stages with distinct morphological characters. Many of them survive best in low soil moistures and temperatures when the rate of metabolism is low and food reserves are consequently depleted slowly. In this way plant nematodes can persist for many months in soil without a host plant. There are some characteristics that appear to aid survival, particularly in dry conditions. Ellenby (1968) showed, for example, that *Heterodera rostochiensis* survives desiccation through its eggshell which is freely permeable to water when wet, but becomes increasingly impermeable when dry, thus providing an effective resistance to loss of water. Wallace (1966a) suggested that the egg sac in *Meloidogyne javanica* is a barrier to loss of water from the contained eggs. His experiments indicated that the gelatinous matrix retained its water while soil pores around the egg sac had drained. Changes in the gaseous composition of the soil atmosphere and water also impose constraints on nematodes. However, nematodes have the capacity for both oxidative and fermentive metabolism, which increases their chance of survival.

Fungi, unlike other soil pathogens, are well known for their resting spores. Oospores and chlamydospores which usually have thick walls are considered to be stages which can survive conditions unfavorable for vegetative growth. Sclerotia have been described as a quiescent viable state which is maintained by many fungi in the absence of a suitable host or of conditions favoring growth (Coley-Smith and Cooke, 1971). With its outer thickened skin and gelatinous matrix that holds the fungal cells together, the sclerotium resembles the egg sac of *Meloidogyne*. In fact they may function similarly in resisting desiccation.

B. Integration of the Preinfective Stages

It would be advantageous to a pathogen if germination or hatch were more likely to occur when a host plant was present and sufficiently close to permit subsequent infection. A good example is provided by the potato cyst nematode, *Heterodera rostochiensis*, which is stimulated to hatch by exudates from the roots of its host plants. In the absence of a host, a spontaneous hatch occurs, but at a slow rate. Consequently eggs of this nematode persist in soil without a host for many years. As far as is known at present few plant nematodes respond in this way. The stimulation of sporangial germination of *Pythium ultimum* and chlamydospore germination of *Fusarium solani* f. *phaseoli* by bean seed exudate is a similar process (Stanghellini and Hancock, 1971) except that with fungal germination emphasis is usually placed on the nutritional status of the exudate. With nematodes, the stimulus is probably sensory in nature.

Less specific in its control of germination is fungistasis, a term used to describe the failure of fungal propagules to germinate in soil under apparently favorable conditions, unless supplied with exogenous nutrients.

Recent research has suggested that fungistasis might be controlled by a few factors (Smith, 1976). The role of ethylene as an inhibitor has been described by Smith (1973) who proposed that, whereas nutrients are the main stimulators of propagule germination, ethylene is the main inhibitor, and that a balance between the two determines germination. However, this mechanism may not apply in all soils. Bristow and Lockwood (1975), for example, have shown that fungistasis in conidia of *Cochliobolus victoriae* in a loam soil is caused primarily by nutrient deprivation whereas they suggest that in a clay loam soil volatile fungistatic substances may play a major role.

There is evidence of a hatching inhibitor of *Meloidogyne javanica* in soil, and hatch is more rapid in the presence of roots (Wallace, 1966b). Thus, the combined effect of inhibitor and stimulator, as yet unidentified, resembles fungistasis.

Thus, germination or hatch may be influenced by the combined effects of an inhibitor and stimulator, by a stimulator alone (*H. rostochiensis*), or by the action of a factor which at one concentration is an inhibitor (nutrient depletion) and at another is stimulatory (adequate nutrition). All these mechanisms probably increase the chance that after germination or hatch a soil pathogen will successfully disperse and find food or infect a host.

Finally, the chance of infection is increased if the various stages in the life history of a pathogen correspond to the growth cycle of the host plant. Thus the pathogen might germinate or hatch at the same time as the plant seeds germinate or form new feeder roots, for example. The plant might have no direct influence on the pathogen, but the coincidence of their activities at least ensures that the host is available at times when the pathogen is active in the soil. Such characteristics are commonplace among soil pathogens and, although they are less striking than fungistasis or stimulation of hatch by root exudates, they certainly have survival value.

C. Renewal and Conservation of Energy Reserves

As the fungal propagule germinates, the nematode egg hatches, or the bacterial resting cell becomes active and the pathogens move, they consume energy. The source of energy is said to be endogenous when the pathogen uses its own energy reserves, or exogenous when the pathogen derives the energy from organic matter or from root exudates. Plant nematodes depend entirely on their own reserves of energy. Once hatch has occurred the problem facing the nematode larva is to find a host before the reserves are too depleted to permit infection. Bacteria are similar;

there is little evidence that they can maintain themselves saprophytically in soil.

Some fungi renew their energy reserves and synthesize new structural materials during growth by saprophytic colonization in the absence of a host. The conditions which influence this ability are many and varied and Griffin (1972) has emphasized the importance of environment, the amount of prior colonization, fungal growth rate, the amount of endogenous energy reserves, and the ability to produce antibiotics and to tolerate antibiosis. Clearly there is no single predominant determinant of saprophytic ability.

The fact that the nutrient requirements for maximum virulence in *Rhizoctonia solani* are greater than those for vegetative growth (Weinhold *et al.*, 1969), and that the root knot nematode, *Meloidogyne javanica*, may be very motile and yet have insufficient energy reserves to invade the host root (Van Gundy *et al.*, 1967) are two examples that support the view that the process of invading the host may require some extra expenditure of energy. Random movement or random growth of hyphae through soil in the vicinity of a host plant are clearly less efficient in conserving energy than if they were to take a more direct route to the host. The orientation to roots at a distance under the stimulus of a chemical gradient emanating from the root is one characteristic that appears to achieve this end. Directed movement to the host not only increases the amount of inoculum or number of pathogens that reach the invasion sites, it also increases the chance that they will get there before their energy resources are insufficient for penetration. Although positive chemotaxis does occur in bacteria, Seymour and Doetsch (1973) consider it to be of little survival value under natural environmental conditions. On the other hand, they consider negative chemotactic responses, which nearly always develop to lethal chemical gradients, to have definite survival value. The same cannot be said of fungal zoospores and nematodes for which there is a wealth of evidence indicating that they move by chemotaxis along gradients of root exudates and eventually aggregate and invade the host just behind the root tip.

VI. DISPERSAL ABOVE SOIL LEVEL

Although soil pathogens are, by definition, largely confined to the soil, some of them occasionally occur aboveground and are thereby subject to dispersal. However, apart from those fungi which form aerial fruiting bodies from which spores disperse, there appear to be few characteristics for aerial dispersal which we can label as adaptations. Nevertheless, as we shall now see, aerial dispersal of soil pathogens is probably a significant factor in accounting for the distribution of pathogens and outbreaks of disease as well as survival in a changing environment. Four types of dispersal will be considered.

A. Sporulation

A good example of a soil-borne fungal pathogen that disperses by sporulation aboveground is *Gaeumannomyces graminis*, the cause of take-all disease in wheat. Gregory and Stedman (1958) recovered ascospores of this fungus from the air following rainfall but no infection of wheat seedling roots occurred when plants were grown in nonsterilized sand (Garrett, 1956). It was subsequently shown that ascospores can infect exposed proximal parts of seminal roots at the surface of unsterilized soil (Brooks, 1965). It seems likely that antagonism by soil microorganisms was responsible for suppression of infection and Brooks concluded that outbreaks of take-all in the newly drained polders of the Netherlands where there was little antagonism could be caused by ascospore infection. Scott (1969) also believes that ascospores of this fungus are only significant in establishing primary infection loci in previously uninfected areas. Root-rotting basidiomycetes disperse in the same way, so, for some soil fungi at least, the problem of dispersal has been solved by emergence from the soil.

B. Transport by Animals and Plants

Species of aquatic phycomycetes, including *Pythium* sp., have been isolated from the feces of several species of birds by Thornton (1971), who suggests that dispersal may also occur through the adhesion of infested soil to birds' feet. Thornton concludes that the widespread distribution of some fungi is associated with their frequent occurrence in the gut of birds. Epps (1971) also stresses the importance of birds in the dispersal of the soybean cyst nematode *Heterodera glycines*. Experiments indicated that such cysts were still infective after passage through the gut so Epps concluded that birds can spread this nematode to noninfested fields over long distances. The work of Warner and French (1970) on the dispersal of fungi by migratory birds indicates that birds can pick up spores, carry them in a viable condition, and effectively deposit them on susceptible host plants. Among the fungi detected were *Fusarium* and *Rhizoctonia*. Above all, the research of Warner and French indicates the way in which future studies should go. Their experimental data carry so much more weight than the usual anecdotal reports that have appeared on this topic.

It is evident from the literature that animals of all kinds are continuously disturbing and transporting soil and plant material from place to place. The fact that spores of the phycomycete, *Endogone*, have been recovered from the stomachs of numerous rodents (Dowding, 1955; Baker-spigel, 1958) indicates that subterranean fungi can be unearthed and transported, but the distances traveled, their ability to survive the journey, and how intensive such dispersal is are unknown. Turner (1967) found that species of *Phytophthora* can survive passage through the gut of the giant African snail, *Achati.* *fulica,* and thus spread from one field to the next. Similarly, observations by Kliejunas and Ko (1976) on the dispersal of *Phytophthora cinnamomi* by feral pigs on the island of Hawaii and those of Ruddick and Williams (1972) on the transport of actinomycete spores by mites, springtails, beetles, and flies, all isolated from soil, suggest that it would be worthwhile to assess by experiment the dispersive influence of animals on soil pathogens in different agricultural regions. Present evidence suggests that their roles may be significant.

That movement of plants is an effective means of dispersing plant pathogens is well recognized by quarantine authorities. If soil accompanies the plants, there is an increased risk of introducing soil pathogens. There is little point in reviewing the multitude of published examples where soil pathogens have been recovered from soil or plants imported from another country. It is sufficient to say that the dispersal by this means is sufficiently frequent, intense, and long range to account for the occurrence of many diseases hitherto unrecorded in a country or part of a country. The spread of *Phytophthora cinnamomi* within Australia in recent years can be attributed to the intensive and widespread movement of infected plants and soil between states. The introduction of the potato cyst nematode from South America to Europe on potatoes is well documented and recent reports that this nematode occurs in New Zealand raises the question of how it can be kept out of Australia.

C. Transport by Wind and Water

Strong wind gusts and occasional major disturbances like tornadoes frequently lift quantities of soil into the air and disperse them. Where a soil surface is smooth and bare, as occurs after fallowing, and when there is a combination of drought and high wind speeds, erosion occurs. Soil particles may travel by a series of jumps, by rolling and sliding over the surface, or they may be lifted as a fine dust into the air. Studies on the physics of erosion indicate that the most erodible particles are about 0.1 mm in diameter, although the size distribution of particles in the aerosol depends on the nature of the parent soil. Furthermore, during wind erosion the organic component becomes preferentially airborne.

It seems likely, therefore, that surface soil is being constantly moved by the wind over long distances (miles). Furthermore, there is some evidence that pathogens contained in such soil are infective when they reach the ground. Roth and Riker (1943) suggest that *Pythium* and *Rhizoctonia* sp. are dispersed from one nursery to another in wind-blown soil. Nematodes have been recovered from dust traps placed 2 m above the ground in Texas. This might explain the wide distribution of such nematodes as *Meloidogyne* sp. (Orr and Newton, 1971). Experiments by Claflin and colleagues (1973), using a wind tunnel, confirmed field observations that wind-blown infested soil is important in Kansas in the epidemiology of bacterial leaf spot of alfalfa and common blight of beans. Although they are not soil pathogens, it is significant to note that cells of *Xanthomonas alfalfae* and *X. phaseoli*, causal pathogens of the diseases, were apparently carried to plant tissue by wind-driven soil particles. Lloyd (1969) found that the number of streptomycete propagules trapped from air above a fallow plot depended on the amount of dust in the air. Most were attached to the surface of airborne soil particles such as are generated by agricultural farm implements or a gusty wind.

Erosion of soil by water is, like wind erosion, a frequent and continuing process. Studies with close-growing vegetation on gradual slopes indicated that about 0.1 to 0.6 ton/acre of soil are removed annually by water (Smith and Stamey, 1965). There was little destruction of essential soil properties and the erosion probably went unnoticed by the farmer, but even at this level a significant number of soil pathogens was probably dispersed. Good evidence has appeared in recent years that irrigation canals are a very effective means of dispersing such soil pathogens as *Phytophthora* sp. and plant nematodes. Such pathogens are carried many miles in a viable state and introduced into those crops which receive the water.

It seems likely, therefore, that the dispersal of soil pathogens by wind and water occurs frequently, intensively, and sometimes over several miles. What is less certain is how successfully they maintain their infectivity especially when exposed to desiccation during wind-blown dispersal.

D. Dispersal and Distribution

Hirst (1965) suggests that ubiquity of some species of soil pathogens indicates their successful disperal over a long time. Baker (1970) makes a similar point when he concludes that the worldwide distribution of

Rhizoctonia and its occurrence in uncultivated desert and forests indicate that it is an ancient species. The inference here presumably is that *Rhizoctonia* has had a long time to disperse thereby compensating for its low dispersive ability. The same argument might be applied to *Agrobacterium tumefaciens, Phytophthora cinnamomi, Heterodera avenae,* and many others. Such a view, however, seems to be contrary to the widely held opinion that the cosmopolitan distribution of a species is associated with an ability to disperse over long distances in a short time. This relationship between ubiquity and high dispersiveness is based on the argument that gene flow limits speciation. Now, although soil pathogens stand a good chance of dispersing to other sites a few miles away within a year, the chance that representatives of a population will disperse across continents and oceans through transport by animals, plants, wind, and water is unlikely, at least in the short term. Hence, it is difficult to reconcile the lack of dispersiveness in ubiquitous soil pathogens with the notion of adequate gene exchange. However, this problem is resolved if, as Ehrlich and Raven (1969) suggest, gene flow is not as important as it was first thought to be; they offer the alternative explanation that widely dispersed populations tend to remain similar because they are subjected to similar selective forces. Such a notion may be useful for soil pathogens if further research indicates that the selective factors in the soil microenvironment have similar characteristics in most agricultural soils over the earth's surface. In other words, some species of soil pathogens are cosmopolitan because they are subjected to the same selective forces. If any differentiation of populations is to occur, selection pressure is most likely to come from the host, hence the occurrence of races sometimes distinguishable only by the species of plant they parasitize.

There is, however, a further explanation of cosmopolitan distribution which must be considered because there is a factor in the dispersal of soil pathogens which has received little attention so far—man. How much man has contributed to the dispersal of soil pathogens can be judged by Zadoks' (1967) review and the paper by Baker (1966). It is safe to assume that given a continuation of man's present movements and his transport of food, fiber, materials, machinery, etc., not to mention plants themselves, many more soil pathogens will have a worldwide distribution over the next 50 years or so. Thus, a ubiquitous distribution of a soil pathogen may be due to a recent, long-range dispersal. Jones' (1969, 1972) accounts of the current world distribution of the round cyst nematodes on potatoes are convincing because dispersal has occurred relatively recently over the last 100 years. Their epicenter in South America, where they originated, is known. The method of dispersal with potatoes is also known.

Whether a particular species of soil pathogen has achieved ubiquity by dispersal over a long time with little difference in selective pressure between populations or by recent and intensive long-range dispersal by man is a problem which cannot be solved by experimentation. In my opinion the second hypothesis is more useful because: (1) we have almost certainly failed to appreciate man's role as an effective agent of dispersal of soil pathogens and (2) it emphasizes the need to control disease by restricting dispersal, especially where man is involved.

VII. CONCLUSIONS

Dispersal of soil pathogens can be considered in three categories: (1) at the level of the soil crumb, (2) within and between fields, and (3) between continents. Each category presents a challenge to the plant pathologist because it is clear from what has been said in this chapter that the subject is complex and technically difficult, hence the lack of data on all aspects.

Study of dispersal at the level of the soil crumb will demand an appreciation of the anisotropic nature of the soil microenvironment and will require new techniques in the measurement of soil properties at microsites as well as the means to measure the numbers and the distribution of pathogens in soil. Information of this kind can then be used to assess the validity of current ecological theory, much of which has been based on studies of larger organisms aboveground.

Dispersal within and between fields is an equally demanding problem. A knowledge of meteorology, hydrology, soil physics, bird migration, and animal behavior will be required to determine the relative importance of wind, water, animals, and plants in the dispersal of soil pathogens. Such research would greatly contribute to our knowledge of how disease spreads in time and may offer useful ways of controlling disease on the farm.

Research into the role of man in the dispersal of soil pathogens over the world will require more than recording further evidence that man carries soil pathogens around on his shoes or in his ships and planes; experiments are needed to assess the extent to which man is responsible for the initiation of disease and for the cosmopolitan distribution of some soil pathogens.

The fact that little progress has been made in the study of dispersal of soil pathogens since Hirst (1965) wrote on the subject 13 years ago may not only reflect the difficult nature of the problem but it may indicate a lack of awareness that dispersal determines the distribution and abundance of soil pathogens which in turn have an important influence on the distribution and intensity of disease. In practical terms control of dispersal at one end of the causal chain may effectively control disease at the other.

References

Baines, R. C. (1974). The effect of soil type on movement and infection rate of larvae of *Tylenchulus semipenetrans*. *J. Nematol.* 6, 60–62.

Baker, G. E. (1966). Inadvertent distribution of fungi. *Can. J. Microbiol.* 12, 109–112.

Baker, K. F. (1970). Types of *Rhizoctonia* diseases and their occurrence. In "*Rhizoctonia solani:* Biology and Pathology" (J. R. Parmeter, Jr., ed.), pp. 125–148. Univ. of California Press, Berkeley.

Bakerspigel, A. (1958). The spores of *Endogyne* and *Melanogaster* in the digestive tracts of rodents. *Mycologia* 50, 440–442.

Bitton, G., Lahan, N., and Henis, Y. (1974). Movement and retention of *Klebsiella aerogenes* in soil columns. *Plant Soil* 40, 373–380.

Blair, I. D. (1943). Behaviour of the fungus *Rhizoctonia solani* Kühn in the soil. *Ann. Appl. Biol.* 30, 118–127.

Bowen, G. D., and Rovira, A. D. (1976). Microbial colonization of plant roots. *Annu. Rev. Phytopathol.* 14, 121–144.

Bristow, P. R., and Lockwood, J. L. (1975). Soil fungistasis: Role of the microbial nutrient sink and of fungistatic substances in two soils. *J. Gen. Microbiol.* 90, 147–156.

Brooks, D. H. (1965). Root infection by ascospores of *Ophiobolus graminis* as a factor in the epidemiology of the take-all disease. *Trans. Br. Mycol. Soc.* 48, 237–248.

Burke, D. W. (1965). The near immobility of *Fusarium solani* f. *phaseoli* in natural soils. *Phytopathology* 55, 1188–1190.

Chantanao, A., and Jensen, H. J. (1969). Saprozoic nematodes as carriers and disseminators of plant pathogenic bacteria. *J. Nematol.* 1, 216–218.

Claflin, L. E., Stuteville, D. L., and Armbrust, D. V. (1973). Wind-blown soil in the epidemiology of bacterial leaf spot of alfalfa and common blight of bean. *Phytopathology* 63, 1417–1419.

Coley-Smith, J. R., and Cooke, R. C. (1971). Survival and germination of fungal sclerotia. *Annu. Rev. Phytopathol.* 9, 65–92.

Dimock, A. W. (1941). The *Rhizoctonia* foot-rot of annual stocks (*Matthiola incana*). *Phytopathology* 31, 87–91.

Dobbs, C. G., and Hinson, W. H. (1960). Some observations on fungal spores in soil. In "The Ecology of Soil Fungi" (D. Parkinson and J. S. Wald, eds.), pp. 33–42. Liverpool Univ. Press, Liverpool.

Dowding, E. S. (1955). *Endogyne* in Canadian rodents. *Mycologia* 47, 51–57.

Ehrlich, P. R., and Raven, P. H. (1969). Differentiation of populations. *Science* 165, 1228–1232.

Ellenby, C. (1968). Desiccation survival in the plant parasitic nematodes *Heterodera rostochiensis* Wollenweber and *Ditylenchus dipsaci* (Kühn) Filipjev. *Proc. R. Soc. London, Ser. B* 169, 203–213.

Endo, B. Y. (1959). Responses of root-lesion nematodes, *Pratylenchus brachyurus* and *P. zeae*, to various plants and soil types. *Phytopathology* 49, 417–421.

Epps, J. M. (1971). Recovery of soybean cyst nematodes (*Heterodera glycines*) from the digestive tracts of blackbirds. *J. Nematol.* 3, 417–419.

Feldmesser, J., Cetas, R. C., Grimm, G. R., Rebois, R. V., and Whidden, R. (1960). Movement of *Radopholus similis* into rough lemon feeder roots and in soil and its relation to *Fusarium* in the roots (Asbtr.). *Phytopathology* **50**, 635.

Garrett, S. D. (1956). "Biology of Root-infecting Fungi." Cambridge Univ. Press, London and New York.

Gray, T. R. G., Baxby, P., Hill, I. R., and Goodfellow, M. (1968). Direct observation of bacteria in soil. *In* "The Ecology of Soil Bacteria" (T. R. G. Gray and D. Parkinson, eds.), pp. 171–192. Liverpool Univ. Press, Liverpool.

Gregory, P. H., and Stedman, O. J. (1958). Spore dispersal in *Ophiobolus graminis* and other fungi of cereal foot rots. *Trans. Br. Mycol. Soc.* **41**, 449–456.

Griffin, D. M. (1972). "Ecology of Soil Fungi." Chapman & Hall, London.

Griffin, D. M., and Quail, G. (1968). Movement of bacteria in moist, particulate systems. *Aust. J. Biol. Sci.* **21**, 579–582.

Hamdi, Y. A. (1971). Soil-water tension and the movement of *Rhizobia. Soil Biol. & Biochem.* **3**, 121–126.

Hepple, S. (1960). The movement of fungal spores in soil. *Trans. Br. Mycol. Soc.* **43**, 73–79.

Hewitt, W. B., and Grogan, R. G. (1967). Unusual vectors of plant viruses. *Annu. Rev. Microbiol.* **21**, 205–224.

Hickman, C. J., and Ho, H. H. (1966). Behaviour of zoospores in plant-pathogenic phycomycetes. *Annu. Rev. Phytopathol.* **4**, 195–220.

Hirst, J. M. (1965). Dispersal of soil microorganisms. *In* "Ecology of Soil-Borne Plant Pathogens" (K. F. Baker and W. C. Snyder, eds.), pp. 69–81. Univ. of California Press, Berkeley.

Hutchinson, S. A., and Kamel, M. (1956). The effect of earthworms on the dispersal of soil fungi. *J. Soil Sci.* **7**, 213–218.

Jensen, H. J. (1967). Do saprozoic nematodes have a significant role in epidemiology of plant diseases? *Plant Dis. Rep.* **51**, 98–102.

Jensen, H. J., and Siemer, S. R. (1971). Protection of *Fusarium* and *Verticillium* propagules from selected biocides following ingestion by *Pristionchus lheritieri. J. Nematol.* **3**, 23–27.

Jones, F. G. W. (1969). Some reflections on quarantine, distribution and control of plant nematodes. *In* "Nematodes of Tropical Crops" (J. E. Peachey, ed.), Tech. Commun. Bur. Helminthol. No. 40, pp. 67–80. Commonwealth Agricultural Bureaux, Farnham Royal, U.K.

Jones, F. G. W. (1972). Plant parasitic nematode pathotypes: Pathotypes in perspective. *Ann. Appl. Biol.* **71**, 296–300.

Kliejunas, J. T., and Ko, W. H. (1976). Dispersal of *Phytophthora cinnamomi* on the island of Hawaii. *Phytopathology* **66**, 457–460.

Lacey, J. (1967). The role of water in the spread of *Phytophthora infestans* in the potato crop. *Ann. Appl. Biol.* **59**, 245–255.

Lloyd, A. B. (1969). Dispersal of *Streptomycetes* in air. *J. Gen. Microbiol.* **57**, 35–40.

Martelli, G. P. (1975). Some features of nematode-borne viruses and their relationships with the host plants. *In* "Nematode Vectors of Plant Viruses" (F. Lamberti, C. E. Taylor, and J. W. Seinhorst, eds.), pp. 233–252. Plenum, New York.

Orr, C. C., and Newton, O. H. (1971) Distribution of nematodes by wind. *Plant Dis. Rep.* **55**, 61–63.

Proctor, J. R., and Marks, C. F. (1974). The determination of normalizing transformations for nematode count data from soil samples and of efficient sampling schemes. *Nematologica* **20**, 395–406.

Redfern, D. B. (1973). Growth and behaviour of *Armillaria mellea* rhizomorphs in soil. *Trans. Br. Mycol. Soc.* **61**, 569–581.

Rode, H. (1962). Untersuchungen über das wandervermögen von larven des Kartoffel-nematoden (*Heterodera rostochiensis* Woll.) in modellversuchen mit verschiedenen bodenarten. *Nematologica* **7**, 74–82.

Roth, L. F., and Riker, A. J. (1943). Life history and distribution of *Pythium* and *Rhizoctonia* in relation to damping-off of red pine seedlings. *J. Agric. Res.* **67**, 129–148.

Ruddick, S. M., and Williams, S. T. (1972). Studies on the ecology of actinomycetes in soil. V. Some factors influencing the dispersal and adsorption of spores in soil. *Soil Biol. & Biochem.* **4**, 93–103.

Scott, M. R. (1956a). Studies of the biology of *Sclerotium cepivorum* Berk. I. Growth of the mycelium in soil. *Ann. Appl. Biol.* **44**, 576–583.

Scott, M. R. (1956b). Studies of the biology of *Sclerotium cepivorum* Berk. II. The spread of white rot from plant to plant. *Ann. Appl. Biol.* **44**, 584–589.

Scott, P. R. (1969). Control of survival of *Ophiobolus graminis* between consecutive crops of winter wheat. *Ann. Appl. Biol.* **63**, 37–43.

Seymour, F. W. K., and Doetsch, R. N. (1973). Chemotactic responses by motile bacteria. *J. Gen. Microbiol.* **78**, 287–296.

Shipton, P. J. (1972). Take-all in spring-sown cereals under continuous cultivation: Disease progress and decline in relation to crop succession and nitrogen. *Ann. Appl. Biol.* **71**, 33–46.

Shurtleff, M. C. (1953). Factors that influence *Rhizoctonia solani* to incite turf brown patch. *Phytopathology* **43**, 484 (abstr.).

Smith, A. M. (1973). Ethylene as a cause of soil fungistasis. *Nature (London)* **246**, 311–313.

Smith, A. M. (1976). Ethylene in soil biology. *Annu. Rev. Phytopathol.* **14**, 53–73.

Smith, A. D. M., and Wallace, H. R. (1976). Fluctuations in the distribution and numbers of *Helicotylenchus dihystera* in Kikuyu turf (*Pennisetum clandestinum*). *Nematologica* **22**, 145–152.

Smith, P. R., Campbell, R. N., and Fry, P. R. (1969). Root discharge and soil survival of viruses. *Phytopathology* **59**, 1678–1687.

Smith, R. M., and Stamey, W. L. (1965). Determining the range of tolerable erosion. *Soil Sci.* **100**, 414–424.

Stanghellini, M. E., and Hancock, J. G. (1971). Radial extent of the bean spermosphere and its relation to the behavior of *Pythium ultimum*. *Phytopathology* **61**, 165–168.

Tarjan, A. C. (1971). Migration of three pathogenic citrus nematodes through two Florida soils. *Soil Crop Sci. Soc. Fla., Proc.* **31**, 253–255.

Thornton, H. G., and Gangulee, N. (1926). The life cycle of the nodule organism, *Bacillus radicicola* (Beij.), in soil and on its relation to the infection of the host plant. *Proc. R. Soc. London, Ser. B* **99**, 427–451.

Thornton, M. L. (1917). Potential for long-range dispersal of aquatic phycomycetes by international transport in birds. *Trans. Br. Mycol. Soc.* **57**, 49–59.

Townshend, J. L., and Webber, L. R. (1971). Movement of *Pratylenchus penetrans* and the moisture characteristics of three Ontario soils. *Nematologica* **17**, 47–57.

Trujillo, E. E., and Snyder, W. C. (1963). Uneven distribution of *Fusarium oxysporum* f. *cubense* in Honduras soils. *Phytopathology* **53**, 167–170.

Turner, G. J. (1967). Snail transmission of species of *Phytophthora* with special reference to foot rot of *Piper nigrum*. *Trans. Br. Mycol. Soc.* **50**, 251–258.

Van Gundy, S. D., Bird, A. F., and Wallace, H. R. (1967). Aging and starvation in larvae of *Meloidogyne javanica* and *Tylenchulus semipenetrans*. *Phytopathology* **57**, 559–571.

Wallace, H. R. (1958). Movement of eelworms. II. A comparative study of the movement in soil of *Heterodera schachtii* Schmidt and of *Ditylenchus dipsaci* (Kühn) Filipjev. *Ann. Appl. Biol.* **46**, 86–94.

Wallace, H. R. (1959). Movement of eelworms. IV. The influence of water percolation. *Ann. Appl. Biol.* **47**, 131–139.

Wallace, H. R. (1966a). The influence of moisture stress on the development, hatch, and survival of eggs of *Meloidogyne javanica*. *Nematologica* **12**, 57–69.

Wallace, H. R. (1966b). Factors influencing the infectivity of plant parasitic nematodes. *Proc. R. Soc. London, Ser. B* **164**, 592–614.

Wallace, H. R. (1968). The dynamics of nematode movement. *Annu. Rev. Phytopathol.* **6**, 91–114.

Wallace, H. R. (1973). "Nematode Ecology and Plant Disease." Arnold, London.

Warner, G. M., and French, D. W. (1970). Dissemination of fungi by migratory birds: Survival and recovery of fungi from birds. *Can. J. Bot.* **48**, 907–910.

Webster, J. M., and Greet, D. N. (1967). The effect of a host crop and cultivations on the rate that *Ditylenchus dipsaci* reinfested a partially sterilized area of land. *Nematologica* **13**, 295–300.

Wehrle, V. M., and Ogilvie, L. (1956). Spread of take-all from infected wheat plants. *Plant Pathol.* **5**, 106–107.

Weinhold, A. R., Bowman, T., and Dodman, R. L. (1969). Virulence of *Rhizoctonia solani* as affected by nutrition of the pathogen. *Phytopathology* **59**, 1601–1605.

Wong, P. T. W., and Griffin, D. M. (1976). Bacterial movement at high matric potentials. I. In artificial and natural soils. *Soil Biol. & Biochem.* **8**, 215–218.

Zadoks, J. C. (1967). International dispersal of fungi. *Neth. J. Plant Pathol.* **73**, Suppl. 1, 61–80.

Zan, K. (1962). Activity of *Phytophthora infestans* in soil in relation to tuber infection. *Trans. Br. Mycol. Soc.* **45**, 205–221.

Zaslavsky, D. (1969). Fluctuations in soil environment and their physical significance. *Soil Sci.* **108**, 326–334.

Zentmyer, G. A., and Richards, S. J. (1952). Pathogenicity of *Phytophthora cinnamomi* to avocado trees and the effect of irrigation on disease development. *Phytopathology* **42**, 35–87.

Chapter 10

Computer Simulation of Epidemics

PAUL E. WAGGONER

I. INTRODUCTION

Without his own Boeing or Concorde, a child can learn how wings, rudder, and thrust affect flight by experimenting with a model airplane. Without the cost, danger, and delays inherent in the real thing, the model will show the holes in the child's knowledge and the function and interrelations of the parts, and by demonstrating the outcome of changes, it will teach the child to forecast the behavior of the real plane. Plant pathologists compose models and simulate epidemics for the same reasons.

Although botanists have built glass models that look like pathogens and diseased hosts, pathologists want models that run like epidemics. One such model is pictured in Fig. 1. From the tank at the top, water that represents healthy foliage drips through a spring-loaded valve into a bucket of water below that represents diseased foliage. The more water in the tank above, the greater the pressure, and the faster the drip that represents conversion to diseased foliage. On the other hand, the more water in the bucket below, the greater the weight, and the further the lever is pulled down, the wider the valve in the tank is opened, and the

203

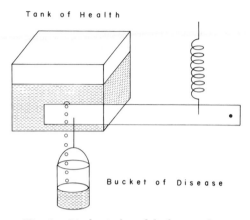

Fig. 1.   Mechanical model of an epidemic.

faster the drip that represents the increase of disease. Experimenting with the tank and bucket, the epidemiologist would learn that when the spring holding up the lever is made only slightly stronger, representing a host that is only slightly more resistant or a pathogen only slightly less virulent, the pull of the bucket of disease is resisted, and there is a remarkable deceleration in the simulated epidemic. In fact, in a short time or simulated season, a little stiffness of spring would practically control the epidemic in the bucket.

The bucket model would simulate the growth of the epidemic, but it would not mimic the pathogenic mechanisms of sporulation, spread, and destruction as the aeronautic mechanisms of wings, rudder, and engine are mimicked in a model airplane. To differentiate the "macro" model that mimics increase in an epidemic but not mechanism from one that mimics mechanisms in the individual plant, one can call the latter "micro."

Perhaps we could assemble enough buckets and springs to make a micromodel of the stages in the life of the pathogen and the stages of the epidemic. We could then experiment with the altered sporulation and flight of spores and the humidity or wind that affect them. With this realistic and revealing micromodel, we could integrate pathology and physiology. Do the tilting levers, stretching springs, and rising water simulate a real epidemic, showing that the essentials of the model are realistic? Does the model show how fungicides will help? Or resistance? Or irrigation? Or arrangements of varieties? Is negligible disease or a severe epiphytotic forecast?

Although making the mechanical model is intriguing, the power of mathematics to mimic mechanisms and the speed of computers to handle the members makes a mathematical model easier. A mathematical simulator run on a computer lets us handle all the hour-by-hour changes in weather, pathogen, and host.

II. USING MULTIPLICATION AND A LIMIT TO COMPOSE MODELS

The whole of a macromodel and the parts of a micromodel, such as the mechanical marvel shown in Fig. 1, operate by the principle shown in Eq. (1).

Rate of increase in the weight of the bucket or in the size of the epidemic $=$ Potential increase with time. Depends on amount of water in bucket or amount of diseased foliage. $\times$ Degree of realization of the potential increase. Depends on amount of water in tank or healthy foliage. (1)

The rate of the entry to the left of the equals sign is dN/dt weight of water or number of lesions per unit time. The potential increase would be bN weight of water or number of lesions per unit time where b is the proportional gain in weight in the bucket or the proportional increase in lesions per unit time. The degree of realization of this potential increase depends upon the fraction $(K - N)/K$ of the original K water or foliage. The principle described in words in Eq. (1) is simply

$$dN/dt = bN(K - N)/K \qquad (2)$$

Integration of Eq. (2) produces the logistic curve solved in 1838 by P. F. Verhulst (Gause, 1934).

In plant pathology the logistic curve has been employed powerfully in a macromodel by van der Plank (1963). He defined the proportion N/K of host diseased as x and wrote b as r [see Eq. (3)].

$$d(N/K)dt = dx/dt = b(N/K)(K-N)/K = rx(1 - x) \qquad (3)$$

Jowett et al. (1974) have reviewed other principles that might have been used, e.g., those of the Gompertz curve. Using the simple logistic curve that embodies the principles of Eq. (1), however, van der Plank has analyzed epidemics and extracted useful results, e.g., the remarkable reduction in epidemics by partial resistance. Others, e.g., Jowett et al. (1974), have shown how r can be used as a summary of epidemic growth to be related to host, pathogen, and environment.

Although the rates of Eq. (3) are easily integrated by the usual methods of calculus to obtain the logistic curve that mimics the sigmoid increase of an epidemic, it is time to arm ourselves for later complications by seeing how a computer would integrate or add the rates by short times. In fact, a computer works from the fundamental differential Eq. (3) that defines the rate. For a computer program, Eq. (3) would be written as shown in Eq. (4).

$$RATEX = R * X * (1 - X) \qquad (4)$$

The computer would calculate the rate of increase in the proportion of foliage diseased at a simulated moment t_0 from the R per time chosen by us and from the proportion X attained up to or by this moment t_0. Then the CSMP (IBM, 1969) command

$$X = INTGRL (XO, RATEX) \qquad (5)$$

would cause the computer to integrate by adding the product of the rate $RATEX$ times an increment of time $DELT$ to the X at t_0 to obtain the larger proportion X diseased at time $(t_0 + DELT)$. XO is the initial disease when calculation began. If $DELT$ were brief enough, the orderly and rapid—if dull—computer would calculate the logistic curve that Verhulst cleverly obtained by integration.

III. DISTILLING A MODEL FROM HISTORY

Since experience shows that host, pathogen, and environment determine the epidemic, we must estimate realistic values of R in different situations if we are to explore which factors influence R and thus control epidemics. The natural vehicles to estimation are observation and then statistics. Since the logistic curve obtained from Eq. (3) is as shown in Eq. (6)

$$\log_e[x/(1 - x)] = rt + \text{constant} \qquad (6)$$

the observer naturally estimates r by observing the logarithmic left-hand member at two times or dates. In fact, the change in disease over a modest range of time can even be assumed to be linear (Last et al., 1969). Then the estimates of r, or even simply change in percentage disease, is related to variety, race, or—especially—environment. Notable examples of such investigations are for wheat leaf rust (Burleigh et al., 1972) and hop downy mildew (Royle, 1973). Essentially they estimated r by multiple regression as functions of environment, and in the case of wheat leaf rust, removing the effect of environment by statistics revealed the effect of variety upon r.

The difficulty with the estimating equations for r is, of course, that they are linear functions of the factors, like wetness and inoculum, and do not appear in logical form as, say, the product of wetness and inoculum (Waggoner, 1974). Nevertheless, such estimation from real epidemics is essential to learn which factors are effective in real fields. Thus macromodels like the logistic curve and statistical estimates from real epidemics set the stage for micromodels that assemble the physiology of individual plants and pathogens logically.

IV. MICROMODELS, ENVIRONMENT, AND PHYSIOLOGY OF PATHOGENS

A. The Grand Model of the Entire Life Cycle of a Pathogen

In chemistry the oxidation of a substrate can be described in the macro by a simple equation, and then the rate parameter examined as a function of, say, environment. Alternatively, respiration can be examined in the micro by a grand chart not unlike Fig. 2 of cycles, compounds, and enzymes and then an attempt made to construct the rate of the whole from the rates of the constituents. Phytopathology is particularly inclined to micromodels because of its long preoccupation with visible components in a pathogen's life cycle. The modern computer permits us to tackle the grand or entire life cycle in a micromodel with components.

The cycle of a typical fungus, e.g., *Phytophthora infestans*, during the growing season is pictured in Fig. 2. It is composed of boxes (or buckets) representing objects and arrows representing passages from one box to another. If the figure is viewed as a flow chart for a computer program, the arrows represent rates like Eq. (4) and the boxes represent integrals

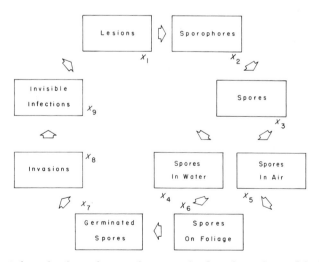

Fig. 2.   Life cycle of a pathogenic fungus or the flow chart of a model of integrals $X_1, X_2, \ldots X_9$.

like Eq. (5). The integrals are increased or decreased by the rates. The generality of the life cycle among pathogenic fungi has evoked a general model (Shrum, 1975).

Having seen the correspondence between the mechanical model of Fig. 1 and equations, and knowing the nature of most life cycles reasonably well, one has few surprises as one writes the equations patiently, arrow by arrow, and box by box. A framework built by Lotka (1924) is both a guide to the work and a reminder of the complex integrations that boggled our predecessors but are made easy for us by a fast computer. The rates of change of the X_i integrals for lesions, spores, etc., are shown in Eq. (7).

$$\frac{dX_1}{dt} = F_1(X_1, X_2, \ldots X_n; P_1, P_2, \ldots P_j)$$

$$\frac{dX_2}{dt} = F_2(X_1, X_2, \ldots X_n; P_1, P_2, \ldots P_j) \tag{7}$$

$$\vdots$$

$$\frac{dX_n}{dt} = F_n(X_1, X_2, \ldots X_n; P_1, P_2, \ldots P_j)$$

This framework says, for example, that in the cycle of Fig. 2 the rate dX_2/dt of change in X_2 or number of sporophores is a function F_2 of X_2 itself and of environmental, varietal, and racial factors P_1 through P_j. If we remember that the r of Eq. (3) is influenced by factors P_1 through P_j, we see that Eq. (7) is more numerous but not different in principle from Eq. (2).

Although the influence of all integrals $X_1, X_2, \ldots X_n$ upon the rates for the integrals in Eq. (7) seems new, an upstream integral, i.e., the $(K - N)/K$ water in the tank, appears in Eq. (2). Thus there are no logical obstacles as we work our way around the cycle and down the rate Eq. (7), economizing by writing only those integrals X and factors P that pathology tells us are influential in each stage.

Equation (7) breaks the project into practical jobs, and the entire model is exemplified by germination of spores. Germination in flow chart Fig. 2 is amplified to Fig. 3 to help us write the rate of change dX_6/dt in the number of ungerminated and germinated spores on healthy foliage. Integral X_6, which is also called census (CSF) of spores on foliage, is the supply for germination, and it is replenished through the rate (RASP) of arrival of spores and depleted by the rate (RK) of killing by desiccation, the rate (REX) of extinction by unfavorable temperatures, and the rate (RG) of germination. Germination, of course, moves the individual into the census (CGT) of germ tubes. In Fig. 3 the integrals are repre-

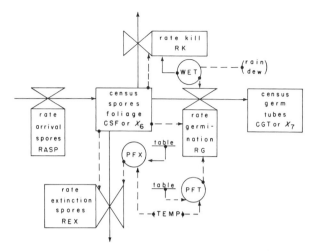

Fig. 3.   Flow chart for germination of spores.

sented by boxes and the rates by valves. The dashed arrows show which of the integrals $X_1, \ldots X_n$ and factors $P_1, \ldots P_j$ control the rates.

The rate of change dX_6/dt in census CSF of spores on foliage is the algebraic sum of four rates, which is shown in Eq. (8).

$$dX_6/dt = \text{RASP} - \text{RK} - \text{REX} - \text{RG spores/hour} \qquad (8)$$

Determining which integrals and factors must appear in Eq. (7) for dX_6/dt becomes an examination of the four rates in the right-hand member of Eq. (8). The most interesting is the rate (RG) of germination. The dashed arrows of Fig. 3 indicate leaf wetness (WET) is a factor, and CSF is an integral affecting RG. An arrow also indicates that RG spores/hr depends upon a proportional rate PFT fraction/hr, which in turn depends upon a table of rates and the temperature. Thus

$$\text{RG} = \text{CSF} * \text{WET} * \text{PFT spores/hour}$$

CSF has dimensions of spores, WET is 0 or 1 and dimensionless, and PFT is a fraction/hour. The other three rates of Eq. (8) are formulated similarly from the pathology of Fig. 3, and all four rates add to dX_6/dt in Eq. (7). Then the other parts of Eq. (7) are similarly obtained.

The reader will have noticed that all the factors in Fig. 3 are environmental and none deals with races of pathogens or varieties of host. In fact, parameters like PFT will vary with race or variety, and a model for more than one race by variety combination must take this into account. While determining the changes in parameters for race or variety, the scientist will at the same time determine the mechanism of virulence or susceptibility.

When all parts of Eq. (7) have been written for the life cycle, the micromodel is essentially composed, embodying our knowledge of the pathogen (Waggoner and DeWit, 1974). Although this model was outlined in 1924 by Lotka, it can be used for simulation today because a fast computer makes numerical integration practical in the grand model of the entire life cycle.

B. Estimating a Parameter in the Model

Numerical integration is possible only if values are given to the parameters like PFT, the portion of the spores on foliage that form germ tubes per hour. Fortunately for our jobs, these parameters are generally unknown, providing something for us to add to the life cycle, to Eq. (7) and to the computer that has been given to us by others.

As an example, PFX for extinction and PFT for germination of wet *Helminthosporium maydis* spores are estimated from observations of the progress of germination at two temperatures (Fig. 4, left side) (Waggoner *et al.*, 1972). It is worth remarking that rate parameters cannot be estimated from observations after a single convenient time.

Since germination is about 50% after a long time at 15°C, extinction and germination rates PFX and PFT are equal. After a shorter time of 9 hr, germination is 24%, and from this we can estimate PFX and PFT to be 0.075/hr. The more rapid and eventually greater germination at 23°C produces estimates of 0.25/hr for PFT toward germination, while PFX toward extinction is only 0.11/hr, little more than at 15°C. We have proceeded as a chemist would to estimate the proportional rates if two reactions were consuming a molecule.

When all spores germinate, PFX is zero, and PFT is simply the reciprocal of the characteristic time, which is the time for the number CSF to decrease to $1/e$ or 37%.

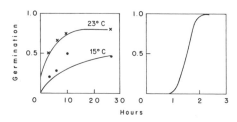

Fig. 4. The course of germination of *Helminthosporium* spores at 15° and 23°C (left) and of *Alternaria* spores at 25°C (right).

The right side of Fig. 4, which depicts the germination of *Alternaria solani* spores, shows, unfortunately, that the simple model of proportional loss from a single integral or box will not always suffice (Waggoner and Parlange, 1974a). Fortunately the delayed onset and then rapid germination increasing as a cumulative normal curve is the behavior of the "break through" at the end of a series of integrals, the more the integrals the steeper the curve. This model of a series of integrals, hooked head to tail like a column of elephants, has been used in a simulator of Southern corn leaf blight (Waggoner and DeWit, 1974). The number of intervening integrals and the rate can be estimated from the curve (Goudriaan, 1973; Parlange, 1974), and this will be discussed in Section III,D.

C. Components or Kits for the Grand Model

If a person boggles at assembling the entire physiology of a pathogen, with all its environmental and host interactions and racial differences reflected in precise parameters, he would be normal. And although I am acclimated to composing models, I am boggled by the thought of thoroughly testing one by catching a robust epidemic underway and measuring the environment and all the integrals of Fig. 2 unfailingly for a season. Fortunately progress can be made by steps.

If we wanted to test the model of the airplane, we could begin with the component of engine and propeller, building this component in a bicycle shop and testing it on a rack in the yard. If it worked, we would not have tested the entire, grand model, but we would know that part of the job was done and could be built upon, step by step.

A plant disease model can be worked at by steps, too. Thus a grand model of *Phytophthora infestans* on potato, including its control by fungicides as well as weather, is a large undertaking, although a realistic model can be written on a couple of pages (Waggoner, 1976). Fortunately the component of germination and infection can be taken separately, like the engine and propeller, and tested against the inoculation of potato foliage in a moist chamber. As it turned out, the model of this component, using rates observed by W. Crosier in New York State in 1934, mimicked the outcome of inoculations by E. Bashi and J. Rotem (1974) in Israel 40 years later. Having designed the model according to the life of the fungus and having tested this component, one can turn to other parts of the model, knowing that one part is tested and waiting for assembly into the whole.

Charles Populer likened the composition of a component for a model to the construction of a carburetor kit for an automobile. This homely analogy makes a good conclusion for it shows the modesty of our accomplishment, suggests that if we have the right kits tested and assembled we shall make a realistic simulation, and emphasizes that infinite perfection of one kit will not carry us far.

D. Integrating Variable Rates in a Variable Environment

Although Nature makes notoriously changeable weather, the practical way of estimating the variation of rates with environment is to create a range of steady, definable environments, as for the germination of spores (Fig. 4). Then the practical way to use those rates in naturally changeable weather is to integrate progress as shown in Eq. (9)

$$f = \int_0^{t_{1/2}} p \, dt \tag{9}$$

where the rate of progress p sums to f by time $t_{1/2}$ when half the individuals have completed the development being considered, e.g., half have germinated, sporulated, or died.

Figure 4 shows two graphs of development. The left graph, showing rapid appearance of the first development, can be modeled by exit from a single integral, i.e., f is 1. When no spores are extinguished, p is 1/(characteristic time). The right graph of Fig. 4 shows a lag in the appearance of the first developments, and I wrote above that this could be simulated by a series of integrals hooked head to tail. In this case, too, f is approximately the number of integrals. Further, the curve is approximately a Normal ogive with variance [see Eq. (10)]

$$\sigma^2 = f/p^2 \tag{10}$$

where p is the rate near $t_{1/2}$ (Parlange, 1974). Equations (9) and (10) allow estimation of f and p from observations like Fig. 4, and a computer model with f integrals and rates p will simulate Fig. 4.

The question is whether rates p from experiments in steady environments will simulate development in variable ones. For example, if spores spend half their time at 15° and half at 23°C, when they are at 23°C will they neither remember their earlier experience nor be affected by the change but instead develop at first p_{15} and then p_{23}, making the sum of $p_{15}t_{15}$ plus $p_{23}t_{23}$ add to f? Although it seems extraordinary luck, the germination of *Alternaria solani* spores does behave in this simple way from 4° to 30°C (Waggoner and Parlange, 1974b).

Heating to temperatures above 30°C complicates this simple model (Waggoner and Parlange, 1974a, 1975), and variable moisture also requires a more complex model (Waggoner and Parlange, 1976). Nevertheless, the studies of *Alternaria* show both that a simple model is realistic over an important range and that simple experiments can establish where matters stand.

Demonstrating that the simple model works in one case provides some foundation for another simplification. A model of relatively few integrals as implied by the diagram of Fig. 3, even when integrals like CSF are subdivided into *f* integrals on the way to the next visible stage, places organisms of different ages but the same degree of development in the same class or integral. One could make a more complicated model with integrals for each age and stage of development, e.g., there would be a separate integral solely for spores that both had been wetted for 2 hr and had passed through a third of the period or the invisible stages between wetting and germination. Fortunately, the success in a fluctuating environment of the model of *f* stages indicates that spores develop according to their stage of development and not their history, encouraging one to neglect age, which is a measure of history, and use only state of development in a relatively simple model.

E. The Problem of Dispersal

The successes and failures of the immigration of pathogenic colonists setting out from the home lesion, tossing on a dangerous medium, and then alighting on the friendly shore of a susceptible host are difficult to quantify and thus to model. Not the least of the difficulties is that, unlike inoculation or fluctuating environment, dispersal is difficult to bring into the laboratory. I begin with aerial dispersal, whose analysis and subsequent modeling is aided by dividing dispersal into takeoff, flight, and landing.

Concentrating upon passive takeoff, one thinks of wind stress, and in a simulator of a blight caused by *Helminthosporium* (Waggoner *et al.*, 1972), we calculated that takeoff increased regularly with wind stress. Fortunately, takeoff can be examined in the laboratory (Waggoner, 1973), and we have been taught that it does not increase regularly with wind.

In fact, since *Helminthosporium* spores are held with a remarkably uniform strength, takeoff is slight until a critical speed is reached and then many blow away (Aylor, 1975a). Thus a factor for passive takeoff, at least for one species, would be BLO = 0 when wind is slower than about 5 m/sec, and BLO = 1 when wind is faster than 5 m/sec at spore height. Spore attachment is not, of course, exactly uniform, and the simple 0 and 1 could be refined to reflect the distribution of strengths.

If this model were employed with average wind speed and with a consideration of the slowing of wind within the boundary layer near the leaf where the little spore dwells, few should take off because 5 m/sec is fast (5 m/sec = 11 miles/hr). As Aylor and Parlange (1975) have found, however, it is not the average but rather the maximum that matters, and the fastest 2 sec wind can be five times the hourly average. Further, when the speed changes abruptly in a short distance in the direction of the wind, the required speed is attained at spore level before the boundary layer grows to that height. Thus the wind used in calculating BLO must be a maximum and not an average, and it is critical whether the change from slow to fast occurs before the boundary layer about the leaf can grow to spore height. It turns out, therefore, that the biological portion of a model of takeoff is relatively simple and depends on a wind speed, but the meteorological portion of the model, which calculates the necessary speed, has yet to be developed.

Turning to the flight to the neighborhood of a new host, one must first decide whether the goal is a model with disease as a function of time alone or a geographical model with disease as a function of location and time.

For change with time alone, the concern need only be how many spores are blown outside the field of susceptible foliage. In EPIMAY (Waggoner et al., 1972) the escape of spores from the system was estimated by division by the wind speed raised to a fractional power that would depend upon the fall of the spores and the turbulence of the air and implicitly upon geographical factors of size of field and height of sporulating lesions relative to target leaves.

If a geographical mode is sought, the observations of diffusion that show that aerial concentrations decrease roughly as the inverse square of the distance from the source (Gregory, 1945) are a good guide. Shrum (1975) has incorporated geography into a disease model and successfully simulated dissimilar increases of *Puccinia striiformis* in seven locations. Zadoks (1976) has made a beginning at the difficult task of composing a geographical model. Although it concerns an actively discharged seed, a model (Strand and Roth, 1976) of mistletoe in a forest incorporates crown structure and positions of source and target as well as horizontal distance between them.

Genetic heterogeneity is looked to for slowing epidemics and is even used on a small scale by mixing varieties in a single field (Browning and Frey, 1969). Clearly a model of this is useful in determining ratios and arrangements of varieties. Change with time alone can be modeled by merely removing a portion of spores that settle on immune foliage (Waggoner, 1976). Alternatively, in a geographical model the gradient of disease dispersal explicitly affects the epidemic (Kiyosawa, 1976).

Finally, the successful spore must alight on a friendly host. Given the usual size of spores, breadth of leaves, and speed of wind, the landing is usually a quiet settling (Aylor, 1975b). Thus it is reasonable to model successful landings simply proportional to the leaf area index (LAI) (Waggoner *et al.*, 1972), perhaps with an adjustment for the horizontal projection of leaves and their overlapping.

Remembering that the potential successful landings are reduced if part of the foliage is diseased, and recurring to the principle of Eq. (2), one would make the landings that may increase infection a fraction (all LAI — diseased LAI)/all LAI of the landings that deplete the aerial load of spores (Waggoner *et al.*, 1972).

Pathogens also move through the soil, and the zoospores of several *Phytophthora* species provide an example. Moving water and saturated soil were required for infiltration of *P. fragariae*, and infection and presumably movement were greater in coarse than in fine soil (Hickman and English, 1951). Similarly, *P. cryptogea* movement required nearly saturated soil (a water potential at least as great as —10 mbars, which would empty pores larger than 150 μm radius), and a porous potting mixture passed zoospores further than a fine sandy loam (Duniway, 1976).

J.-Y. Parlange and I suggest how this knowledge could be transformed into a computer simulation of spores infiltrating and lodging in a column of soil as water percolates at VEL cm/hr. First, assume that motile zoospores move with the water, but still zoospores do not. Then divide the soil into strata thin enough to be essentially uniform. Finally, calculate the number of motile MV and still ST zoospores per stratum by integrating rates of spores per hour entering, leaving, and becoming still in the stratum.

The number of spores in a stratum would be

$$MV = INTGRL (IMV, RMV), \text{spores}$$
$$ST = INTGRL (IST, RST)$$

where IMV and IST are the initial numbers in the stratum. The rate RMV of change in motile spores/hour in the stratum would be the algebraic sum of the rate RIN motile spores per hour entering, ROUT leaving, and RST becoming still in the stratum.

$$RMV = RIN - ROUT - RST \text{ spores/hour}$$

Assuming that movement of spores in still water is negligible compared to the movement of the water itself and remembering that spores do not move when pores at least 150 μm large are not filled with water, we can write spores/hour

$$ROUT = O$$

when water columns are less than 150 μm radius,

$$ROUT = (MV/VOL) * A * VEL$$

when water columns are 150 μm in radius or larger. VOL is the volume, and A is the area of upper or lower faces of the stratum. The RIN of the top stratum would be given while the RIN of other strata would be the ROUT of the stratum above. Finally stilling would be RST = MV/TIMEM, spores/hour where TIMEM is the characteristic time of motility.

The computer program would include the above equations for each stratum, a subprogram for determining the radius of water columns, and such parameters as RIN to the top stratum. For saturated soil, the model would proceed simply, simulating the effects of different durations of motility or soil textures upon the zoospores lodged at each depth or escaping from the bottom of the column. For unsaturated soil, the disease simulator would have to be coupled with a model of water movement.

A recently observed characteristic of movement of water in soil lends an intriguing aspect to pathogen movement. We assumed above that water moves evenly downward. In fact, when fine soil lies over coarse or when the solution at the top is heavier with solute than that beneath, fingers of water move downward, here and there, much faster than expected from the simple view of an average advance of a uniform wetting front (Hill and Parlange, 1972; Starr and Parlange, 1976). This rapid penetration of a few fingers could carry a few spores rapidly to considerable depths and obviously deserves investigation and then modeling. In the meantime, just learning whether motility really affects movement of zoospores with water in the soil or coupling a model of disease to a model of the infiltration of water will provide useful work because the movement of zoospores in soil "is urgently in need of further analysis" (Hickman and Ho, 1966).

F. Coping with Microclimate, Winter, and Races

Future tasks include coping with microclimatic differences that make the hospitality for a pathogen different on an upper leaf cooled by the wind, a lower leaf parched near the soil, and an inner leaf shaded from the sun. Incorporating this diversity may explain the occurrence of late blight in the seemingly inhospitable potato fields in Israel (Rotem et al., 1970), a paradox that an average environment cannot explain (Waggoner, 1976).

Varied temperatures, humidities, and evaporation rates near and on a leaf (Raschke, 1960) or within a canopy of leaves (Waggoner, 1975) can be calculated from the characteristics of leaves and macroenvironment, which are familiar parameters. Thus when the refinement is merited, the tools are at hand to calculate the environment of a pathogen in different locations within the canopy, sometimes causing it to perish in one location, persist in another, and then be renewed throughout when the environment becomes wholly favorable. As a first attempt, I would compose a model of two compartments, one representing the outer leaves with an environment the same as the macroenvironment and the other with the calculated microenvironment of shaded, sheltered leaves. The pathogen would grow or perish in the separate compartments by the rules of two EPIDEMS with different environments but connected by dissemination between the two compartments.

The winter as well as the summer surely affects epidemics as Stevens (1934) realized in his forecasting scheme for Stewart's wilt of corn. The overwintering of a pathogen has yet—so far as I know—to be simulated by a model of the sort that is my subject here, but when the time is ripe it might be done by adding in parallel to lesions, X_1 (Fig. 2) integrals that represent the fall of infected host organs onto the soil, the survival of a few, and then their sporulation.

Overwintering provides an example of how race or virulence can be incorporated into a model. Different races of the pathogen of tobacco black shank survive the cold of winter soil differently (McIntyre and Taylor, 1977), and a model of overwintering would have different mortality rates of the pathogen in the same cold according to racial characteristics. In general, of course, virulence or resistance could affect the rates for other arrows in Fig. 2, a correct model simulating the effect of virulence or susceptibility by incorporating the true mechanism as in more abundant sporulation or less frequent invasion.

V. HOST AND VECTOR

Up to this point, I have written about models of the pathogen that are really "mycological" and in concluding I must write a few words about "phytopathological" models, remembering that it takes a host and sometimes a vector to make a disease. The present writing improves my modesty because a classic model of disease (Gause, 1934) presented years ago concerns the interdependent changes in malaria in mosquitoes and in people.

A rudimentary model of host growth, a vector, and a disease has been designed by Saul Rich, Don Aylor, Gary Heichel, and myself and provides an illustration from plant pathology. The model represents the growth, infection with *Erwinia stewartii*, and death of an established population P of corn plants and the increasing infestation of a constant population B of the vectors, flea beetles, which have overwintered. In this model, Fig. 5, will be found integrals for the host and vector, but no integral for the pathogen.

The increase in the fraction A of flea beetles infested and infective is set proportional to K_2 bites/day, K_8 infestation/bite, and the fraction $(1 - A)$ of beetles still clean. The increase in A is also set proportional to the fraction $I/(H + I)$ of the living leaf area that is lesion; this could mean that the flea beetles are being infested by biting lesions or by biting an infectious area proportional to lesion area [see Eq. (11)].

$$dA/dt = K_2 K_8 (1 - A) I/(H + I) \qquad (11)$$

The decrease in the number P of healthy plants is set proportional to K_2 bites/day, K_3 infections/bite, the number AB of infested beetles, and the fraction $P/(P + Q)$ of plants still healthy [see Eq. (12)].

$$dP/dt = -K_2 K_3 A B P/(P + Q) = L_5 \qquad (12)$$

The increase in diseased plants Q is L_5 minus the rate L_3 of death, which we shall call extinction. Extinction L_3 of plants is set proportional to the infected plus dead leaf area $(I + D)$ on the Q plants relative to all the foliage estimated for the Q plants $[Q(H + I + D)/(P + Q)]$. Thus the extinction rate [Eq. (13)] L_3 is

$$L_3 = K_4 Q(I + D)/[Q (H + I + D)/(P + Q)] \qquad (13)$$

where the factor K_4 is 1/day.

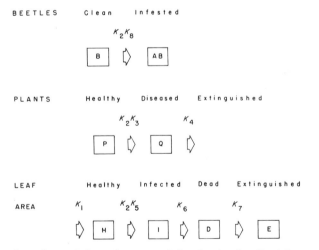

Fig. 5. Flow chart of the infestation of flea beetles by *Erwinia stewartii*, the infection of maize plants, and the growth, infection, and death of leaf area. (Leaf area is extinguished when a plant dies.)

Finally, we come to the leaf area that grows, becomes infected, dies on a living plant, and is extinguished when an entire plant dies. The healthy area H grows in proportion to H and to the portion $[M - (H + I + D + E)]/M_1$ of the potential growth still unrealized. M_1 is a maximum leaf area attained by a healthy crop, while M is smaller because of disease, as will be formulated later. Healthy area is decreased by bites L_6, by expansion L_1 of lesions, and by the extinction L_4 of healthy area stranded upon a dying plant. Thus, as shown in Eq. (14),

$$dH/dt = K_1 H [M - (H + I + D + E)]/M_1 - L_6 - L_1 - L_4 \qquad (14)$$

where a realistic value of K_1 is 0.1/day and of M_1 is 5 ha of leaf/ha of land.

The increase in infected area I is, of course, L_6 and L_1. The L_6 infected/day at the moment of biting is set at an area/bite times the ABK_2 bites/day by infested flea beetles times the area $H/(H + I)$ still healthy. The expansion L_1 of lesions is set proportional to the diseased area I and to the portion of the area on the diseased plants that is still healthy. The infected area I is decreased by the death L_7 of lesion area on healthy plants, which is simply proportional to I, and by the extinction L_2 of infected area stranded on dying plants. Thus

$$dI/dt = L_1 + L_6 + L_7 - L_2 \qquad (15)$$

Next an equation must be written for the change in dead area D on living plants, which is L_7 minus the estimated area of D that is extinguished when plants die. Also, an equation must be written for the change in E, which is the area of healthy, diseased, and dead foliage that is extinguished when an entire plant dies.

The final rate equation is written for M, the maximum leaf area. In a healthy crop, leaf area increases sigmoidally to a maximum M_1. Disease, however, decreases the maximum because early death of foliage reduces potential growth. In the model, therefore, M is decreased by the three losses $(L_1 + L_4 + L_6)$ of healthy foliage area, and the decrease in M is weighted by dividing it by the relative size $(H + I + D + E)/M$, at the time of loss of the healthy area.

VI. CONCLUSION

The complexity of the interacting environment, host, and pathogen has frustrated us. For example, one could measure the effect of temperature upon the single stage of germination of *Phytophthora*, as Crosier (1934) did, but linking this to the effect of temperature upon lesion growth, except in a general way, was not easy. Or, one could draw a sigmoid curve of the entire epidemic growth, as van der Plank (1963) did, but varying this with environment, except in a general way, was not easy. Now fast computers have made it easy to deal with complexity, and new work begins. The concluding model of wilt of corn illustrates that beginning. While using the old foundations of limit and multiplication and of observations of a single stage, the model shows how a computer numerically integrates several integrals pertaining to different stages or factors. Typically the model reveals the lack of knowledge of rates and evokes their observation as when one wonders how height or age of leaf affect the K_2 bites/day of a flea beetle. Especially, this model begins the essential task of calculating the loss of yield to disease. The easy computation makes the hard pathology.

References

Aylor, D. E. (1975a). Force required to detach conidia of *Helminthosporium maydis*. *Plant Physiol.* **55**, 99–101.

Aylor, D. E. (1975b). Deposition of particles in a plant canopy. *J. Appl. Meteorol.* **14**, 52–57.

Aylor, D. E., and Parlange, J.-Y. (1975). Ventilation required to entrain small particles from leaves. *Plant Physiol.* **56**, 97–99.

Bashi, E., and Rotem, J. (1974). Adaptation of four pathogens to semi-arid habitats as conditioned by penetration rate and germinating spore survival. *Phytopathology* **64**, 1035–1039.

Browning, J. A., and Frey, K. J. (1969). Multiline cultivars as a means of disease control. *Annu. Rev. Phytopathol.* **7**, 355–382.

Burleigh, J. R., Eversmeyer, M. G., and Roelfs, A. P. (1972). Development of linear equations for predicting wheat leaf rust. *Phytopathology* **62**, 947–953.

Crosier, W. (1934). Studies in the biology of *Phytophthora infestans* (Mont.) DeBary. *N.Y., Agric. Exp. Stn., Ithaca, Mem.* **155**, 1–40.

Duniway, J. M. (1976). Movement of zoospores of *Phytophthora cryptogea* in soils of various textures and matric potentials. *Phytopathology* **66**, 877–882.

Gause, G. F. (1934). "The Struggle for Existence." Williams & Wilkins, Baltimore, Maryland.

Goudriaan, J. (1973). Dispersion in simulation models of population growth and salt movement in the soil. *Neth. J. Agric. Sci.* **21**, 269–281.

Gregory, P. H. (1945). The dispersion of air-borne spores. *Trans. Br. Mycol. Soc.* **28**, 26–72.

Hickman, C. J., and English, M. P. (1951). Factors affecting the development of red core in strawberries. *Trans. Br. Mycol. Soc.* **34**, 223–236.

Hickman, C. J., and Ho, H. H. (1966). Behavior of zoospores in plant-pathogenic Phycomycetes. *Annu. Rev. Phytopathol.* **4**, 195–220.

Hill, D. E., and Parlange, J.-Y. (1972). Wetting front instability in layered soil. *Soil Sci. Soc. Am., Proc.* **36**, 697–702.

IBM. (1969). "System/360 Continuous System Modeling Program Users Manual," H20–0367–3. IBM. White Plains, N.Y.

Jowett, D., Browning, J. A., and Haning, B. C. (1974). Non-linear disease progress curves. *In* "Epidemics of Plant Diseases. Mathematical Analysis and Modeling" (J. Kranz, ed.), pp. 115–136. Springer-Verlag. Berlin and New York.

Kiyosawa, S. (1976). A comparison of simulation of disease dispersal in pure and mixed stands of susceptible and resistant plants. *Jpn. J. Breed.* **26**, 137–145.

Last, F. T., Ebben, M. H., Hoare, R. C., Turner, E. A., and Carter, A. R. (1969). Build-up of tomato brown rot caused by *Pyraenochaeta lycopersici* Schneider and Gerlach. *Ann. Appl. Biol.* **64**, 449–459.

Lotka, A. J. (1924). "Elements of Physical Biology." Williams & Wilkins, Baltimore, Maryland.

McIntyre, J. L., and Taylor, G. S. (1978). Race 3 of *Phytophthora parasitica* var. *nicotianae*. *Phytopathology* **68**, 35–38.

Parlange, J.-Y. (1974). Analytic solution to model passages through phenophases. *In* "Phenology and Seasonality Modeling" (H. Lieth, ed.), pp. 323–326. Springer-Verlag, Berlin and New York.

Raschke, K. (1960). Heat transfer between the plant and environment. *Annu. Rev. Plant Physiol.* **11**, 111–126.

Rotem, J., Palti, J., and Lomas, J. (1970). Effects of sprinkler irrigation at various times of the day on development of potato late blight. *Phytopathology* **60**, 839–843.

Royle, D. J. (1973). Quantitative relationships between infection by the hop downy mildew pathogen, *Pseudoperonospora humuli,* and weather and inoculum factors. *Ann. Appl. Biol.* **73**, 19–30.

Shrum, R. (1975). Simulation of wheat stripe rust (*Puccinia striiformis* West.) using EPIDEMIC, a flexible plant disease simulator. *Pa. State Univ., Prog. Rep.* **347**, 1–81.

Starr, J. L., and Parlange, J.-Y. (1976). Solute dispersion in saturated soil columns. *Soil Sci.* **121**, 364–372.

Stevens, N. E. (1934). Stewart's disease in relation to winter temperature. *Plant Dis. Rep.* **18**, 141–149.

Strand, M. A., and Roth, L. F. (1976). Simulation model for spread and intensification of western dwarf mistletoe in thinned stands of ponderosa pine saplings. *Phytopathology* **66**, 888–895.

van der Plank, J. E. (1963). "Plant Diseases: Epidemics and Control." Academic Press, New York.

Waggoner, P. E. (1973). The removal of *Helminthosporium maydis* spores by wind. *Phytopathology* **63**, 1252–1255.

Waggoner, P. E. (1974). Simulation of epidemics. *In* "Epidemics of Plant Diseases" (J. Kranz, ed.), pp. 137–160. Springer-Verlag, Berlin and New York.

Waggoner, P. E. (1975). Micrometeorological models. *In* "Vegetation and the Atmosphere" (J. C. Monteith, ed.), Vol. 1, pp. 205–228. Academic Press, New York.

Waggoner, P. E. (1976). Predictive modeling in disease management. *In* "Modeling for Pest Management" (R. L. Tummala, D. L. Haynes, and B. A. Croft, eds.), pp. 176–186. Michigan State University, East Lansing.

Waggoner, P. E., and DeWit, C. T. (1974). Growth and development of *Helminthosporium. In* "Simulation of Ecological Processes" (C. T. DeWit and J. Goudriaan, eds.), pp. 99–123. Pudoc, Wageningen, The Netherlands.

Waggoner, P. E., and Parlange, J.-Y. (1974a). Mathematical model for spore germination at changing temperature. *Phytopathology* **64**, 605–610.

Waggoner, P. E., and Parlange, J.-Y. (1974b). Verification of a model of spore germination at variable, moderate temperatures. *Phytopathology* **64**, 1192–1196.

Waggoner, P. E., and Parlange, J.-Y. (1975). Slowing of spore germination with changes between moderately warm and cool temperatures. *Phytopathology* **65**, 551–553.

Waggoner, P. E., and Parlange, J.-Y. (1976). Germination of *Alternaria solani* spores in changing osmotic pressures. *Phytopathology* **66**, 786–789.

Waggoner, P. E., Horsfall, J. G., and Lukens, R. J. (1972). EPIMAY. A simulator of southern corn leaf blight. *Conn., Agric. Exp. Stn., New Haven, Bull.* **729**.

Zadoks, J. C. (1976). Role of crop populations and their deployment, illustrated by means of a simulator, EPIMUL 76. *N.Y. Acad. Sci. Annals.* **287**, 164–190.

Chapter 11

Forecasting of Epidemics

ROBERT D. SHRUM

I. INTRODUCTION

A forecaster is a prophet and a prophet has his troubles. This has been true throughout history. A fragment from an Egyptian papyrus of the second or third century AD says that "a prophet is not acceptable in his own country." This same idea was expressed a couple of centuries earlier by St. Matthew and 18 centuries later by Victor Borge, the famed Danish comedian, who put his own twist on it by saying that forecasting is a difficult business, especially forecasting the future.

In this chapter I hope to avoid the hazards of being a prophet by discussing the procedures for forecasting rather than the actual making of forecasts. It is interesting that the word prognosis is used in the German and Russian literature of plant pathology. In English the word prognosis is used mainly to describe the expected progress of disease in humans, but taken literally it means simply "knowing beforehand." In English the word forecasting means "to foresee or to calculate beforehand." Thus, the calculation of probabilities is implicit in the meaning of the word.

223

Although I shall mention a few of the earlier, simpler methods for forecasting, major attention will be devoted to a discussion of more modern procedures that are based on systems analysis. Despite major achievements in this field (Waggoner *et al.*, 1968, 1969, 1972; Bourke, 1970; Tschumakov, 1973; Berger, 1977), plant pathologists generally have not recognized the full value of forecasting in translating our substantial knowledge of disease dynamics into practical disease management. Thus, this chapter has been developed with two parallel purposes and audiences in mind: (1) to introduce methods by which epidemics can be modeled by the many plant pathologists who can contribute to forecasting but have been inhibited from doing so by the confusing jargon of this specialized branch of biology; and (2) to encourage plant pathologists already conversant with modeling to increase forecasting accuracy by selecting models which are as ideal as possible for the purpose intended and for the biological systems to be modeled. Although mathematics and computer science are beguiling in themselves, reliable forecasting requires that the mathematical tools used be appropriate to the biological system being modeled, and not the other way around.

As Bourke (1970) has said so well, modeling must not be "a technique and a jargon which can be made to serve as a figleaf to hide the shame of barren thought." Bourke refers to A. S. C. Ehrenberg as the inventor of the word "sonk," an acronym for the "scientification of nonknowledge." I hope to show that mathematical tools can be used properly in forecasting plant disease and that "sonk" need not be involved. Fortunately, I do not stand alone in this effort (Brockington, 1971; Conway and Murdie, 1971; Hanson, 1971; Headley, 1972; Hyre, 1954, 1957; James, 1974; Jeffers, 1971; Kranz, 1974, also Chapter 3, this volume; Krause and Massie, 1975; Large, 1955; McNew, 1972; van der Plank, 1963; Waggoner, 1960, also Chapter 10, this volume; Zadoks, 1971, also Chapter 4, this volume).

II. THE PURPOSE OF FORECASTING

Every rational person indulges in forecasting. We expect visitors this weekend; therefore, we had better buy enough food for the extra people. I expect snow this winter; therefore, I will buy a new pair of skis. We expect the cost of gas and oil to increase; therefore we will purchase a smaller car and add more insulation in the attic.

Every rational farmer also indulges in forecasting. I expect that drought may hurt my corn crop; therefore I will buy some irrigation equipment this year. Labor prices may continue to increase during these next few years; therefore I will buy a mechanical tree shaker for my cherry orchard. Late blight attacks my potatoes 3 years out of 4; therefore I must spray my crop every year even though I may not need it in some years.

Obviously, every farmer has the chance to increase his profits with more accurate forecasts of dry weather, increasing labor costs, the probability of late blight, and many other uncertainties in his risky business.

Plant pathologists are interested in forecasting for three major reasons: (1) disease forecasts are useful to farmers in the practical management of disease in their crops; (2) research aimed at accurate forecasting helps us identify the gaps that remain in our knowledge; and (3) modern research on modeling the complexities of epidemics is fun.

The ultimate purpose of modeling, the "raison d'etre" of plant pathology if you will, is to minimize economic losses due to plant disease (see Main, 1977; Chapter 4, Vol. I; Ordish, 1969; Headley, 1972; James, 1974). As Main points out, the farmer must balance three factors in disease management—risk, cost, and benefit. Here forecasting can be immensely useful. If the forecast suggests that the risk of disease is low, there is no sense in spraying or purchasing an expensive resistant variety. If the forecast suggests that the risk of disease is great, however, there may be a big advantage in spraying or using a resistant variety. These decisions about alternative strategies or tactics must be made "down on the farm" since the probabilities, costs, and benefits will vary from farm to farm and from time to time. The forecasts must be accurate, adaptable to different locations, and convenient to apply in a timely fashion.

Forecasting provides a means for determining if, when, and where a given management practice should be applied. For example, regular application of fungicides is a widely used tactic for management of foliage diseases; but with reliable forecasts, applications of fungicides can be timed to correspond with disease development so that fewer sprays will provide the same degree of control.

Forecasting provides other critical information as well. Even in cases where a disease has been studied experimentally over a long period of time, many aspects of the disease turn out to have been overlooked so that critical bits of information remain undiscovered. Forecasting provides a systematic way to assess our knowledge of environmental factors and their influence on the progress of disease.

Finally, forecasting is also important because it is intellectually stimulating. Forecasting allows a pathologist to test his understanding of disease and his ability to integrate its pieces into a dynamic, functioning system. The challenge can include biological, climatological, mathematical, and/or computational lines of action—all are formidable and challenging and all must function together if the forecast is to work. It is somewhat intimidating, but always challenging, to make the inevitable end-of-season comparison of forecast with actual occurrence of disease. This is the acid test of the quality of our thinking and understanding. Sometimes we celebrate our success; and other times we go back to the drawing boards for another try next year.

III. SOME EARLY FORECASTING PROCEDURES

Apple farmers in the scab belt, potato farmers in the late blight belt, and pine nurserymen in the rust belt have all learned the hard way how frequently their crops become diseased. Therefore they forecast disease and decide whether to use fungicides or resistant varieties. In some years the hazard of disease may be low and the sprays unnecessary. Then the farmer or nurseryman must decide if it is better to expect the worst and buy the sprays anyway or to accept the risk and thus increase his profits. Farming has always been a risky business and some of our early plant pathologists devised some very useful systems by which these risks could be reduced.

A. Forecasting Based on Inoculum Potential

Perhaps the most workable of the early efforts to forecast disease were based on inoculum potential—how much inoculum is available? Heald (1921) helped farmers forecast wheat bunt a year in advance by counting the number of spores on the grain. Knowing the rate of spread through the soil, Taubenhaus and Killough (1923) were able to tell their farmers fairly accurately where *Phymatotrichum omnivorum* would attack their cotton the following year. Similarly, Hepting (1935) devised a system by which foresters could predict the amount of heartrot that would be expected 10, 20, or even 30 years after hardwood trees were damaged by logging wounds or by fire. Wilhelm (1950) estimated the amount of *Verticillium alboatrum* on cotton by growing tomatoes in the greenhouse in samples of soil from the farmers' fields. Leach (1938) developed a system for forecasting decay in sugar beets by *Sclerotium rolfsii* by counting the sclerotia left in the soil from the previous crop.

These forecasts based on inoculum potential contain a hidden assumption—that the weather will continue to be favorable for disease. If the weather changes, of course, the forecast may be in error. If the impact of the change is great, the error can be unacceptably large; if the impact is small, the error in the forecast may be correspondingly small.

B. Forecasting Based on the Weather

Some of the most popular early forecasting methods were based on analyses of weather patterns. Late blight of potatoes is a classic case. The disease is highly dependent on rainy weather and local temperature. Beaumont (1947) in Britain devised this type of a forecast for late blight. He assumed that one or two periods of favorable weather were critical.

Bourke (1957) in Ireland improved the system by using synoptic weather charts. Low barometric pressure passing over Ireland creates the weather pattern needed for late blight. The passage of the "low" can be forecast with some accuracy. A third development involved negative forecasts. Farmers need to know when weather will be unfavorable for blight as well as when it will be favorable. Negative forecasting systems have been perfected in Germany by Burckhardt and Freitag (1969), and in the United States by Hyre (1954) and Wallin (1962).

Of course, forecasting based on the weather carries its own hidden assumption—that inoculum potential will be high enough to explode into an epidemic before the season ends. Sometimes this is a very safe assumption; sometimes it is not.

IV. USE OF MATHEMATICAL MODELS IN FORECASTING

The kind of forecasting system we need for most diseases is a procedure that can handle and integrate all significant features of an epidemic—the pathogen, the host, the weather, the time, and perhaps the space. Neither the procedures nor the calculating machines needed for doing this were available to the pioneers in plant pathology. The tools became available with the advent of "operations research" and capacious computers during and after World War II. The approach is called systems analysis and the tool is called mathematical modeling or simulation. The first application to botanical epidemiology was made by Waggoner (1968) using the old classic disease—late blight of potatoes. The procedure was amplified and improved by Waggoner and Horsfall (1969) for the early blight disease of potato and tomato.

In recent years, epidemics of other diseases have been simulated by computer. A few examples are given below: wheat rust by Zadoks et al. (1972) and by Shrum (1975); southern corn leaf blight by Waggoner et al. (1972) and Massie (1973); downy mildew of hop by Butt and Royle (1974); apple scab by Kranz et al. (1973 and Analytis (1973); and diseases of various crops as discussed by Berger (1977).

A. Some Definitions

Let us begin this discussion of systems forecasting with a few definitions and descriptions. We shall consider that an epidemic is a system which progresses with time and space and varies with the weather. An epidemic model is an abstract representation of the epidemic system. A holistic model is one that deals with the epidemic as a whole, as van der Plank (1963) did in his now-classic book. A systems analytic model breaks the system into its component parts, treats each one separately, and accounts for the weather inputs at successive stages in the progress of the epidemic.

Forecasting models can take many forms: (1) a set of decision-making guidelines or rules; (2) mathematical or statistical formulas, as characterized by van der Plank (1963); and (3) a computer-based simulator that includes all significant parameters, as discussed more fully by Waggoner in Chapter 10 of this volume.

B. The Epidemic as a System

A typical epidemic is depicted in Fig. 1 as a system of sequential steps by which disease progresses. This diagram has many features in common with the disease cycles we all have learned in beginning courses in plant pathology. It is only slightly different from the one Waggoner has used in Chapter 10 to illustrate how a computer simulator of disease can be constructed. There is an undeniable sequence to the processes which propels the epidemic.

The epidemic system depicted in Fig. 1 is comprised of several states. These states represent the stages through which the pathogen passes as the disease progresses, for example, the propagules, the invasions, and the lesions. We shall call each of these important stages a state variable. Each state variable is a biologically significant unit that can be represented by some measurable attribute or group of attributes which we shall call a component of the epidemic system.

As one state gives way to another during disease development, a rate is involved. For example, spores on the foliage comprise a state. Germinated spores comprise another state. As spores on the foliage become germinated spores, a rate is involved. We call this the rate of germination. The rates are shown as arrows in the diagram.

These various rates of transformation from one state to another are all influenced by what we shall call external factors or external variables. These comprise the factors in the environment that impinge on the various states of the epidemic system. An external variable may either slow down or accelerate the rate of change from one state to another.

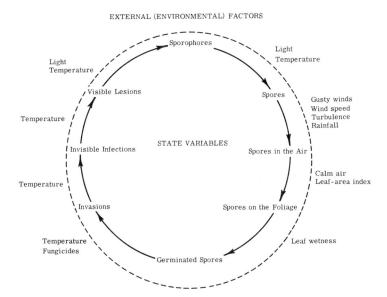

Fig. 1. Schematic diagram depicting the dynamics of a foliar pathogen. The broken line encloses the major state variables (the epidemic system). Factors outside the broken line are environmental factors that determine the rate of change from one state to another (system's environment).

For example, with *Alternaria solani,* leaf wetness caused by dew or rain slows the rate of change from spores to spores in the air. Leaf wetness also accelerates the rate of change from spores on the foliage to germinated spores (Waggoner, 1969). Similarly, gustiness of wind accelerates the change from spores to spores in the air, but may decrease the rate of change from spores in the air to spores on the foliage. A single external factor can exert different degrees of influence or even opposite influences at different parts of Fig. 1. This is a major difference between systems analytic and holistic modeling procedures. For a more complete description of such influences see Chapter 8 by Aylor in this volume.

In Fig. 1, all state variables are shown within a dotted line to emphasize that they are internal variables, as distinct from the external variables which though they act on it are themselves not part of the system.

All of these aspects of the epidemic have to be dealt with in a workable model, because in systems analysis the comprehensiveness and accuracy of description of the parts will determine the accuracy of the forecast and its sensitivity to the many factors, both biological and physical, which affect the dynamics of the epidemic as a whole.

In common with van der Plank (1963) I shall use the letter *r* to denote the apparent rate of increase of disease, or stated another way, the apparent rate at which the disease cycle turns. In mathematical terms this is simply the slope of the line on a disease progress curve.

V. ANALOGY WITH A ROAD MAP

"Models are used to frame concepts and to organize knowledge to the end that the right questions may be asked" (Patten, 1972). Models are summaries of the important features of the real system and depend on the judgment of the forecaster as an epidemiologist to determine the proper balance between realism and abstraction. Stated in Bourke's (1970) inimitable style: "The main problem is to steer a safe course between the Scylla of stripping the model to a stark simplicity which throws out the baby with the bath water, and the Charybdis of clogging the machinery with irrelevancies which stop just short of the kitchen sink."

The complexity of an epidemic viewed as a whole, the time, the space and the intricate interactions, will boggle the mind, but it can be both easy to understand and easy to model one component at a time. Let us examine this reduction of complexity to simplicity by using the analogy of a road map.

This modeling process can be compared to constructing a national highway map with complex details such as road structure and condition, county lines, topography, and historical points of interest. Although the whole job may seem mind-boggling at first, the job becomes manageable if it is done by components. Instead of mapping all features simultaneously, map the roads first, then the county lines, and finally the topography. Or, divide the nation into states and provinces, develop maps for each component, and then assemble the separate units into a national map.

Similar processes can be used to model complex epidemics. As shown in Fig. 1, the real system can be dissected into its various states, and the pertinent environmental factors can be identified and their influence on rates can be measured experimentally. Each such unit can be constructed independently. Then the whole system can be assembled into a descriptive model whose predictive value can be tested by trying it out on next year's crop.

When we plot a disease progress curve, we map disease development. We can use this map to determine the time it takes to increase disease from amount A to amount B (say, from 1 to 40% of plants infected). Even though the disease progresses along a fixed path, i.e., by successive turns of the disease cycle, the time it takes to get from percentage A to percentage B may vary with environmental conditions just as the time to drive from city A to city B may be affected by the topography or road conditions.

A map is an abstract model of the real world. A road map can be full of complex detail or it can be very simple. A map cluttered with too many details may be useless to the traveler seeking direction. Likewise, no single model of an epidemic should try to represent all aspects of the real system at once. For this reason, the objective(s) of the model should be defined before the model is built.

With this in mind let us assume that our objective is to compare the relative travel time from city A to city B and from city A to city C on a road map. If we assume that the topography along the two paths has no effect on the apparent rate of travel (r), our objective could be accomplished by comparing the length of straight lines drawn between the two cities. If, on the other hand, r is not the same for both, other factors must be modeled. Perhaps the lengths of actual roadways are not proportional to the straight lines used in the model, or perhaps one route is a superhighway and the other a gravel road. If the values of r for AB and AC differ, then r must be calculated for each route using parameters that correct for actual length of road, travel speed, etc.

Each of the above would be a holistic model because in each case the whole trip was treated as a unit, and the value of r did not change during different parts of the trip—an average value of r was applied over the entire trip.

If a different kind of environmental factor, such as rainfall or snow, has a differential effect on the speed of travel on gravel roads and paved roads, inclusion of parameters to adjust r for the amount of time that those factors operate will give more accurate forecasts for each route; but still the holistic model is appropriate for the job. If the route becomes more complex, part gravel and part paved highway for each route, a holistic model becomes less useful because it is unable to accommodate the differential effect of rain or snow on different parts of the trip. If it rained for half the trip, one would need to know whether it was the gravel half or the paved-highway half. The holistic model would not permit this kind of discrimination; thus it would be unable to predict differences in rate due to timing of rain or snow. This analogy illustrates a fundamental deficiency of holistic models in representing complex systems—sequence is ignored.

An analogous situation in disease dynamics is the case where, for instance, dew is necessary for spore germination but inhibits spore dissemination. Since the relative abundance of spores in the infection court versus spores waiting to be disseminated can vary, the impact of the event "dew" will vary. Sometimes the epidemic as a whole will be enhanced, sometimes it will be suppressed. Consistent with the need to keep track of the status of these variables, the systems analytic model monitors each state variable as the disease cycle turns. On the other hand, the holistic class of models would calculate the impact on the basis of what the average response to dew had been in the past—without knowledge of the specific biological situation leading to it.

The primary difference between holistic and systems analytic models is that the systems analytic model "knows" where it is along the trip, and evaluates the influence of specific environmental conditions against that background, while the holistic model does not. The latter deals with the epidemic cycle as though it were a single process of "average" sensitivity to an "average" environment. Individual processes (the undeniable sequences) within the system are not recognized.

Averages are compelling in science, and arguments can be made both ways about which modeling approach is best; but forecasters should always be reminded that an experienced farmer or researcher can predict "normal" epidemics as well as, and perhaps better than, models. At the same time one should be reminded that it is the unusual or unexpected epidemic, not the expected one, that is most costly to the farmers. Models which accurately forecast the atypical epidemic will be more valuable, and the more gratifying to construct.

To recap, ability to distinguish between specific combinations and permutations of environment by process interactions allows the systems analytic class of models the capability to consider the sequence of the disease cycle as it relates to the environmental events which propel it. Such sensitivity is especially important for multiple-cycle (compound interest) diseases where the influence of small perturbations is magnified in each subsequent turn of the disease cycle.

With this brief background, let us move to a consideration of timing in the use of forecasts based on either holistic or systems analytic models.

VI. PREPLANTING FORECASTS

Some strategies and tactics for management of disease must be implemented before planting—crop rotation, resistant varieties, soil fumigation, seed treatments, disease-free seeds, etc. Management decisions before planting have to be based on disease potential which in turn is often based on estimates of inoculum potential—how much of the pathogen is lying in wait for the crop to be planted? What is the abundance of overwintering vectors? Several examples based on inoculum potential have been cited earlier in this chapter. Stevens (1934) has published another excellent example. Having found that severe winter kills-off the beetles that transmit Stewart's wilt of corn, he could predict disease by knowing the winter weather. Preplant forecasts are helpful mainly for "simple interest" diseases in the sense of van der Plank (1963). Such diseases usually have only one infection cycle per season.

VII. AFTER-PLANTING FORECASTS

After-planting forecasts are essential for the explosive "compound interest" diseases in which several generations of the pathogen (turns of the disease cycle) occur each year. This type of forecast is also useful for diseases that are highly sensitive to external environmental factors so that the rate of the epidemic fluctuates widely with the weather.

The management strategy for compound interest diseases must be changed for these explosive diseases as the environment changes during the growing season. Most of the current mathematical models for forecasting deal with this type of disease because they are volatile, and thus most difficult to anticipate without such aids. After-planting management is discussed by Chiarappa (1974) as supervised plant disease control.

VIII. HOLISTIC VERSUS SYSTEMS ANALYTIC MODELS

Most holistic models treat the epidemic as though it were a single biological entity. Holistic models possess only a single dependent variable (usually some measure of disease intensity) which is expressed as a function of one or more rate-determining environmental factors. These models assume that environmental factors (temperature, leaf wetness, etc.) will always have the same effect on the epidemic (no state-times-environment interactions). Measurements of disease intensity from laboratory or field experiments can be used to identify critical environmental factors but only data on the epidemic as a whole can be used in the construction and verification of holistic models.

Systems analytic models, on the other hand, subdivide the epidemic into its various disease-cycle components, individually model the response of each state variable to their own set of pertinent environmental parameters, and then resynthesize the whole epidemic system by linking the components in a "bio-logical" manner. Greenhouse and field data can be freely used in the development and verification of any component, and thus of the whole model. Systems analytic models reserve the possibility that any given environmetal factor (for example, dew) may simultaneously result in positive, neutral, and negative impacts on disease development, the net effect of which is determined by the current relative states of the specific components being impacted. They permit the mapping of highly complex sets of interactions and responses and thus can be readily adapted from one location or environment to another.

With modern computer programs, one can easily fit holistic models to data with little concern for the specifics of how the operation is being accomplished—simply select the basic form of the model, identify the combination of parameters to be used, and let the computer do the rest. Likewise, it is also a simple operation to reverse the process and use the resultant "best fit" model for forecasting. But this often is done without realizing the restrictive biological and mathematical implications of that action. Historic data, from which such "best fits" are attained, often are not adequately inclusive. The greater the number of external factors that differentially influence the individual processes of the disease cycle, the greater the possibility that the model will fail because certain combinations and/or ranges of environmental factors were not represented in the historic data.

Multiple regression analyses provide one of the more adaptive types of holistic models yet any time one uses regression care must be taken to be sure that each independent variable is linear or linearized with respect to the dependent variable, and that criteria are met for uniform variance. These requirements are easy to violate, and extrapolating outside of the range of the original data is dangerous but usually unavoidable, when applying the model to real situations (Butt and Royle, 1974).

Regression models provide estimates of the net effect for some dependent variable, Y, based on the continuous operation of each in a set of independent variables $(x_1, \ldots, x_n)$. The relative influence of each x on y is measured by the respective partial regression coefficients $(b_1, \ldots, b_n)$. Each of these b's is a *constant* which represents an average rate and functions continuously in the operation of the model (thus a holistic model). Even where the x's in the regression might each have a biological counterpart, the real world counterparts of the b's are not likely to be either continuous or constant, and those representing sequential processes would change depending on the sucess or failure of the previous process. Butt and Royle (1974) discuss this time–series problem and give suggested solutions.

Another important limitation of holistic models is their inability to account for the critical dependence of some events on the prior occurrence of other events in a required sequence. For example, infection cannot begin until spore germination occurs, spore germination cannot occur until after spores are produced, etc. I know of no method by which holistic models can be made to adequately account for the large numbers of such mandatory sequences of events in epidemics. This is the proverbial "fly in the ointment" which limits the applicability of empirically derived decision rules, regression models, or mathematical models of whole systems, i.e., holistic models.

In considering which of these two classes of models will be best for a given application, the following factors are usually pertinent. Holistic models usually are easier and therefore less time-consuming and less expensive to constuct than systems analytic models. They also require less sophisticated computers. Of course, holistic models are applicable only within the range and combination of variables that operated to generate the data on which they are based. If changes are made in crop varieties, pathogen biotypes, management practices, and other environmental conditions, holistic models will have to be reworked. And the only way to adjust such models to accommodate the new situation is to collect new data when one (or preferably several) epidemics have occurred under the new set of conditions. The old model cannot be adjusted.

Properly constructed systems analytic models, on the other hand, can accommodate such perturbations more quickly and easily. The impact of a change to a more aggressive pathogen, to a more effective antisporulant spray, etc., can be incorporated immediately into the proper component(s) to provide a tentative evaluation of its impact on the epidemic as a whole. Specific components of the epidemic can be studied under controlled conditions and the new information then incorporated into an updated (adjusted) model as soon as it becomes available. Verification of the model, however, can be carried out only as the epidemic develops under the new conditions.

In some cases, it may be desirable to develop both holistic and systems analytic models simultaneously—a holistic model to be used as a temporary tool while the more comprehensive analytic model is being perfected.

In the long run, however, I believe that systems analytic models will predominate for most compound interest diseases of intensively managed, high-value crops, as they can cope with the vagaries of sequential events and differential environmental effects. They can provide information on a broad array of state variables and are generally more flexible than holistic models. But, of course, all these advantages are not without their price in terms of time, energy, computer time, and all the other costs of sophisticated research.

IX. FORECASTING FOR FARMERS

Where does all this discussion of holistic and systems analytic models lead? If we are to meet the objectives stated at the outset—to aid agriculture in solving its disease problems—we must put forecasting among the tools the farmer can use. The variable in the farmer's field or orchard that changes hourly is the weather and so the practical use of a forecasting model will depend on being able to record the weather in each field and orchard, to enter the data into a computer, and to get the forecast back again. Krause *et al.* (1975) have developed such a system at Pennsylvania State University for late blight of potatoes. Every participating farmer has a hygrothermograph in his potato field to monitor changes in the weather. He reads the instrument and then calls in the data on the telephone to the university. There it goes into the computer which gives back three messages: the weather is bad—spray immediately; the weather is not bad—do not spray today; or the weather is marginal—call back tomorrow and we will reassess the situation.

With the present rapid increase in practitioners in agriculture, it is likely that commercial firms will soon rise up to create these necessary computer services. I can only conclude that the sky looks bright ahead for the commercial use of mathematical models and computers for forecasting.

References

Analytis, S. (1973). Methodik der analyse von epidemien dargestellf am Apfelschorf [*Venturia inaequalis* (Cooke) Aderh.]. *Acta Phytomedica* **1**. 76 pp.

Beaumont, A. (1947). The dependence on the weather of the dates of outbreak of potato blight epidemics. *Trans. Brit. Mycol. Soc.* **31**, 45–53.

Berger, R. D. (1977). Application of epidemiological principles to achieve plant disease control. *Annu. Rev. Phytopathol.* **15**, 165–183.

Bourke, P. M. A. (1957). The use of synoptic weather maps in potato blight epidemiology. *Irish Dept. Industry and Commerce, Met. Services Tech. Note* **23**, 1–35.

Bourke, P. M. A. (1970). Use of weather information in the prediction of plant disease epiphytotics. *Annu. Rev. Phytopathol.* **8**, 345–370.

Brockington, N. R. (1971). Summary and assessment: An agricultural scientist's point of view. *In* "Mathematical Models in Ecology" (J. N. R. Jeffers, ed.), pp. 361–365. Blackwell, Oxford.

Burleigh, J. R., Eversmeyer, M. G., and Roelfs, A. P. (1972). Development of linear equations for predicting wheat leaf rust. *Phytopathology* **62**, 947–953.

Butt, D. J., and Royle, D. J. (1974). Multiple regression analysis in the epidemiology of plant diseases. *In* "Epidemics of Plant Diseases: Mathematical Analysis and Modeling" (J. Kranz, ed.), pp. 78–114. Springer-Verlag, Berlin and New York.

Chiarappa, L. (1974). Possibilities of supervised plant disease control in pest management systems. *FAO Plant Prot. Bull.* **23**, 65–68.

Conway, G. R., and Murdie, G. (1971). Population models as a basis for pest control. *In* "Mathematical Models in Ecology" (J. N. R. Jeffers, ed.), pp. 195–213. Blackwell, Oxford.

Eversmeyer, J. G., Burleigh, J. R., and Roelfs, A. P. (1973). Equations for predicting wheat stem rust development. *Phytopathology* **63**, 348–351.

Hanson, A. A. (1971). A directed ecosystem approach to pest control and environmental quality. *J. Environ. Qual.* **1**, 45–54.

Headley, J. C. (1972). Defining economic threshold. *In* "Pest Control Strategies for the Future," pp. 100–108. Natl. Acad. Sci., Washington, D.C.

Heald, F. D. (1921). The relation of spore load to the percent of smut appearing in the crop. *Phytopathology* **11**, 269–287.

Hepting, G. H. (1935). Decay following fire in young Mississippi Delta hardwoods. *U.S., Dep. Agric., Tech. Bull.* **494**, 1–32.

Hyre, R. A. (1954). Progress in forecasting late blight of potato and tomato. *Plant Dis. Rep.* **38**, 245–253.

Hyre, R. A. (1957). Forecasting downy mildew of lima beans. *Plant Dis. Rep.* **41**, 7–9.

James, W. C. (1974). Assessment of plant diseases and losses. *Annu. Rev. Phytopathol.* **12**, 27–48.

Jeffers, J. N. R. (1971). The challenge of modern mathematics to the ecologist. *In* "Mathematical Models in Ecology" (J. N. R. Jeffers, ed.), pp. 1–11. Blackwell, Oxford.

Kranz, J. (1974). Comparison of epidemics. *Annu. Rev. Phytopathol.* **12**, 355–374.

Krause, R. A., and Massie, L. B. (1975). Predictive systems: Modern approaches to disease control. *Annu. Rev. Phytopathol.* **13**, 31–47.

Krause, R. A., Massie, L. B., and Hyre, R. A. (1975). Blitecast: A computerized forecast of potato late blight. *Plant Dis. Rep.* **59**, 95–98.

Large, E. C. (1955). Methods of plant-disease measurement and forecasting in Great Britain, *Annu. Appl. Biol.* **42**, 344–354.

Leach, L. D. (1938). Determining the sclerotial population of *Sclerotium rolfsii* by soil analysis and predicting losses of sugar beets on the basis of these analyses. *J. Agr. Research* **56**, 619–631.

Massie, L. B. (1973). Modeling and simulation of southern corn leaf blight disease caused by race T of *Helminthosporium maydis* Nisik. and Miyake. Ph.D. thesis. The Pennsylvania State Univ., University Park, Pa. 84 pp.

McNew, G. L. (1972). Concepts of pest management. *In* "Pest Control Strategies for the Future," pp. 119–133. Natl. Acad. Sci., Washington, D.C.

Ordish, G., Dufour, D. (1969). Economic base for protection against plant diseases. *Annu. Rev. Phytopathol.* **7**, 31–50.

Patten, B. C. (1972). A primer for ecological modeling and simulation with analog and digital computers. *In* "Mathematical Modeling and Simulation in Ecology" (J. N. R. Jeffers, ed.), pp. 3–75. Blackwell, Oxford.

Shrum, R. D. (1975). Simulation of stripe rust of wheat using EPIDEMIC—a flexible plant disease simulator. *Prog. Rept.* **347**. Agric. Expt. Sta., The Pennsylvania State Univ. 81 pp.

Stevens, N. E. (1934). Stewart's disease in relation to winter temperatures. *Plant Dis. Rep.* **12**, 141–149.

Taubenhause, J. J., and Killough, D. T. (1923). Texas root rot of cotton and methods for its control. *Texas Agric. Expt. Sta. Bull.* **307**.

Tschumakov, A. E. (1973). Scientific principles of prognosis of sick plants. Central Science-Research Institute of Information and Research in Technology and Economics in Agriculture. Ministry of Agric. of USSR. pp. 1–59.

van der Plank, J. E. (1963). "Plant Diseases: Epidemics and Control." Academic Press, New York.

Waggoner, P. E. (1960). Forecasting epidemics. *In* "Plant Pathology: An Advanced Treatise" (J. G. Horsfall and A. E. Dimond, eds.), Vol. 3, pp. 291–312. Academic Press, New York.

Waggoner, P. E. (1968). Weather and the rise and fall of fungi. *In* "Biometeorology" (ed. W. P. Lowry), pp. 45–66. Corvallis: Oregon State Univ. Press.

Waggoner, P. E., and Horsfall, J. G. (1969). EPIDEM: A simulator of plant disease written for a computer. *Conn., Agric. Exp. Sta., New Haven, Bull.* **698**, 1–80.

Waggoner, P. E., Horsfall, J. G., and Lukens, R. J. (1972). EPIMAY: A simulator of southern corn leaf blight. *Conn. Agric. Expt. Sta. Bull.* **729**, 1–84.

Wallin, J. R. (1962). Summary of recent progress in predicting late blight epidemics in the United States and Canada. *Amer. Pot. J.* **39**, 306–312.

Wilhelm, S. (1950). Vertical distribution of *Verticillium albo-atrum* in soils. *Phytopathology.* **40**, 368–376.

Zadoks, J. C. (1971). Systems analysis and the dynamics of epidemics. *Phytopathology* **61**, 600–610.

Chapter 12

Changes in Host Susceptibility with Time

C. POPULER

I. INTRODUCTION

Plants vary in susceptibility to disease with age. Except when they are grown in artificially stable conditions, their susceptibility also responds over time to the influence of changing environmental factors. In the field and in the glasshouse, these factors vary according to a daily cycle and a yearly cycle.

The effects of varying environmental conditions with time of day on susceptibility are easy to dissociate from effects of plant age because plants do not age appreciably during a single day. Such effects on susceptibility to disease are of minor practical interest and are dealt with briefly at the end of this chapter.

Changes in susceptibility with age and time of year are much more important. In annuals, the effects of the yearly environmental cycle on susceptibility are closely associated with effects of age of the plant or its parts, since all these effects are on the same time scale. This association also holds in perennials between the yearly environmental cycle and the age of the herbaceous shoots and new vascular increments; but

239

here susceptibility may also change over the years on successive generations of annual shoots and vascular layers, and thus plant age becomes
a factor clearly dissociated from the yearly environmental cycle. In
evergreen perennials, the age of the leaf can be separated in two time
components: age expressed as a whole number of years, and age in the
context of the yearly progression of seasons. The latter component is
associated with the yearly environmental cycle in its effects on leaf susceptibility to disease, while the former component is not. These effects
of the age of the plant or its parts on susceptibility, with the possible
interference of the yearly environmental cycle, make the bulk of our
subject.

This subject has somewhat been neglected by plant pathologists and
a good part of the evidence to be found in the literature is of an inci-
dental kind. Sometimes it is also based on insufficiently controlled experimental conditions. This fact has already been noted earlier by Yarwood (1959) and Schein (1964) and is still true. Since experimental
avenues and results in this subject are more than ordinarily influenced
by methodological and semantic constraints, let us first examine the
questions of methods and terminology in some detail.

II. CHOICES AND CONSTRAINTS IN METHODOLOGY

A. Measuring the Components of Susceptibility

Susceptibility is the antonym of resistance and expressions like more
resistant and less susceptible are commonly used as equivalent. There
is no harm in this. But we are interested in expressing quantitatively
how much the host supports the multiplication of the pathogen and how
much it is diseased. Expressing this on a resistance scale is embarrassing
because it is difficult to define zero resistance. Zadoks (1972) suggested
that any disease reading be expressed as a fraction of the highest observed reading and to call this fraction "relative disease." He defined
"relative resistance" as the difference between the unit and relative
disease. This amounts to defining zero resistance as the state of the most
severely diseased host. If a still more severely diseased host is found
later on, there is no other alternative than to discard it and forget it or
to change the scale. On a susceptibility scale, zero susceptibility is for a
completely healthy plant, and the upper part of the scale can be left
open to accommodate any supersusceptible host. Whenever there is an
implication of a measurement scale, it is thus preferable to use the term
"susceptibility" rather than the term "resistance."

In studies on changes in host susceptibility with time, it is desirable to measure susceptibility precisely and during time intervals that are as short as possible. This implies working inside a single multiplication cycle of the pathogen. Consequently, the effects of susceptibility components are not integrated as in an overall damage assessment at the end of many multiplication cycles in a crop, and they must usually be evaluated separately. As Johnson and Taylor (1976) pointed out for the spore-yield component of susceptibility, variation in susceptibility may occur with both race-specific and non-race-specific host reactions and is not restricted to the latter.

1. Lesion-Causing Diseases

a. Fungal Pathogens. Higher susceptibility of a host to a fungus disease may be caused by any one or more of four components of susceptibility: (1) a larger number of infections resulting from a given amount of inoculum; (2) larger lesions (for diseases which produce nonexpanding lesions) or lesions enlarging more rapidly; (3) a shorter latent period, i.e., a shorter time from infection to sporulation; (4) a higher yield of spores per unit area of tissue or per lesion.

Ideally, all four of these components should be measured when studying changes in susceptibility of a plant with time. In fact, experimenters often concentrate on one or two components because it is felt that these components are the more important ones or because they are more convenient to work on. The problems specific to the measurement of each of these four components for studies of changes in host susceptibility with time are discussed below.

In *infection* studies with spores, inoculum may either be applied by spraying with a spore suspension, by dusting with dry spores, or by depositing individual calibrated drops of a spore suspension.

When inoculum is applied as a spray or dust, the number of lesions formed per unit area of host surface may increase proportionally to the number of spores deposited. But the proportionality will hold only up to a certain point of deposition density beyond which the additional number of lesions for each increment of deposition density becomes progressively smaller (Schein, 1964; C. Populer, unpublished data, 1973). This dose–response curve must be established in a preliminary trial with host material of uniform susceptibility, such as leaves of the same age and position. (In fact, the individual susceptibility of such leaves is variable and uniformity is only achieved between the average susceptibility of adequate samples.) A spore suspension density may then be chosen which gives a lesion density suitable for counting either with the naked eye or under a magnifying glass. Schein (1964) has proposed

"infection efficiency" as a measure of susceptibility; this is the ratio between the number of lesions formed and the number of spores deposited
on the same area. This definition of infection efficiency, which implies
proportionality, permits experimentation with an inoculum of varying
deposition density, provided that this density is known. Either with this
method or when working with a constant deposition density, care must
be taken to stay well within the basal linear part of the response curve.
This is true even though greater precision in measurements of susceptibility can be obtained by applying large numbers of spores per unit area
of tissue as well as by increasing the number of sampled areas.

The number of lesions formed also varies—usually along a sigmoid
curve—with the duration of exposure to conditions favorable to infection,
such as duration of wetness. The duration is usually chosen long enough
to be in the upper flat part of the curve. In the case where plants are
screened for disease resistance, the experimenter may, however, prefer
to use a shorter duration so that the less susceptible plants are not infected at all (see Fig. 1 in Umaerus, 1969).

With inoculum applied as calibrated drops, McKee (1964) found with
potato blight that the response to varying spore doses in the drops was
approximately linear if the percentage of drop sites that became infected
was transformed to probits and plotted against log dose. This response,
which is similar to the familiar relationship between the percentage of
killed spores and the fungicide dose in fungicide tests, implies that
spore drop sites in a plant sample differ in susceptibility according to a
normal distribution when the number of spores per drop (the drop dose)
is on a logarithmic scale. On this basis, a concise expression of the susceptibility of a plant sample is the drop dose at which 50% of the drop
sites become infected within a given wetness duration (Lapwood and
McKee, 1966). This is equivalent to the LD_{50} in fungicide assays. When
susceptibility to infection is measured with a constant spore dose in the
drops and expressed as percentage infected drop sites (Warren *et al.*,
1971), the 100% mark is reached in plant samples with only moderate
susceptibility. In such cases, it is not possible to distinguish among
samples with susceptibilities higher than moderate. Variation of the
response with duration of wetness in drop inoculation follows a sigmoid
curve similar to that observed with spray inoculation (Lapwood, 1968).

For very accurate infection trials, drop inoculation is much more time-
consuming than spray inoculation because of the need to use a series of
dose dilutions with each plant sample. A criterion for choosing between
the two methods might be to use spray inoculation with spores normally
airborne and to restrict the drop method to spores dispersed in groups
by splashing or trickling water in the field.

When the *size of lesions* is expressed as a diameter, a measure is obtained of the susceptibility of the host tissue to colonization by the pathogen. When the size of lesions is expressed as an area, however, an epidemiological measure of susceptibility is obtained. There is presumably no drawback here to inoculation with drops, provided they are much smaller than the lesions produced naturally in the field. The borderline between successful infection and growth of a lesion is not always clear. When counting late blight lesions on potato, for instance, it may be difficult to decide whether minute specks should be discarded as abortive infections or counted as successful infections which did not enlarge because of the resistance of the host tissue to colonization.

The length of the *latent period* following drop inoculations may decrease with increasing spore load (Lapwood and McKee, 1966). In diseases in which the natural dispersal process is by single spores, methods based on single-spore infection are preferable. This can be achieved by dilute spraying or dusting or with highly diluted drops.

The measurement of *spore yield* is often laborious. Despite its obvious epidemiological importance, it has been much less used than measurements of the former three components of susceptibility to disease. Since sporulation may extend over several days or weeks, repetitive sampling may be required. Useful indications on the measurement of spore yield and its expression are to be found in the review paper by Johnson and Taylor (1976).

With some diseases, visual assessment of the abundance of lesions per unit area of host surface by comparison with standard diagrams can give sufficient precision (Jones and Hayes, 1971). This technique is reduced to its simplest form in diseases where susceptibility abruptly drops to or rises from nil as time proceeds (Populer, 1972a; Lapwood and Adams, 1973). Visual evaluation is also adequate for measurements of the size of lesions or when changes in susceptibility of the host tissues to colonization are expressed by the host as qualitative differences in lesion type. Visual methods for estimating sporulation are acceptable only when objective standards can be set.

The component most often used in studies of changes in host susceptibility with time has been the number of lesions formed. There may be reasons for this other than its convenience. For example, the number of lesions usually varies with time and host much more than lesion size; also spore yield may be considered unimportant when the inoculum in the field is known to come predominantly from other host plants or varieties or exclusively from an alternate host.

b. Viruses. Local lesions are not a particularly important class of symptoms in the field, where systemic infection is the usual condition. The interest in local lesions is tied to their utility for assaying virus preparations in the laboratory. In these assays, virus preparations are applied to leaves that are injured mechanically and counts are made of the ensuing local lesions. Most studies on changes in susceptibility to viruses with time have consequently aimed at determining the conditions of plant age, leaf age, and leaf position which ensure high and reproducible counts of local lesions.

When the term "susceptibility" is used by virologists in connection with the production of local lesions, it refers only to the tendency of the leaf to produce local lesions and it does not imply that a "nonsusceptible" leaf is not infected. When leaves are inoculated mechanically, viruses may indeed produce local lesions without further progress in the host, or multiply within the leaf without developing local lesions, or produce both local lesions and systemic symptoms (Bawden, 1964; de Bockx, 1972).

Studies of changes in susceptibility to viruses also require preliminary dose–response tests to establish a suitable dilution of the virus preparation. For practical details on this and for a comparison between mathematical models of the dose–response relationship with viruses and experimental reality, Bawden (1964, pp. 155–159) and then Brakke (1970) are enlightening.

2. Vascular, Systemic, and Root-Rotting Diseases

Fungal diseases in this group, which comprise disorders such as *Verticillium* or *Fusarium* wilts, systemic smuts and rusts, and extensive root or stem rots, affect the whole plant or a major part of it. In principle, susceptibility to these diseases may be separated into the same components as for lesion-causing fungal diseases. In diseases which are predominantly transmitted by root contacts or by mycelium growing through the soil, susceptibility to initial infection and to subsequent spread of the pathogen inside the host may, however, arouse more interest than the spore yield on the infected plants.

In systemic virus diseases, the logic is distinct from that of fungal diseases and susceptibility can be separated into three components: (1) infection, (2) virus multiplication, and (3) virus translocation or spread, followed by further multiplication. Viruses do not sporulate and the term "latent period" is used by virologists to denote the time between inoculation and detection of newly produced virus. Thus it varies with the sensitivity of the detection method and is not equivalent to the term "latent period" as used for fungal diseases.

Studying the growth of a vascular or systemic fungus inside its host is much more time-consuming than in the lesion-causing group since it entails taking samples of tissues from different places on the plant and testing them for presence of the pathogen (Howell *et al.*, 1976). A similar technique is necessary for virus translocation studies (Beemster, 1972).

Just as with lesion-causing diseases, simple visual evaluation may be adequate for defining and expressing changes in susceptibility as various grades of wilt, growth depression, abnormality, etc. It is especially adequate for handling cases of tolerance where the host is not affected in proportion to the development of the pathogen inside it. These simple and convenient methods must not be looked down on. They have great utility so long as their limitations are recognized.

B. Quantitative Inoculation

In most instances, changes in host susceptibility to infection can be discovered only by quantitative inoculation. Exceptions to this are cases of on–off susceptibility changes. An example of this is the work by Lapwood and Adams (1973) on common scab of potatoes, which they carried out with natural infection under field conditions. With susceptibility components other than infection, quantitative inoculation is only necessary when they are sensitive to the initial inoculum load. Inoculation methods, irrespective of their qualification, have been reviewed by Waterston (1968).

Quantitative inoculation requires an inoculum of fixed dosage, applied on the host under standard conditions as close as possible to field conditions. The applied dose and the infectivity of the inoculum must be reproducible.

Numbers of spores in suspensions are commonly determined with a hemacytometer or an automatic cell counter, and then adjusted to the required concentration. With the hemacytometer, uneven sedimentation may be the source of large errors. This can be avoided by adding a thickener to the suspension sample. Spores dry-deposited onto the host are either weighed before or counted after deposition. Checks of spore deposits and spore viability on agar-coated slides at intervals during lengthy spraying operations may also be necessary (Schein, 1964). Virus preparations are brought to an empirical dilution (Matthews, 1953a; Crowley, 1967) or to a determined titer, in cases where this is feasible (Helms and McIntyre, 1967).

While inoculum concentration is usually determined with some care, the deposition technique is sometimes less satisfactory, especially with spraying. When the inoculum is stated without further details to have been "sprayed uniformly," this can usually be interpreted to mean "nonuniform hand spraying." A quantitative spray inoculator delivering timed target shots has been described by Schein (1964). A spray distribution line where detached leaves travel past a continuously spraying nozzle has been used by Populer (1972b, 1973); systems similar to the latter but with attached leaves on whole plants are worthless because of the varying leaf angles. For dry deposition of rust spores, which are hydrophobic, several generations of settling towers have been evolved (Petersen, 1959; Kulik and Asai, 1961; Melching, 1967); later models contain turntables. Spores are injected into the tower with a blast of compressed air to avoid clumping. For the soft and sticky spores of powdery mildew, Kirby and Frick (1963) have described a settling tower where spores are blown in by a fan from sporulating plant material. Simple hand shaking of mildewed plants over detached leaves is, however, still used in studies of susceptibility changes (Jones and Hayes, 1971). Accuracy with deposition techniques other than spraying or dusting can be achieved with comparatively simple equipment.

Inoculum deposition methods either result in mass inoculation, where large numbers of plant parts are inoculated simultaneously, or in discrete inoculation, where one site is inoculated at a time. Examples of mass inoculation are settling-tower deposition or spray-line distribution on detached plant parts, dipping tubers or root systems of uprooted plants in inoculum or transplanting into soil homogeneously mixed with inoculum. Discrete inoculation can be achieved by such techniques as: target spray shots, deposition of drops or of inoculum in pieces of filter paper, growth medium or diseased tissue, and injection or insertion into plant or soil.

The advantages of mass inoculation are rapidity and uniformity. With a spray distribution line traveling at 15 cm/sec, for instance, detached 5-cm leaves are inoculated three at a time in one-third of a second, while Schein (1964) reports 5 sec/leaf with the spray-shot inoculator.

A drawback of mass inoculation is the necessary use of detached plant parts or uprooted plants, although this is commonly accepted in studies on susceptibility changes. Warren et al. (1971) observed with drop inoculation that hypersensitivity to *Phytophthora infestans* was shown less frequently (or not at all) by detached as opposed to attached potato leaves. But C. Populer (unpublished data, 1973) found no such reduction of the hypersensitive reaction on detached potato leaves with spray inoculation (see Fig. 1). Nicholson et al. (1973) report susceptibility reactions to *Venturia inaequalis* similar on detached as on attached apple leaves. Differences in numbers of virus local lesions related to time of leaf detachment were quantified by Nakagaki and Matsui (1971); similar work with some fungal pathogens would be instructive.

When spores are sprayed or dusted onto leaves which are not yet fully

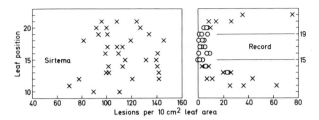

Fig. 1. Leaf susceptibility to infection by *Phytophthora infestans* in potato cultivars Sirtema and Record. Both cultivars were planted in the field on April 18 and spray-inoculated on July 11, 1972. Leaf positions were counted upward. The numbers of normal (X) and hypersensitive (O) lesions are given for each of the leaves of three stems. On the cultivar Record, leaf growth was still active on positions above 19; it was finished during the week ending on inoculation day for positions 19 to 15 and finished for more than a week on lower positions (C. Populer, unpublished data, 1973).

grown, the number of spores deposited per unit area of leaf decreases between the time of inoculation and the time when lesions are counted because of the leaf expansion. Correcting the lesion counts per unit area to allow for this expansion is difficult. It can be avoided by using an inoculator such as Schein's (1964), with which a known amount of spores is deposited on a target area. The problem is also avoided when inoculating detached leaves if these are incubated under conditions where their growth is arrested.

Spore suspensions prepared by washing plant material or cultures may contain inhibitors or nutrients which do not accompany the spores when they are dispersed naturally. Rinsing the spores after collection to eliminate these inhibitors or nutrients is not necessarily appropriate since it can also extract spore material essential for germination (Shepherd and Tosic, 1966). In order to avoid interferences of this kind with spores of the dry-dispersed type the following methods can be used: (1) suction collection of spores, as reviewed by Johnson and Taylor (1976); or (2) direct inoculation from sporulating material in a settling tower, as described by Kirby and Frick (1963).

Reproducibility of inoculum dosage and infectivity is indispensable in determining changes in susceptibility at intervals over time. How this is achieved is often not specified. With fungal spores, careless hemacytometer counts can conceal appreciable variations in the amount of inoculum applied. The proper procedure is to repeat counts in sufficient numbers to reach an adequately small confidence interval around the mean. One way of avoiding changes in concentration or infectivity of the inoculum is to infect plants or plant parts of staggered ages simultaneously.

C. Field versus Growth Room

Changes in host susceptibility with time can be tested with plants produced in the field or in growth rooms. Plants in a glasshouse are raised in intermediate conditions, where watering can be controlled and most unwanted pests or pathogens avoided, but where temperature can hardly be regulated and light still less so.

There is something of a philosophical option in the choice between plants grown in a growth room or in the field. Barring equipment failure, the growth room provides the control of environment necessary for investigating the effects of age and environmental factors one by one with various constant combinations of the other factors. This explanatory dissociation of factors, dear to the reductionist, and their recombination, is no simple task, as plants not only respond instantaneously to environmental changes but also accumulate past environmental experience. Until recently, susceptibility changes with age have mostly been studied on growth room plants of a very early age and in one set of constant environmental conditions; young plants take up little space and their environmental experience can hopefully be neglected. The problem of inoculation uniformity over time is easily solved in growth rooms, since plants of staggered ages can then be produced under constant conditions and be inoculated all at the same time (Schein, 1965).

The alternative choice is to test susceptibility changes on plants growing in the field or in the glasshouse as they do in agricultural practice. This may include more than one growth period in the year. Replication over a few years may be necessary (at least from a purist's viewpoint) to reduce the effects of deviations from mean seasonal climatic trends. Here, the association between age and yearly environmental cycle, as discussed in the introduction, cannot be separated as with growth room plants. Results of this kind need to be verified, though with a simpler procedure, before being extended to other climatic areas. But then, experiments in a growth room also would need to be completed by an examination of the host varieties grown in other climatic areas. To the pragmatist, the field approach has the appeal of immediately exploitable knowledge. The main problem with field-grown plants, but not with glasshouse plants and not a serious one with perennial plant parts, is to avoid their being damaged naturally by pests or diseases, including the one under study. There is also a risk that undetected virus infections in the field interfere with host susceptibility to the disease under study. Spraying is to be looked on with distrust, even with insecticides, which can have a side effect on fungal infection (Populer, 1972b). The solution lies in isolation and luck.

The choice of this pragmatic approach is also a matter of experimental feasibility: beyond the seedling stage, plants are often too large for growth rooms and dwarf varieties may not always be available or acceptable. Experimenters on susceptibility changes have mainly used potted plants from the glasshouse for annuals and field plants for perennials, rarely the opposite.

Potted plants and detached material are best tested in the growth room under standard environmental conditions. Field inoculation with lesion-causing diseases on aerial herbaceous plant parts is objectionable because the pathogen is subjected to changing weather conditions and changes in the microclimate brought on by growth of the crop itself. But it is much less objectionable for bark diseases, systemic or vascular diseases, and root rots, where short-term environmental variations are more or less smoothed out.

III. SUSCEPTIBILITY CHANGES WITH AGE AND TIME OF YEAR

A. Age of Plant and Age of Plant Part

Susceptibility varies with the age of the plant part affected by the pathogen, not only in lesion-causing diseases but also in systemic and vascular diseases, as examples in the next two sections will show. Susceptibility may also vary with plant age when changing susceptibility is displayed by plant parts, e.g., leaves, that are of the same type and physiological age but are produced at different times in the life of the plant.

Another necessary distinction is that plant parts such as leaves vary in age not only with passage of time but also with position on the plant. These two age sequences are not comparable, because leaves in different positions on a stem do not necessarily go through the same changes in susceptibility with age, just as they do not grow to the same size or develop the same form. The reason for this is not only that the plant or stem grows older as it produces its successive leaves, but it is also that lower leaves are gradually affected by different environmental conditions due to their shading by the upper leaf stories. Complete coverage of susceptibility changes is obtained by sampling leaves of different positions either at successive times in the season or at the same time on staggered plantings, or both.

A puzzling detail is how to identify leaves on a shoot. Counting leaves from the shoot top downward gives the same number to leaves of similar size and development stage, while counting upward gives the same number to a given leaf position throughout the shoot growth period. In Fig. 1, the top leaf of the three stems of cultivar Record bore respectively numbers 21, 21, and 24 counted upward, i.e., a mean position of 22. Reducing arbitrarily each of these three top leaves to the number 22 and correcting accordingly the numbers on the other leaf positions has been found a satisfactory compromise.

B. Changes in Susceptibility to Fungi with Age

In a paper on the more general subject of predisposition, Yarwood (1959) briefly reviewed some cases of susceptibility changes with age under the heading of ontogenetic predisposition. He recognized four classes of such changes: both decreasing and increasing susceptibility with age, and susceptibility in middle life either more or less than in youth and later age. He also stated that few generalizations could be made in this regard.

The approach to a more biological frame proposed hereafter is based on distinctions made earlier in this paper which will be apparent in the discussion.

1. Susceptibility to Lesion-Causing Fungi

a. *Diseases on Herbaceous Plant Parts.* The available data suggest that changes in susceptibility to infection are either concentrated during the growth period of the plant part, or are stretched over its entire life span.

Infectibility changes restricted to the growth period of herbaceous plant parts are found both in annual and perennial hosts. On potato tubers, Lapwood and Adams (1973) observed that susceptibility to infection by *Streptomyces scabies* on any internode was strictly limited to when it was in the third and fourth position while growing away from the tuber apex, an interval lasting about 6 days in the field. With *Uromyces phaseoli* on growth room bean plants, Schein (1965) demonstrated that infectibility of the first three leaves was low at unfolding, maximum at 20% area expansion, and low again at the end of leaf growth; only one leaf at a time was highly susceptible on a plant. A similar variation of lesion numbers during the growth of apple leaves was reported by Aldwinckle (1975) with *Gymnosporangium juniperi-virginianae* on glasshouse plants and by Sys and Soenen (1969) with *Venturia inaequalis* on orchard trees. In the latter two references, there is no question about the decrease in susceptibility during the later part of the growth period. But the few lesions reported during the early part of the growth period can be due, at least in part, to counting lesions per leaf concurrently with the small size and furled state of these leaves at inoculation time, instead of making allowance for the expanding leaf area as explained in Section II,B. On leaves of rubber trees, susceptibility to infection by *Oidium heveae* is high from the earliest age but stops completely with a cuticle change occurring while the leaf passes from 30 to 55% of its adult length. Resistance is also achieved at the same time against *Helminthosporium heveae, Gloeosporium albo-atrum, Dothidella ulei,* and even mite damage (Populer, 1972a). Susceptibility during early leaf age clearly decreases with tree age for *Helminthosporium,* and seems to increase for *Oidium* and to remain constant with the other two diseases.

When infectibility changes extend over the whole life of a plant part, a common pattern is that of susceptibility increases with increasing age of the plant part. For *Erysiphe graminis* on glasshouse oat plants, Jones and Hayes (1971) reported that infectibility increased with the age of leaves after their full expansion. On leaves at the same physiological age produced on plants of staggered ages, however, susceptibility to infection decreased with plant age, a typical case of "adult plant resistance." Increased susceptibility to lesion growth with age of the plant part is also known for various fungal diseases. A special case is that of "latent infections" on fruit (Verhoeff, 1974), where the host tissues are penetrable at an early age but allow lesion growth only when ripening.

Another frequent pattern of infectibility changes over the life of plant parts is juvenile susceptibility, followed by some measure of resistance in early adult age, and later susceptibility again. In potato tubers, Boyd (1967) observed that susceptibility to infection by *Fusarium caeruleum* declined steadily during the two months of tuber growth in the field, was nil shortly after natural death of haulm or its premature removal, and increased again during winter storage. Plant age as such obviously cannot be involved any longer after lifting time since the tuber is then surviving as a detached plant part. Warren *et al.* (1971) demonstrated on glasshouse and outdoor potted plants of the potato cultivars King Edward and Arran Victory that all the leaves were highly susceptible to infection by *Phytophthora infestans* on very young plants. During both the pre- and postflowering stages, however, top and bottom leaves were also highly susceptible while the intermediate leaves were partly resistant or hypersensitive. At flowering time itself the greater susceptibility of the top zone was less apparent. Lesion growth on King Edward followed the same trends. Among 24 field-planted potato cultivars—all without *R* genes—12 late or midseason ones were found to behave similarly, while the other midseason or early ones were highly susceptible on all leaves at all times of the season (C. Populer, unpublished data, 1973). This difference between cultivars may explain earlier conflicting reports. On the late or midseason cultivar group, hypersensitivity and low infection numbers were observed on leaves which had just reached adult size while leaves either more mature or still growing had numerous normal lesions (see Fig. 1). An effect of plant age was not apparent, and seasonal variations in susceptibility to late blight earlier attributed to plant age (Grainger, 1968) are in fact explained by the varying distribution of leaf age classes on the plant.

A tentative system for explaining the changing patterns of susceptibility to infection or to lesion growth in relation to plant part age is given in Fig. 2; the special case where no susceptibility changes were observed is not illustrated. Herbaceous stems on perennials would probably come only under Type I since they become lignified a short time after growth ceases and presumably become more resistant. Leaf-gall or leaf-curl diseases such as those caused by some *Taphrina* and *Exobasidium* species should *a priori* also come under Type I since leaf distortions are produced during the growth process. Experimenting on very early growth stages of Type Ib is necessary to know whether it is fundamentally different from Type Ia. Similarly, observations begun too late on Type III may truncate it into Type II. All three of these patterns typically proceed upward along growing stems on successive internodes, leaves, etc. More data would be necessary to determine whether susceptibility to lesion growth usually follows a pattern similar to infectibility in the same disease.

Decreasing susceptibility with plant age is known as adult plant resistance. It is obvious in some of the examples discussed earlier. Increased susceptibility with plant age, as in the case of *Oidium* on rubber, would need to be substantiated. Grainger (1968) has proposed a number of other patterns of susceptibility changes with plant age. His inferences were drawn from the changing ratio between total plant carbohydrate and residual shoot dry weight. Effects of plant age versus plant part age on specific diseases would, however, need to be directly measured and separated on these plants before further considering these patterns.

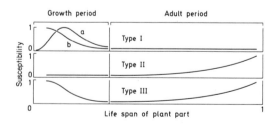

Fig. 2. Patterns of susceptibility to infection or to colonization by the pathogen related to age of the affected plant part. In the horizontal sections of the curves, susceptibility is low but not necessarily nil.

b. Bark and Cambium Diseases on Stems. This group of diseases includes lesion-causing diseases which develop in the cambium and surrounding living layers of xylem and phloem. Sewell and Wilson (1973) found that 1- to 3-year-old field apple trees inoculated at cambium level in a cork-borer hole above the graft union were infectible by *Phytophthora cactorum* only from March until early May, i.e., from bud swelling to onset of shoot extension. On 35-year-old trees, the infectible period was longer, extending from March to June, which points to an increased susceptibility to infection with plant age. With *P. syringae,* young apple trees were only infectible from October to December, i.e., during tree dormancy. Lesions of *P. cactorum* grew throughout the year with a maximum growth occurring during May to July, while growth of *P. syringae* was maximal in March and April but was checked in June–July. There was no effect of tree age on the growth rate of *P. cactorum* lesions.

Similar annual patterns of susceptibility have been reported for other bark diseases. In these diseases, susceptibility to infection or to lesion extension generally coincides with spring and early summer, i.e., during the first months of cambial activity and the main period of shoot growth, or with the late autumn months, when the tree is dormant. There is something in this which is suggestive of susceptibility Types I and II respectively. Before taking this suggestion seriously, however, more would need to be known about the seasonal activity of the cambium, the growth periods of phloem and xylem layers, the nature of the tissues in which the susceptibility to initial infection and colonization is located, and the effect of the ambient temperature on the activity of the fungus.

Bark infections through leaf scars are a special case because they are restricted to stems less than 1 year old on deciduous species and are possible only during a brief time. On mature apple trees, for instance, Dubin and English (1974) found that the percentage of leaf scars infected by *Nectria galligena* fell from 30–90% to 5% and finally to zero when inoculation took place, respectively, 1, 10, and 20–30 days after leaf removal.

2. Susceptibility to Root-Rotting, Vascular, and Systemic Fungi

a. Root Rots and Vascular Diseases. Most of these diseases are soilborne or transmitted by root contacts, but in some of them infection also occurs through wounds in the aerial parts.

Working on growth room seedlings of snapdragon, Mellano *et al.* (1970) found that *Pythium ultimum* mainly penetrated the elongation and maturation regions of the root tips, but rarely the mature root portions and never the meristem or the differentiation region just behind it. Seedlings inoculated when 15 days old or earlier were killed, growth of the pathogen being rapid in the root tissues—except the meristem and differentiation region—and in the stem and leaves. On seedlings inoculated when they were more than 20 days old, the fungus easily colonized tertiary and quaternary roots but its growth was restricted in the cortex of secondary and primary roots, and the plants survived. Glasshouse experiments with *Fusarium oxysporum* on peas by Nyvall and Haglund (1976) also suggest that the decreasing severity of wilt with plant age at the time of infection is due to restricted fungal growth in the plant vessels rather than to decreased infectibility of roots.

In both of these examples, susceptibility to pathogen growth decreases with the age of the affected plant part, i.e., vascular tissue for *Fusarium* and root cortex or other tissues for *Pythium,* according to a pattern similar to Type Ib. The pattern of root infectibility revealed in the *Pythium* example above is strictly Type Ia (Fig. 2), the maximum susceptibility being coincident with the period of maximum growth of the plant part. This suggests that the infectibility patterns found for fungal lesions on aerial plant parts may also be applicable to roots.

A somewhat unexpected kind of relationship with age of plant parts was discovered by Howell *et al.* (1976), who reported that young leaves at the top of cotton plants were less susceptible than older leaves to vascular invasion by *Verticillium dahliae* when syringe-inoculated through their petioles. The development stage of these leaves was not indicated.

On 2-year-old glasshouse elms inoculated with *Ceratocystis ulmi* on stem wounds, Smalley (1963) found that vascular discoloration extending from the wounds and that wilt symptoms developed mainly from the time of bud burst to the end of the terminal growth. This changing susceptibility to growth of a vascular fungus in a woody host is reminiscent of the annual susceptibility patterns described above for bark diseases. It suggests that such patterns might be common to all diseases affecting the vascular layers of woody stems.

   b. Systemic Smuts, Rusts, and Downy Mildews. In contrast with the root rots and vascular diseases discussed above, penetration by systemic smuts, rusts, and downy mildews occurs through aerial parts. In oat seedlings, *Ustilago avenae* infects the coleoptile when it is 2–3 cm long, but not when it is older. On wheat and barley, *Ustilago nuda* penetrates the pericarp wall only at flowering time (Manners, 1971). These are cases of an infectibility pattern of Type I (Fig. 2); similar situations may be suspected with some systemic rusts.

Some downy mildews cause systemic infection of leaves, stems, and roots on young seedlings. When the plant is more than a few weeks old, however, these pathogens are seldom able to produce more than local lesions on the new young leaves, a typical case of decreasing susceptibility with plant age. With *Bremia lactucae* on growth room lettuce, Dickinson and Crute (1974) have shown that both systemic fungal growth and sporulation decreased as inoculation receded from the third to the sixth week of seedling age, with no change in infectibility.

C. Changes in Susceptibility to Viruses with Age

For bean leaves inoculated mechanically with TMV, Schein (1965) obtained an infectibility curve identical to Type Ia (Fig. 2). Similarly, Crowley (1967) found that leaves of *Nicotiana glutinosa* were most susceptible to lettuce necrotic yellows virus when one-half to three-fourths expanded. In other reports on the variation in numbers of local lesions with leaf position, the evidence is made precarious by the lack of data on the growth stages of leaves (Bawden, 1964; de Bockx, 1972).

In experiments which are probably the most extensive yet conducted on changes in host susceptibility with age, Beemster (1972) inoculated attached potato leaves with four potato viruses (PVX, PVY, PVS, and PLRV). He found that all four viruses multiplied progressively more slowly as leaf age at the time of inoculation increased, and that translocation to the tubers was also delayed. Topping old plants and inoculating the newly formed leaves resulted in quick tuber infection. This demonstrated that both the decreasing virus multiplication in aging leaves and the delay in virus translocation were attributes of the age of the leaves and not attributes of the age of the plant to which the leaves were attached. In the field, leaf growth stopped in the first days of July and this marked the transition from high to low tuber infection in inoculation experiments.

D. Relating Susceptibility to Ontogenetic Drift

The examples in the former two sections make it clear that experimentation on changes in host susceptibility with time necessitates an exact knowledge of the age of the plant and the age of each of its parts. Age in days from sowing coupled with an indication of ambient temperature may be sufficient with vascular and root-rotting diseases on young plants, and age in years or calendar time in the season may be sufficient with bark diseases. In other cases, however, precise data are needed on the physiological age of both the plant and the plant parts. Stating that an individual leaf is young or that it is 10 days old is nearly useless information. Physiological age covers not only the period of growth, but also the adult period, senescence, and death. Physiological aging is better known as the "ontogenetic drift," which makes the subject more interesting.

For herbaceous species, plant age is appropriately described by "growth stages," for which scales have been established for a number of crops (Chiarappa, 1971). The date in the season should accompany each growth stage as this furnishes indications on the course and effects of day length and temperature during the plant life cycle; concurrent measurement of plant height may also be useful for knowing how stem growth proceeds and when it stops.

The physiological age of plant parts such as leaves or stems is measured by their size and development stage during growth, and afterward by age in days, with due reference to time in the season. Since size is useful as a guide mark to physiological changes during growth, it is best quantified by expressing it as percent of the final adult size, i.e., measuring relative growth. Growth curves expressed in percentages are easily averaged for plant parts receiving a similar treatment by sliding them either graphically or mathematically along the time abscissa until their 50% points coincide; taking this common origin in the middle of the sigmoid growth curve minimizes time errors because of the steep slope in that area. Giving the mean final size in real value for a treatment will be useful for readers who may want to use this value for modeling calculations. Morphologically distinct development stages of the leaf may in some species be useful in delimiting parts on the relative growth curve (Populer, 1972a).

Relative growth of stems, leaves, or other plant parts can be expressed in percentages of their final lengths, widths, areas, or conceivably also of other dimensions such as final volume or weight. Susceptibility related to percentage of final length is easy to visualize; when related to percentage area, it may be more attractive to the model maker. The transformation of leaf length measurements into leaf areas requires an empirical calibration between the two (Schein, 1965), but squaring relative length gives a fair approximation of relative area. If this is tried on Schein's calibration graph for first formed bean leaves, the calculated relative areas are found to be within 3% of the experimental values, except for the single measurement he gave under 20% of final length, i.e., 4% area. In this case, the discrepancy between estimated and measured values is much larger but of little significance. The latter figures show, by the way, that measurements started at 20% relative length give an impression of better coverage of growth when transformed into 4% relative area.

Graphically, susceptibility can be plotted against relative growth. This method is limited to the growth period and requires that relative growth be separately plotted against time. Plotting susceptibility and relative growth simultaneously against time is adequate over the whole life span of the plant part.

Growth should not be measured on herbaceous plant material which is to be inoculated, but on distinct and comparable material, since even careful handling, when repeated, can modify its susceptibility.

E. Susceptibility Changes in Plant Populations

In an ideally uniform plant population, the growth curve for a single plant could be used to represent the growth of the plant population, and the family of curves attached to successive leaves on a shoot would also adequately describe leaf growth in the shoot population. In real populations, however, plants or plant parts do not grow synchronously; they go through a given stage at different times which roughly follow a normal distribution. This can be represented by plotting against time the percentage of plants or plant parts which are at different stages, or by cumulating against time those which are or have been through these stages (Fig. 3). In the latter representation, the horizontal distance between two successive curves represents the time a plant or plant part stays in one stage before entering the next stage. Thus, a susceptible stage also can be recorded easily on such a diagram. For instance, in Fig. 3, the shaded area on the cumulative graph covers the leaf-growth period, which is the susceptible period in Type I of Fig. 2. Obviously, different degrees of susceptibility could also be separated by a set of supplementary curves parallel to the phenological curves. Susceptible stages can be treated as phenological stages.

The time and duration of a susceptible stage in a plant population varies like any phenological stage with year and climate. This variation may be surveyed and mapped to serve as a basis for adapting control

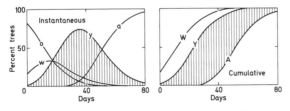

Fig. 3. Phenological curves for a clonal rubber tree population about wintering time. The graphs show the percentages of trees with old leaves (o), of defoliated, wintering trees (w), of trees with new growing leaves (y), and trees with new adult leaves (a). The same data are plotted on the left as instantaneous values and on the right as cumulative values (Populer, 1972a).

tactics to different climatic areas; for example, this has been done for the rubber tree with respect to *Oidium* leaf fall (Populer, 1972a). Epidemics are dependent on the effects of the climate on the host just as much as on the pathogen.

IV. SUSCEPTIBILITY CHANGES WITH TIME OF DAY

Daily changes in living organisms are known to be related to the photoperiod and the thermoperiod. These daily biological rhythms may be exogenous, i.e., directly controlled by the physical environment. They may also be endogenous, i.e., under control of a so-called biological or circadian clock, presumed to have evolved as an adaptation to the physical environment.

In experiments on glasshouse plants, Matthews (1953a,b), working with tobacco necrosis virus on beans, TMV on *Nicotiana glutinosa*, and lucerne mosaic and turnip mosaic viruses on tobacco, found that the number of local lesions produced by mechanical inoculation varied continuously with time of day, being highest for afternoon and least for dawn inoculations. Crowley (1967) reported the same trend with lettuce necrotic yellows virus on *N. glutinosa*. Other data from Matthews (1953b) and from Helms and McIntyre (1967) suggest that these daily variations in lesion numbers are related more to the photoperiod than to the thermoperiod and that they may persist for some time when the plants are switched from alternating day and night to continuous light or darkness. In all these experiments, the authors explicitly refer to the changes in lesion numbers as expressing changes in host susceptibility. The assumption, which seems rather reasonable in the case of viruses, is that the infectivity of the inoculum is not influenced by time of day.

This assumption, however, cannot be made for fungal spores. Fungi grown on artificial media exhibit a daily rhythm in germination, growth, sporulation, and spore release. The state of predisposition of a fungal inoculum and its response to ambient conditions at the time of inoculation may thus be expected to change with time of day, as does the host susceptibility to infection. Separating these factors does not seem easily feasible. A special case where the daily variation in host susceptibility to a fungal pathogen can be shown is when it is related to the rhythm of stomatal opening and closing, since this rhythm can be induced experimentally by chemicals while leaving the physical environment constant (Royle, 1976).

If the conditions of plant growth and inoculum production are similar to field conditions, simply measuring the changes in infection efficiency with time of day, without any ambition to separate the parts of host infectibility and pathogen infectivity, may be sufficient for all practical purposes, as for instance in forecasting work and in modeling. For studies of changes in susceptibility with age and time of year, and for precise work in resistance testing, it is easy to avoid the possible interference of changes in host infectibility and pathogen infectivity with time of day by always inoculating at the same hour.

V. CONCLUSIONS

A survey of one of the most recent volumes of the *Review of Plant Pathology* shows that 5% of all references were devoted to the effect of environmental factors on the pathogen and only 0.5% were devoted to changes in host susceptibility with environmental conditions and time. Epidemiologists have plainly been busier with the pathogen than with the diseased plant. Data like those shown in Fig. 1, however, demonstrate that variations in host susceptibility with time are substantial and must not be neglected.

Epidemiologists have grown used to thinking about time in the multiplication of a pathogen as a financier envisages time in the increase of money at interest. This is the viewpoint that time involves rates of multiplication among pathogens. The viewpoint that time involves physiological changes in the host plants should now also be incorporated in the concept of the epidemic process.

Some general patterns of changes in plant-part susceptibility with time are suggested in this chapter for various groups of diseases caused by fungi or viruses, but these patterns only concern susceptibility to infection or colonization by the pathogen (Fig. 2). Very little is known about changes in the length of the latent period or changes in spore yield with age of the plant part, and these subjects would be worth investigating. More data are also needed on the effect of plant age on susceptibility changes, correctly dissociated from the effect of plant part age. While examples of changes in host susceptibility to diseases caused by bacteria, seed plants, and air pollutants can be found in the literature, they have been omitted because they are too few at present to allow a generalization.

Practical applications of knowledge about susceptibility changes with time are tactical and bear on the timing of control. In breeding work, this knowledge is basic for soundly testing resistance and for identifying some kinds of resistance genes. It also is an essential prerequisite of studies on the host–parasite relationship involved in resistance and susceptibility reactions and it forms part of the material to be incorporated in modeling, but these are applications once removed and their practicality depends on the ultimate use of the studies they support.

References

Aldwinckle, H. S. (1975). Effect of leaf age and inoculum concentration on the symptoms produced by *Gymnosporangium juniperi-virginianae* on apple. *Ann. Appl. Biol.* **80**, 147–153.

Bawden, F. C. (1964). "Plant Viruses and Virus Diseases." Ronald Press, New York.

Beemster, A. B. R. (1972). Virus translocation in potato plants and mature plant resistance. *In* "Viruses of Potatoes and Seed Potato Production" (J. A. de Bockx, ed.), pp. 144–151. PUDOC, Wageningen, The Netherlands.

Boyd, A. E. W. (1967). The effects of length of the growth period and of nutrition upon potato-tuber susceptibility to dry rot (*Fusarium caeruleum*). *Ann. Appl. Biol.* **60**, 231–240.

Brakke, M. K. (1970). Systemic infections for the assay of plant viruses. *Annu. Rev. Phytopathol.* **8**, 61–84.

Chiarappa, L., ed. (1971). "Crop Loss Assessment Methods. FAO Manual on the Evaluation and Prevention of Losses by Pests, Diseases and Weeds." Commonw. Agric. Bur., Farnham Royal.

Crowley, N. C. (1967). Factors affecting the local lesion response of *Nicotiana glutinosa* to lettuce necrotic yellows virus. *Virology* **31**, 107–113.

de Bockx, J. A. (1972). Test plants. *In* "Viruses of Potatoes and Seed Potato Production" (J. A. de Bockx, ed.), pp. 102–110. PUDOC, Wageningen, The Netherlands.

Dickinson, C. H., and Crute, I. R. (1974). The influence of seedling age and development on the infection of lettuce by *Bremia lactucae*. *Ann. Appl. Biol.* **76**, 49–61.

Dubin, H. J., and English, H. (1974). Factors affecting apple leaf scar infection by *Nectria galligena* conidia. *Phytopathology* **64**, 1201–1203.

Grainger, J. (1968). C_p/R_s and the disease potential of plants. *Hortic. Res.* **8**, 1–40.

Helms, K., and McIntyre, G. A. (1967). Light-induced susceptibility of *Phaseolus vulgaris* L. to tobacco mosaic virus infection. II. Daily variation in susceptibility. *Virology* **32**, 482–488.

Howell, C. R., Bell, A. A., and Stipanovic, R. D. (1976). Effect of aging on flavonoid content and resistance of cotton leaves to verticillium wilt. *Physiol. Plant Pathol.* **8**, 181–188.

Johnson, R., and Taylor, A. J. (1976). Spore yield of pathogens in investigations of the race-specificity of host resistance. *Annu. Rev. Phytopathol.* **14**, 97–119.

Jones, I. T., and Hayes, J. D. (1971). The effect of sowing date on adult plant resistance to *Erysiphe graminis* f.sp. *avenae* in oats. *Ann. Appl. Biol.* **68**, 31–39.

Kirby, A. H. M., and Frick, E. L. (1963). Greenhouse evaluation of chemicals for control of powdery mildews. I. A method suitable for apple and barley. *Ann. Appl. Biol.* **51**, 51–60.

Kulik, M. M., and Asai, G. N. (1961). Use of a portable inoculation tower in laboratory, greenhouse and field tests of fungicides to control rice blast. *Plant Dis. Rep.* **45**, 907–910.

Lapwood, D. H. (1968). Observations on the infection of potato leaves by *Phytophthora infestans*. *Trans. Br. Mycol. Soc.* **51**, 233–240.

Lapwood, D. H., and Adams, M. J. (1973). The effect of a few days of rain on the distribution of common scab (*Streptomyces scabies*) on young potato tubers. *Ann. Appl. Biol.* **73**, 277–283.

Lapwood, D. H., and McKee, R. K. (1966). Dose-response relationships for infection on potato leaves by zoospores of *Phytophthora infestans. Trans. Br. Mycol. Soc.* **49**, 679–686.

McKee, R. K. (1964). Observations on infection by *Phytophthora infestans. Trans. Br. Mycol. Soc.* **47**, 365–374.

Manners, J. G. (1971). Cereals: Rusts and Smuts. *In* "Diseases of Crop Plants" (J. H. Western, ed.), pp. 226–253. Macmillan, London.

Matthews, R. E. F. (1953a). Factors affecting the production of local lesions by plant viruses. I. The effect of time of day of inoculation. *Ann. Appl. Biol.* **40**, 377–383.

Matthews, R. E. F. (1953b). Factors affecting the production of local lesions by plant viruses. II. Some effects of light, darkness and temperature. *Ann. Appl. Biol.* **40**, 556–565.

Melching, J. S. (1967). Improved deposition of airborne uredospores of *Puccinia graminis tritici* and *P. striiformis* on glass slides and on wheat leaves by use of a turntable. *Phytopathology* **57**, 647.

Mellano, H. M., Munnecke, D. E., and Endo, R. M. (1970). Relationship of seedling age to development of *Pythium ultimum* on roots of *Antirrhinum majus. Phytopathology* **60**, 935–942.

Nakagaki, Y., and Matsui, C. (1971). Effect of bean leaf detachment on susceptibility to tobacco mosaic infection. *Phytopathology* **61**, 354–356.

Nicholson, R. L., Van Scoyoc, S., Kuć, J., and Williams, E. B. (1973). Response of detached apple leaves to *Venturia inaequalis. Phytopathology* **63**, 649–650.

Nyvall, R. F., and Haglund, W. A. (1976). The effect of plant age on severity of pea wilt caused by *Fusarium oxysporum* f.sp. *pisi* Race 5. *Phytopathology* **66**, 1093–1096.

Petersen, L. J. (1959). Relations between inoculum density and infection of wheat by uredospores of *Puccinia graminis* var. *tritici. Phytopathology* **49**, 607–614.

Populer, C. (1972a). Epidemics of powdery mildew of rubber and phenology of its host in the world. *Publ Inst. Natl Etude Agron. Congo, Ser. Sci.*, **115**.

Populer, C. (1972b). Infection of potato leaves by *Phytophthora infestans* inhibited by malathion sprays. *Meded. Rijksfac. Landbouwwet., Gent* **37**, 507–510.

Populer, C. (1973). Quantitative inoculation of potato leaves with late blight. *Ann. Phytopathol.* **5**, 109.

Royle, D. J. (1976). Structural features of resistance to plant diseases. *In* "Biochemical Aspects of Plant-Parasite Relationships" (J. Friend and D. R. Threllfall, eds.), pp. 161–193. Academic Press, New York.

Schein, R. D. (1964). Design, performance and use of a quantitative inoculator. *Phytopathology* **54**, 509–513.

Schein, R. D. (1965). Age-correlated changes in susceptibility of bean leaves to *Uromyces phaseoli* and Tobacco Mosaic Virus. *Phytopathology* **55**, 454–457.

Sewell, G. W. F., and Wilson, J. F. (1973). Phytophthora collar rot of apple: Seasonal effects on infection and disease development. *Ann. Appl. Biol.* **74**, 149–158.

Shepherd, C. J., and Tosic, L. (1966). The roles of riboflavin and inhibitors in conidial germination in *Peronospora tabacina* Adam. *Aust. J. Biol. Sci.* **19**, 335–337.

Smalley, E. B. (1963). Seasonal fluctuations of young elm seedlings to Dutch elm disease. *Phytopathology* **53**, 846–853.

Sys, S., and Soenen, A. (1969). Onderzoek naar de invloed van faktoren die van belang zijn bij het bepalen van het ritme der behandeling tegen *Venturia inaequalis* (Cke) Aderh. *Agricultura* **17**, 3–19.

Umaerus, V. (1969). Studies on field resistance to *Phytophthora infestans*. II. A method of screening young potato seedlings for field resistance to *P. infestans*. *Z. Pflanzenzuecht.* **61**, 167–194.

Verhoeff, K. (1974). Latent infections by fungi. *Annu. Rev. Phytopathol.* **12**, 99–110.

Warren, R. C., King, J. E., and Colhoun, J. (1971). Reaction of potato leaves to infection by *Phytophthora infestans* in relation to position on the plant. *Trans. Br. Mycol. Soc.* **57**, 501–514.

Waterston, J. M. (1968). Inoculation. *Rev. Appl. Mycol.* **47**, 217–222.

Yarwood, C. E. (1959). Predisposition. *In* "Plant Pathology: An Advanced Treatise" (J. G. Horsfall and A. E. Dimond, eds.), Vol. 1, pp. 521–562. Academic Press, New York.

Zadoks, J. C. (1972). Reflections on disease resistance in annual crops. *In* "Biology of Rust Resistance in Forest Trees" (R. T. Bingham, R. J. Hoff and G. I. McDonald, eds.), pp. 43–63. *U.S. Dep. Agric., For. Serv., Misc. Publ.* **1221**.

Chapter 13

The Genetic Basis of Epidemics

P. R. DAY

I. PERSPECTIVE

The cultivation of crop plants by man involves several significant departures from a natural ecosystem. One of the most important of these is the replacement of a diversified natural vegetation, having many component species, with uniform stands made up of a single species. In such stands each species is generally represented by a single variety or, for some crops, by a single clone composed of genetically identical individuals. In some parts of the world the scale of this replacement is enormous and a contiguous stand of a single variety may cover areas measured in millions of hectares or thousands of square miles. In natural ecosystems the components interact to produce what appears to be a relatively stable or balanced condition. In fact the organisms that compose them undergo constant evolutionary change.

263

A current theory concerning the evolution of ecosystems, known as the Red Queen hypothesis, suggests that "each evolutionary advance made by one species in an ecosystem is experienced as a deterioration of the environment by other species, and consequently if a species is to survive it must evolve continuously and rapidly" (see Maynard Smith, 1976).

It is useful to consider a farm crop from this point of view. Although man destroys natural ecosystems by the operations of clearing land for farming and ploughing, the crops he plants are subject to strong competition from weeds and parasites. These may be components of the displaced system or colonizing species from other systems. Man has had to enter the fray himself by wielding hoe and applying all sorts of pesticides and also by genetically manipulating crop plants through breeding to maximize yield and introduce resistance to pests and diseases. If he gives up the struggle the land rapidly reverts as the crop plants for the most part disappear and are replaced by a succession of plants from the natural vegetation eventually leading back to climax vegetation.

The Red Queen hypothesis contains a zero-sum assumption: "an increase in fitness in any one species is exactly balanced by the sum of the losses of fitness of all the others" (Maynard Smith, 1976). In fact some of the early breeders for disease resistance confidently predicted that extinction of certain disease organisms would follow the widespread use of resistant cultivars. However, the increase in crop fitness resulting from introduction of a gene for resistance was soon negated by the evolution of virulent forms of pathogens that were no longer held in check by genes for resistance. By the same token, large areas of unprotected and relatively uniform crop plants invited destruction by epidemics of pests and diseases able to spread rapidly and without hindrance.

II. GENETIC UNIFORMITY

Examination of past epidemics reveals that host plant uniformity is an important cause (see Horsfall, this volume, Chapter 2). The operational character is of course uniformity of susceptibility. Three examples illustrate how the hazard of genetic uniformity comes about.

A. Hybrid Maize

The primary cause of the 1969–1970 epidemic of southern corn leaf blight in North America brought about by race T of *Helminthosporium maydis* was the widespread introduction of Texas male sterile cytoplasm (cms–T) among hybrid maize cultivars grown at that time. Unfortunately a pleiotropic effect of this cytoplasmic determinant, now known to be carried by mitochondrial DNA (Levings and Pring, 1976), was sensitivity to a pathotoxin excreted by *H. maydis* race *T*. It has been estimated (Horsfall, 1972) that by 1968 from 75 to 90% of commercial maize hybrids were produced by the use of this single source of male sterility. Labor costs and losses of seed caused by damage from hand detasseling the plants in fields used to produce hybrid seed were avoided. Although United States breeders knew that maize hybrids containing cms–*T* had been reported as unusually susceptible to *H. maydis* in the Philippines (Mercado and Lantican, 1961), tests with local isolates had revealed no threat in North America. We now assume that, at the time the tests were made, race *T* had not appeared, was too infrequent to be recovered in sampling, or was not present in a form capable of causing epidemics. Race *O* does not discriminate between normal and cms–*T* plants. It is important to note that the range of maize hybrids available, bearing different names from different seed companies, obscured their mitochondrial uniformity and fatal flaw. At the time of this writing at least two other sterile cytoplasms (cms–*C* and cms–*S*) are in use for preparing hybrids. When these are unsuitable, or ineffective, detasseling is carried out by machine or by hand. The seed industry still hopes to find an effective and economical chemical male gametocide to provide yet another way of controlling pollination in fields used for the production of hydrid seed corn.

B. Crown Rust-Resistant Oats

An earlier plant disease epidemic in the United States provides another example of the major role of genetic uniformity. The Victoria oat cultivar was introduced in Iowa in 1942 primarily to control crown rust (*Puccinia coronata*). It carried a gene *Pc–2* that conferred resistance to most races of crown rust known at the time. This gene was discovered in a sample of *Avena byzantina* originally collected in Uruguay in 1927. By 1945 varieties carrying *Pc–2* made up 97% of the land area planted to oats. In 1946 a seedling blight caused by *Helminthosporium victoriae* appeared. In 2 years the Victoria blight had reached epidemic proportions in Iowa. *Helminthosporium victoriae* was subsequently found to be a widely distributed minor parasite of native grasses. The epidemic occurred because of a pleiotropic effect of the dominant gene *Pc–2* for crown rust resistance. Plants carrying this gene are extremely sensitive to the pathotoxin (victorin) produced by *H. victoriae*. Dilution end-point tests showed that oats carrying *Pc–2* are some 400,000 times more sensitive to Victorin than the recessive homozygote (Wheeler, 1975). The cure for Victoria blight was simply to abandon varieties with the Victoria gene for crown rust resistance and turn to other sources of resistance that were available.

C. Chestnut Blight

Genetic uniformity also may occur in nature. *Castanea dentata*, the American chestnut, was devastated by the blight fungus *Endothia parasitica* that first appeared in North America around the turn of this century. There were three theories about the origin of the disease. Clinton (1908) thought it endemic and that several very cold dry winters had weakened the trees. Murrill (1906) also believed the disease to be endemic but that it had suddenly acquired unusual virulence. Metcalf (1908) suggested that the disease had been introduced from the Far East. According to this last theory, which now seems the most likely, the pathogen encountered a host population that had never been exposed to blight infection. In the absence of the pathogen there was no selection in the natural population for resistance and so none accumulated. To this date there has been no demonstration of useful levels of resistance to blight among surviving trees and sprouts (Jaynes *et al.*, 1976). Chestnut species with moderate to high levels of resistance are found in Japan and China where the pathogen also occurs. These have been used as sources of resistance in breeding programs to produce resistant forms of American chestnuts. Other examples of substantial genetic uniformity resulting from lack of natural selection for resistance in natural populations include the susceptibility of the European chestnut (*C. sativa*) to chestnut blight and of European and North American elms to Dutch elm disease (*Ceratocystis ulmi*).

Natural plant populations normally exist in a balanced equilibrium with their parasites. Harlan (1976) recently drew attention to a striking example from the North American wheat belt. This is a region that covers some 18 or 19 million hectares and produces about 60 million metric tons of grain each year. Three native grasses (*Bouteloua gracilis, Andropogon scoparius*, and *Agropyron smithii*) have a comparable range, although they vary in abundance from place to place within this region. All three grasses are, like wheat, subject to rust diseases and most individual plants will carry a few pustules at some time during the year. However, rusts are not a serious problem on these wild populations of grasses largely because each species has a great deal of genetic diversity. While the genetic diversity within wheat and its relatives is no less complex, the range deployed among the cultivars grown in the wheat belt of North America is extremely restricted. Consequently, rust epidemics are an ever-present threat.

D. The Extent of Genetic Uniformity

The fate of maize with T cytoplasmic male sterility and oats with the Victoria gene for crown rust resistance shows how plant breeders may unwittingly lay the foundations for plant disease epidemics. It is very likely that some current plant breeding activities and anticipated developments in crop improvement will have similar consequences in the future. The dangers of genetic uniformity in modern agriculture were recognized long before the southern corn leaf blight epidemic (see references in discussions by Day, 1973; Marshall, 1977). However the economic repercussions that followed it were severe enough to raise concerns about the vulnerability of other major crops. Recently a committee of the National Academy of Sciences examined the extent of genetic uniformity among major crops in the United States (Horsfall, 1972). Thirteen major crops were examined. They occupied from 53,000 to 26.8 million ha. From a maximum of seven to as few as a single cultivar were planted on the bulk of these huge land areas. For example five F_1 hybrid cultivars of sugar beet accounted for 73% of the 770,000 ha grown in 1970. The balance was made up of 11 others that each occupied less than 5% of the land area. The situation was even more hazardous than these figures suggest since all hybrid seed of sugar beet contained a single source of cytoplasmic male sterility and one gene for monogerm seed. Monogerm seed ensures that only one viable embryo is present in each seed cluster and so avoids the need for thinning after sowing and germination in the field.

In 1976 the situation was essentially unchanged in the United States (Duvick, 1977). Reports from the USSR (Krivchenko, 1977), India (Safeeulla, 1977), and Britain (Moore, 1977) showed that extensive uniformity also exists in the major crops of these countries. Genetic uniformity is a consequence of improved technology and economic incentives. Modern farmers grow the highest yielding and most profitable cultivars they can and as a consequence in any one region they all tend to concentrate on the same few cultivars.

The improvements in technology are beguiling but have a built-in risk. Each year the International Spring Wheat Yield Nursery tests 50 entries in more than 60 different spring wheat regions of the world. The assessments resulting from these tests are provided rapidly to all cooperating breeders. In the early 1970's, the daylength-insensitive dwarf Mexican varieties proved to be so outstanding in yield and so broadly adaptable in many different locations that few breeders could afford to exclude such outstanding material from their own programs. Such an efficient testing and information-sharing service has the potential to bring about global uniformity for spring wheats unless breeders deliberately introduce diversity into their programs to satisfy local needs. Jennings (1974) has stressed the importance of adapting and modifying both traditional varieties and the products of the Green Revolution for marginal lands, and of the developing countries producing many forms, each designed to fill a particular need.

The evolution of agriculture during the last several thousand years has resulted in increasing concentration on a few food crops. Man has used at least 3000 plant species as food (Mangelsdorf, 1966). Of these at least 150 are cultivated for marketing but only 15 now feed most of the world's people. Clearly within these 15 intensively cultivated crops the same processes of concentration and narrowing have occurred. Each step in this progression has moved agriculture further away from the natural dynamic equilibrium between plants and their parasites and in the direction of monoculture.

III. THE GENETICS OF SUSCEPTIBILITY

Epidemics of plant diseases can only develop on crops that are susceptible. Thus the genetic basis of epidemics is really the genetics of disease susceptibility. Since plant breeders do not deliberately breed crop plants that are disease susceptible and in fact wherever possible strive for resistance, we must ask why our major crops are sometimes susceptible and thus subject to epidemics. One reason has already been mentioned. When a host is geographically isolated from its pathogens there can be no selection pressure for resistance. Chance introduction of a pathogen into a host population that is uniformly susceptible to disease dramatically increases the probability of an epidemic. This happened with chestnut blight and with potato late blight in Ireland in the 1840's. Plant quarantine services are designed to intercept and prevent the introduction of exotic pests and diseases primarily for this reason.

Soon after discovery of genetically controlled resistance to plant diseases early in this century, plant breeding began to incorporate genes for disease resistance in a wide range of crops. In the beginning, the resistance of choice appeared to be single, dominant genes with clearcut and easily observable effects. In such cases, the difference between resistance and susceptibility was determined by the allele present at a single host locus. Other sources of resistance were more difficult to work with either because they were less dramatic, allowing appreciably more disease development, or the inheritance of resistance was complex and dependent on a number of genes with individually small, usually undetectable, effects on the phenotype.

The increasingly widespread use of major gene resistance soon revealed the occurrence of pathogen variants, called physiologic races, that were virulent or able to produce as much disease on "resistant" plants as on susceptible plants. These variants temporarily restored the balance between virulence and resistance. As we shall see later they sometimes produced more disease on the resistant variety. The difference between virulence and avirulence on a given host plant was also found to be determined by alleles at a single locus in the pathogen. In pathogens, such as rusts and smuts, where the infecting mycelium is dikaryotic, avirulence is dominant. Host and parasite each exhibit one of two phenotypes and the phenotype of one defines that of the other. When the host is resistant, the parasite is by definition avirulent; when the parasite is virulent the host is susceptible.

It is not surprising that the early plant breeders and plant pathologists were bewildered by the variety of physiologic races they began to find. Each time a new gene for resistance was introduced to combat a plant disease, two new pathogen races could be distinguished—one avirulent and the other virulent. Each new gene for resistance doubled the number of races of the pathogen since n genes for resistance define 2^n physiologic races. The very activity of breeding for resistance seemed to encourage ultimate susceptibility. A treadmill of activity resulted; a succession of new resistant cultivars from the breeders that selected new physiologic races of the pathogens.

A. Genetics of Resistance and Virulence

Our understanding of the genetic basis of host–parasite interaction owes much to the work of Harold Flor. His studies of the inheritance of virulence in flax rust (*Melampsora lini*) and rust resistance in flax (*Linum usitatissimum*) led to the formulation of the gene-for-gene hypothesis (Flor, 1946, 1947). The hypothesis implies that the products of a resistance gene interact with the products of a specific gene in the pathogen to determine the phenotype of host resistance and pathogen avirulence. "Low infection type" is a convenient shorthand for this tautology that was suggested by Loegering (1966). Each gene for resistance is thus matched with a corresponding pathogen gene. Plants with a given resistance gene are resistant to pathogen strains with the corresponding avirulence allele but are susceptible to strains with the corresponding virulence allele. The latter interaction can be called a "high infection type."

Among the better known systems it is common to find multiple allelism at the loci in the host controlling its resistance (see Day, 1974). The alleles, although dominant, are expressed independently. Multiple allelism also has been reported for avirulence genes of the flax rust pathogen

(*Melampsora lini*) (Lawrence, 1977). A further complication arose from the detailed analysis carried out by Lawrence. One single or several closely linked dominant inhibitor genes in *Melampsora* interfere with certain interactions expected to be avirulent. Thus four different interactions involving the flax resistance genes M^1, L^1, L^7, and L^{10} and their corresponding rust avirulence genes expected to give low infection types are affected by the inhibitor(s) and give high infection types instead. Evidently the range of interactions between hosts and parasites includes other levels of complexity than can be accounted for by the gene-for-gene hypothesis.

A still more subtle complication was recently reported by Martin and Ellingboe (1976) in studies of *Erysiphe graminis tritici* on wheat. The gene *Pm4* for mildew resistance in wheat determines a low infection type in combination with mildew strains that carry the gene *P4*. The mildew allele, *p4*, determines a high infection type. Superficially the extent of pathogen development in the association of *p4/Pm4* is equivalent to that in *P4/pm4* and *p4/pm4* where *pm4* represents an allele for mildew susceptibility in wheat. All three interactions have high infection types with abundant mildew development. However, while the early stages of development in the latter two interactions are similar, *p4/Pm4* shows a significant lag over the period 22–26 hr after infection. This lag disappears after 28 hr. These observations suggest that the mutational change from avirulence (*Px*) to virulence (*px*) in mildew does not necessarily make *px/Pmx* exactly equivalent to *Px/pmx* or *px/pmx*. Martin and Ellingboe (1976) went on to compare three *p4/Pm4* reactions which differed in that three different and presumably unrelated strains of mildew were employed. All three gave high infection types of *Pm4* but their infection efficiencies on *Pm4* were 80, 35, and 19%, respectively, and yet all three had infection efficiencies of 80% on *pm4*. Infection efficiency was expressed as the percentage of spores that formed secondary hyphae longer than 10 μm 28 hours after inoculation. Whether the differences observed on *Pm4* were due to different alleles of *p4* or to modifiers at other loci is not known.

B. Complex Resistance

The interactions defined by major genes for resistance and virulence can be obscured or complicated by other genes whose effects are revealed only by detailed analysis. Much resistance to disease in plants in fact is determined by genes having little effect individually. This polygenic resistance can be handled by plant breeders in the same way as yield, maturity date, product quality, and other characters that are similarly inherited. It is not surprising that when dramatic and simply inherited (oligogenic) resistance was available it was nearly always used and polygenic resistance was neglected. Breeding was easier, and while it remained effective this kind of resistance afforded more protection. A survey of more than 900 papers describing disease resistance published from 1912–1970 found that only 60 (6.5%) described polygenic resistance (Person and Sidhu, 1971). Some 80% of these papers were published prior to 1950 before the gene-for-gene hypothesis had focused attention on host specificity and the need for precise control of races used in screening.

Polygenic resistance has not been eroded by virulent forms of pests or diseases. Van der Plank (1963) introduced the term "horizontal resistance" to draw attention to this character of stability and absence of race-specific interactions. In contrast, vertical resistance is oligogenic, unstable, and differential; it is effective against some pathogen races but not others. Selection for vertical resistance resulted in loss of horizontal resistance, as shown by the extreme susceptibility of vertically resistant crop cultivars to virulent pathogen strains. The new variety was often more badly damaged than the original susceptible variety that was the point of departure for resistance breeding. Simultaneous selection for horizontal and vertical resistance requires the use of a virulent pathogen race for the former and an avirulent race for the latter.

The treatment of breeders' nurseries with protective applications of insecticides and fungicides will also result in a loss of horizontal resistance. For example, cotton breeders have maintained insect-free plant breeding nurseries and, under the insecticide shroud, produced high-yielding cultivars that were highly vulnerable to insect pest attack and required many insecticide treatments to produce high yields (Bottrell and Adkisson, 1977). Similar criticisms can be leveled at all breeders who "dress up their plots" by controlling insects, diseases, and water stresses and thereby lose the opportunity to detect useful segregants (Jennings, 1974).

C. Vertical and Horizontal Distinctions

Most discussions of the genetic basis of host–parasite interactions tend to regard vertical and horizontal systems as distinct entities. While this is convenient and simplifies the discussion of examples it neglects the fact that in nature both systems coexist and merge in a spectrum of genetic effects that ranges from one extreme to the other. For example, a wild relative of a cultivated plant may show a substantial amount of resistance to a pathogen in a screening test. However, a breeder who evaluates resistance among segregants of this selection crossed to a susceptible domesticated form will usually only look for oligogenic effects.

Given the time and resources he could no doubt expect to find a series of oligogenes of smaller and smaller effect. In the absence of oligogenes with major effects the breeder may make do with those of lesser effect. He may also deliberately avoid the use of major gene resistance. Maize breeders in the United States have made little use of oligogenic resistance. Much of their success in accumulating polygenic resistance to *Puccinia sorghi* and *Ustilago maydis* is due to the fact that selection for resistance takes place constantly during evaluation of inbreds and hybrids.

In the meantime, apparently small but sometimes epidemiologically significant differential effects are first observed as a haziness obscuring the sharp edges of major gene interactions. A good example is Johnson and Taylor's (1976) finding that slight differences in the high infection type of *Puccinia striiformis* on wheat can not only be race specific but also are associated with large differences in the amount of sporulation. The example came to light as a result of widely scattered and sometimes severe outbreaks of yellow rust reported for the first time on the wheat Joss Cambier in England in 1971. The differences were revealed by weighing the spores formed by the two types of lesions. Although this technique has great epidemiological value, so far it has been beyond the resources of most plant breeders.

It is not surprising that detailed analyses will reveal significant differences that a breeder would either fail to detect or assume were part of horizontal polygenic resistance. The finding that such effects are differential, or race and host specific, has led to the suggestion that "nonspecific or horizontal resistance is that resistance which hasn't yet been shown to be specific" (Ellingboe, 1975). The implication that all resistance can be overcome as a result of genetic change in the pathogen is probably too sweeping. The extremely limited host range of most pathogens shows that the vast majority of potential host plants they encounter are so uncongenial as to be beyond their capacity to cause high infection types. I would not expect to find a clear distinction between such "non-host resistance" which is evidently very successful and very stable, and many elements of horizontal resistance that a pathogen encounters on an otherwise congenial host (see Day, 1974).

D. Virulence Analysis

The effective life of a resistance gene in a crop depends on the frequency of pathogen strains that carry an appropriate virulence allele. Surveys of physiologic races record the occurrence of particular pathogen genotypes defined in terms of the reactions they produce when inoculated to a set of differential host cultivars. The fact that many pathogens could be maintained in culture as clonal populations which reproduced by conidia or by uredospores led many early investigators to think of physiologic races in nature also as fixed clonal entities. The fallacy of this assumption is now being recognized in race surveys which are moving in the direction of describing the frequencies of virulence genes in particular pathogen populations.

Wolfe and Schwarzbach (1975) proposed this approach for cereal mildews and have pointed out the improvement in efficiency. For example, if surveys show that the frequencies of four different virulence genes in a pathogen population are 0.75, 0.35, 0.50, and 0.09, respectively, then the frequency of any specific race (i.e., specific combination of genes) can be calculated as a simple product. The frequency of races carrying the first two virulence genes is $0.75 \times 0.35 = 0.26$, but the frequency of the race carrying only those two genes is $0.75 \times 0.35 \times (1 - 0.50) \times (1 - 0.09) = 0.12$. In fact, the first frequency, 0.26, is more important to a breeder or farmer growing a host cultivar with the two corresponding resistance genes than is the second frequency. The second, 0.12, is included in the first. The frequency 0.26 also avoids the complication that it is the sum of the frequencies of four different races all of which carry the first two genes.

Where these predicted frequencies have been tested good agreement has been found between the predicted and the observed frequencies. However, examples where the observed frequency was lower than the predicted frequency are of interest. These suggest that there is a selective disadvantage associated with a specific combination which could be exploited by breeding resistant cultivars with the two corresponding resistance genes (Wolfe et al., 1976).

Direct analysis of the virulence of pathogen populations can also be accomplished by alternative sampling techniques. One method involves the use of mobile nurseries consisting of trays of barley seedlings with known genes for resistance to barley mildew. The seedlings are exposed for 24 hr in the field and then returned to a greenhouse for incubation and counts of barley mildew colonies (Wolfe and Minchin, 1976). Comparisons of the frequencies of different virulences determined by this method and by direct isolation from colonies on the field crop showed they gave similar results. The chief disadvantage of the mobile nursery is its limitation to field crops within easy reach of a base laboratory or greenhouse for preparing the seedlings and incubating them after exposure. This limitation can be overcome if spores can be trapped, transported to a test center, and then inoculated to seedlings.

E. The Origin of Virulent Alleles

The frequency of any given allele for virulence in a pathogen population is a function of the mutation rate that gives rise to the allele and the effect of the allele on the survival of the pathogen carrying it. If the allele reduces fitness, its frequency in the population will be close to the mutation rate. If the allele increases fitness, however, its frequency will approach 1.0. Several recent analyses of mutation rates in plant pathogens make it clear that where the population is sufficiently large it is reasonable to expect that most virulence genes already occur and are maintained principally by mutation.

In Sweden, Leijerstam (1972) examined the spontaneous production of mutants of *Erysiphe graminis* f. sp. *tritici* arising within small populations raised on wheat seedlings in a greenhouse. Each population was maintained on seedlings of one of 10 different mildew-resistant cultivars of wheat together with seedlings of the mildew-susceptible cultivar Thatcher. The wheat plants were grown in pots covered with plastic cylinders that were put in place prior to emergence of the seedlings. Each pot of seedlings was heavily inoculated with spores from a race of mildew which was avirulent on all of the resistant varieties under test. At the same time the seedlings were thinned to one plant of each variety so that each pot contained one resistant and one Thatcher seedling. Each week the resistant plants were examined for mildew colonies and the cylinders were shaken. This treatment produced uniform infections covering 50% of the leaf area of Thatcher. When the Thatcher plants began to die from infection a new set of seedlings was inoculated. Noninoculated seedlings were used to monitor contamination. The mildew populations were maintained for periods of either 241 or 438 days. Four mutants were recovered. Three mutants were virulent on hosts with single genes; one on *Pm1*, another on *Pm2*, and the third on *Pm3a*. These appeared in populations kept for 438 days. The fourth mutant, recovered on *Pm4*, was also virulent on hosts with the genes *Pm1* and *Mle* and appeared in a population kept for 241 days.

Leijerstam converted these rates to field mutation frequencies by assuming the spring wheat crop had a density of approximately 5×10^6 plants/ha and that approximately 5% of the plant surface area carries mildew in a typical crop. The number of mutations per locus per hectare per day was thus 1141 for the three single loci and 2074 for the triple mutant. These field rates are somewhat lower than estimates based on assumed mutation rates of 10^{-8} to 10^{-7} per locus and spore production rates of 10^{13} per hectare per day (Day, 1974; Person *et al.*, 1976).

Virulent mutations in a pathogen population are recombined by all of the mechanisms that produce variation. Combinations of two or more different virulent mutations can arise by sequential mutation in one strain, and by sexual recombination, and parasexual processes involving different strains. For prokaryotes (bacteria and Actinomycetes) plasmid-borne virulence factors are transferred among different organisms. No doubt other systems including sexduction, transduction, and transformation will all be found capable of effecting reassortment of virulence genes in prokaryotes.

Deliberate attempts to restrict sexual recombination and hence control pathogen variation have been few in number and not markedly successful. For example, eradication of *Berberis* spp., the alternate host of *Puccinia graminis*, largely eliminated infection foci and rust overwintering in North America, and probably reduced variation, but had no effect on mutation, mitotic recombination, or movement of rust from other parts of the continent.

F. Stabilizing Selection

Virulent mutants that are a threat to introduced resistant varieties will arise in a pathogen population maintaining itself on susceptible hosts (Person *et al.*, 1976). Hence their fitness on susceptible hosts is important in determining the frequency of such mutants. There has been much speculation about the effect of unnecessary genes for virulence in a pathogen population. Some investigators believe that possession of these unneeded genes will reduce the fitness of pathogen races. The end result of such reduced fitness would be to eliminate either races with many unnecessary virulence genes or races with unnecessary virulence genes that impose an unusually large loss of fitness. Van der Plank (1963) drew attention to the fact that this elimination results in stabilizing selection and later described genes for resistance in the host that could only be countered by the latter kind of virulence as epidemiologically "strong" (van der Plank, 1968).

Unfortunately it is not easy to measure these effects. Many attempts to demonstrate such differences have been unsuccessful probably because they were overshadowed by other effects. For example, Osoro and Green (1976) compared the rates of survival of seven Canadian races of *Puccinia graminis tritici* inoculated in pairs to seedlings of several different susceptible wheat varieties in a greenhouse or growth cabinet, and in the field. The seven races each possessed from one to four virulence alleles with respect to five different genes for resistance. The races were related inasmuch as they could be arranged according to origin in a temporal sequence of single mutations at specific loci. In the majority of comparisons, the races with the largest number of virulence genes predominated in mixtures. Osoro and Green concluded that "selection . . . favored the most aggressive races, which by chance were also the most virulent. . . ." Such experiments suggest that reduction in fitness associated with increased virulence is generally slight, or that it is often compensated by other genetic effects described by the terms aggressiveness, or horizontal pathogenicity.

Leonard (1977) has discussed other situations that explain why pathogen populations may maintain virulence genes that are apparently unnecessary. These include low frequencies of undetected resistance genes in the host population, and the necessity for maintaining virulence genes to allow a pathogen to overwinter in another part of its range that is remote from the sampling area.

The effects of alleles for virulence and avirulence on pathogen fitness can be measured without the confounding effects of other genes by comparing near-isogenic lines of the pathogen on a susceptible host. Leonard (1977) compared lines of race O and race T of *Bipolaris maydis* (*Helminthosporium maydis*) prepared by 12 generations of backcrossing to a recurrent parent isolate of race T. From the increase in frequency of race O in mixtures of near-isogenic lines of races O and T on maize with normal cytoplasm, he calculated a cost of virulence (k) in race T of 0.12, where relative fitness of the pathogen is equal to $1 - k$. In another comparison, 50 different isolates of race O were combined with 50 different isolates of race T and the bulked population tested on maize with normal cytoplasm. The large samples of each race were intended to provide enough genetic diversity to cancel out genetic differences other than virulence. In this experiment k was estimated at 0.30. In fact this measurement of k is probably more useful than the first which has the advantage of comparing isogenic lines but the disadvantage of determining the effect of only one of possibly several different forms of the allele for virulence.

In Leijerstam's (1972) example of the spontaneous origin of virulent mutants of *E. graminis tritici*, two independent mutants were virulent on wheat hosts carrying *Pm1*. The first was virulent only on this host but the second was virulent on two others in addition. A simple explanation for this result is that the first mutation was a point change and the second a deletion that included the two other avirulence loci. Removal of a dominant allele for avirulence by deletion of a segment of the chromosome can have the same effect as a mutation to virulence by a point change within the locus. Admittedly, without data to confirm the implied close linkage among the three loci, other explanations are possible. However, several lines showing such deletion of genes were obtained by Flor and others among induced mutants of flax rust (*Melamp-sora lini*) (see Day, 1974). The point is simply that it would not be at all surprising to find that Leijerstam's two mutants had quite different k's. In the absence of the two other differential hosts there would be no way to distinguish between them other than by k. In sampling 50 different T strains Leonard covered the possibility of including different copies of the virulence gene which although having the same phenotype may not have identical k values. As Leonard (1977) has pointed out much of the confusion over stabilizing selection has arisen because of failure to distinguish effects of individual virulence genes on fitness from the effects of the rest of the genes in the genotype.

Person *et al.* (1976) and Groth and Person (1977) have developed models that take into account reduced fitness to predict selection effects in mixed populations. They concluded that: (1) there is a limit to the number of virulence genes accumulated through natural selection; (2) this limit will be reached when the advantage of infecting more hosts is balanced by loss of fitness in individual disease reactions on the expanded host population that can be infected; and (3) the overall reproductivity, or fitness, of the pathogen population will depend on the number of genes for resistance in the host population. However, they point out there still is too little information to use these models to assess the various strategies available for disease management. If the strategies employ new genes for resistance, then virulent races are either unknown or occur in such low frequencies that measurement of k values by Leonard's methods is impossible or unreliable. At the present time there is no way to predict k, or, in van der Plank's terms, measure the strength of a resistance gene.

IV. CHOOSING STRATEGIES

Genetic uniformity offers many advantages for modern agriculture (Marshall, 1977). Large-scale farming operations require uniform stands of plants that germinate, emerge from the soil, flower, and ripen for harvest at the same time. Farm machines cannot cope with much variation in time of emergence, plant size, habit, vigor, or maturity date. Machines allow very large-scale operations with their attendant increases in efficiency and resulting lower costs. Food processors and vendors also require uniform quality and appearance to simplify and speed their work. Apples, tomatoes, potatoes, lettuce, asparagus, and other produce must be of uniform size and appearance, be easy to ship and to store, and have a long shelf life for the food chains that market them. Millers and maltsters have stringent requirements for wheat and barley that have discouraged blending plants in the field rather than blending harvested grain during processing. Even the laws to protect plant breeder's rights require uniform, homogeneous lines as cultivars so that they may be precisely defined and standardized at registration.

All these requirements are abhorrent to those who urge a return to nature and a simple life sustained by subsistence farming. Diversity is much easier to sustain in smaller scale, labor-intensive, agriculture although of course even genetically diverse crops are liable to some loss due to disease. Breeders who deal with the problems of marginal lands and improving food production in developing countries have unparalleled opportunities to avoid many of the pitfalls of "factory farming" (Jennings, 1974). However, industrialized societies with large urban populations must have cheap, abundant, and nutritious food. A measure of genetic uniformity therefore has to be accepted and the genetic strategies chosen as defenses against plant disease epidemics must take this into account.

Large-scale agriculture has three major strategies for disease control: (1) uniform, vertically resistant cultivars; (2) multiline, vertically resistant cultivars; and (3) uniform, horizontally resistant cultivars. Below is presented a brief account of the advantages and disadvantages of these methods.

A. Uniform Vertical Resistance

The search for disease resistance and incorporation of single major genes conferring resistance into crop plants has been the most successful and most widely used genetic strategy for management of plant diseases. In a few cases the resistance has been permanent or sufficiently effective to give control for many years. Examples of single-gene horizontal resistance include loose smut of barley (*Ustilago nuda hordei*) and milo disease of sorghum (*Periconia circinata*) (van der Plank, 1968) resistance to woolly aphid (*Eriosoma lanigerum*) in apple (Robinson, 1976), and *Cercospora* blight in cucumber (Day, 1974). More often farmers and breeders have experienced "boom and bust cycles." A newly introduced resistant cultivar that is resistant to a prevalent disease is more and more widely grown each year until it is subject to an epidemic caused by a virulent race. As a result of this "bust" the variety is abandoned in favor of a new one with a different gene for resistance that, in due course, suffers a similar fate. Where breeding is simple and straightforward, as in most small grains, breeders produce a series of resistant cultivars. As the current one builds to its "bust" phase its successors are at different stages of development ranging all the way from unimproved new sources, through early crosses, to the final large-scale trials needed before release of the variety. The limitations of this method include dissatisfaction with the temporary nature of the solution in the face of the inevitable appearance of virulent races, and the problem of locating new genes for resistance.

Many breeders have advocated development of lines with many genes for resistance. They reason that, if virulent races arise by mutation, then the frequency of mutants virulent on combinations of two or more resistance genes will be the product of the individual mutation rates. Since multiply virulent races will be much more rare than singly virulent races, the resistant combinations should remain useful much longer. If the frequency of singly virulent mutants is greater than that maintained by mutation, as is suggested by the paucity of evidence for reduced fitness associated with virulence, then multiply virulent mutants may be a good deal more common than expected. Unfortunately there is almost no information on the fitness of virulent mutations. The competition experiments referred to earlier beg the question of fitness. The reason is that in general only successful virulent races will ever be used. Since they have already demonstrated their ability to survive and be detected in nature, they clearly have great fitness. During reproduction on selective resistant hosts they have undergone selection to improve fitness. This will not be true of new virulent mutants.

The reader must bear in mind that the question of fitness here is only relevant on nonselective or susceptible host plants or, in the case of pathogens that can exist as saprophytes on some other substrate. Once a virulent mutant produced by the population maintained on the nonselective host alights on and infects a resistant host, its fitness is now very much greater than that of any avirulent individual. The only exception to this is the interference effect that a large number of adjacent low-infection types sometimes has on the establishment of a single high infection type during the first critical hours of infection.

The widespread use of F_1 hybrids in a variety of crops lends itself to the synthesis of multiply resistant hybrids by combining different dominant genes for resistance in the parent lines. The genetic constitution of these hybrids can be rapidly changed to meet the threat of new races. The method was suggested for tomatoes (Darby and Day, in Day, 1968) and cereals (Done, 1973).

The problem of maintaining a constant supply of new resistance genes is referred to in the next section.

B. Multiline Vertical Resistance

This strategy is based on the premise of reintroducing variation and employing it against the pathogen. A multiline cultivar consists of a set of from 8 to 16 lines that are as near as possible isogenic except that they carry different genes for resistance to a given pathogen. For small grains the composition of the blend is controlled by mechanically mixing seeds of the component isolines. These are selected on the basis of agronomic performance, their reactions to specific pathogen races, and pathogen population trends (Frey *et al.*, 1977). In practice the blends can be changed from year to year and the proportion of each component in the blend can also be varied. Multilines are designed to use stabilizing selection to maintain simple pathogen races and prevent selection of super races. A second important effect is that inoculum buildup on susceptible components of the host population should be delayed by the trapping, or so called "fly-paper," effect of surrounding resistant plants. Field tests of oat multilines designed to control *Puccinia coronata* (reviewed by Frey *et al.*, 1977) have shown delays in rust buildup and, in artificially induced massive epidemics, have produced grain yields comparable to those of resistant isolines and nearly double those of susceptible isolines.

Two criticisms of multilines are that varietal improvement is 8–16 times as laborious as in a single resistant line and that the variation present is polarized, that is directed only toward a single pathogen. The labor in varietal improvement is increased because each change of genetic background for improved yield or quality requires the introduction of all the resistance genes that are to be used into that background to produce an improved set of isolines. The polarization of resistance would be unacceptable in a crop like rice where current programs are assembling resistance to as many as eight different pests and diseases in single lines (Ou, 1977; Khush, 1977). To date there is no field data on whether multilines will prolong the effective, useful life of vertical resistance genes. A computer simulation compared the effects of introducing strong resistance genes in three different ways; in multilines, as a single cultivar with a combination of all genes available, or as a sequence of cultivars each with one gene (Trenbath, 1977). The rate of "breakdown" of the resistance genes in the simulation was approximately the same in all cases. However, the conclusions reached in a computer simulation will be no better than the estimates and data fed to it. Multilines, and to some extent multigene combinations also, depend heavily on the effects of stabilizing selection. In view of the difficulty of measuring these effects it is probably premature to judge these methods.

C. Horizontal Resistance

Given that all plant breeders have to work with polygenic systems for many agronomic characters other than disease resistance, it is surprising that more have not turned to polygenic horizontal resistance. The obvious reasons include the availability of satisfactory and sufficiently stable vertical resistance, inadequacy of the horizontal resistance available, and more recently, the belief that this kind of resistance is really not horizontal at all but will eventually be eroded by pathogen variation.

The recent history of breeding potatoes for resistance to late blight (*Phytophthora infestans*) is of some interest in this connection. Vertical resistance failed and the story of its failure was complete with the observation that in Mexico even the wild potato sources of vertical resistance were susceptible to local late blight races. Under these conditions the plants survive epidemics as tubers below the soil and resprout in weather conditions not favorable to late blight development. Although the level of horizontal resistance in potatoes bred for the exacting markets of North America and Europe has not been great, plants with high levels of horizontal resistance are available in Mexico. For example the Mexican variety Atzimba was developed for areas such as the Toluca Valley where the blight hazard is great. In North America and Europe the blight hazard is not as great. Hence progress in developing horizontally resistant varieties has been slower.

Other crops in which adequate levels of horizontal resistance have been found and are widely used include: maize to *Puccinia sorghi, P. polysora*, and *Ustilago maydis*; wheat to *P. striiformis*; and tobacco and peanuts to the bacterial wilt caused by *Pseudomonas solanacearum* (Day, 1974, Robinson, 1976). Since all of these are still examples of horizontal resistance, it is perhaps superfluous to add that they have not been eroded by pathogen virulence. This is not for want of looking for evidence (see Day, 1974). The demonstration that quite small effects may be vertical (see Section III, C.) while providing a model of how erosion may appear does not affect the premise that the horizontal effects which make up the major component of the resistance are beyond the adaptive capacity of the pathogen.

Several years ago in two review papers R. A. Robinson formulated some useful rules for the use of vertical and horizontal resistance in disease control. These rules take into account the breeding system of the crop, whether it is annual or perennial, and how it is propagated. The rules also consider salient features of the pathogen, such as how it is spread, what plant organs it attacks, how it survives from one crop to the next, and so on. Recently Robinson (1976) presented his ideas in the form of a systems concept applied to crop improvement and disease management. While the treatment is largely theoretical and cloaks some old ideas and concepts under new names (for example, physiologic races are "vertical pathotypes" and a host cultivar is a "polyphyletic pathodeme") it presents some bold suggestions and deserves careful study and evaluation. All too often plant breeders and plant pathologists are unaware of the genetic patterns in their breeding work. A rigorous systems approach may help to remedy this through consideration of facts and principles that would otherwise be ignored.

One of these important but often ignored ideas is that plant breeders do not require new sources of horizontal resistance. For many crops the germ plasm available in commercial cultivars is sufficient. All that is necessary in selection is to avoid the confounding effects of vertical resistance genes that may be present. This is most conveniently done by using virulent pathogen races for screening the most resistant individuals from as a broadly based a random polycross as can be made. In West Africa, 10–15 generations were sufficient to accumulate enough horizontal resistance in maize to *Puccinia polysora* without importing any germplasm. A number of FAO projects are presently engaged in breeding for horizontal resistance (Robinson, 1977).

V. CONCLUSIONS

Perhaps the clearest conclusion from examining choices among strategies in breeding for disease resistance is that no one strategy for controlling epidemic disease is superior to all others. The advantages to modern agriculture from high levels of genetic uniformity also will continue to put crops at risk for the foreseeable future perhaps to a greater extent than now. Uneconomic genetic diversity, except on a small scale in areas or social systems where large-scale monoculture is impractical, is ruled out by the overwhelming pressures on fewer farmers to feed and clothe more people. If the availability of widely adaptable germplasm continues to expand through international testing programs for other crops to match that for wheat and rice then the introduction of local diversity must be stressed. The National Academy Report (Horsfall, 1972) pointed to the need for backup potential in the event that epidemics occur in spite of our efforts to control them. Possibly the most satisfactory form of back up for multiline oats and wheat, and multigenic rice would come from equally strenuous, but local, efforts to develop horizontally resistant equivalent lines. This kind of backup will not run out of genetic resources since it is not dependent on a continuing supply of newly discovered resistance genes.

The conservation of genetic resources is unquestionably the most important form of long term protection. The erosion of genetic diversity in crop plant germplasm has been well documented (see Frankel, 1977). There is an urgent need to collect and preserve valuable and irreplaceable materials now before future breeding programs are jeopardized by their loss.

References

Bottrell, D. G., and Adkisson, P. L. (1977). Cotton insect pest management. *Annu. Rev. Entomol.* **22**, 451–481.

Clinton, G. P. (1908). Report of the botanist for 1908. *Conn., Agric. Exp. Stn., Rep.* pp. 879–890.

Day, P. R. (1968). Plant disease resistance. *Sci. Prog. (Oxford)* **56**, 357–370.

Day, P. R. (1973). Genetic variability of crops. *Annu. Rev. Phytopathol.* **11**, 293–312.

Day, P. R. (1974). "Genetics of Host-Parasite Interaction." Freeman, San Francisco, California.

Done, A. C. (1973). Implications of the introduction of hybrid cereals on disease patterns. *Ann. Appl. Biol.* **75**, 144–149.

Duvick, D. N. (1977). Major United States crops in 1976. *Ann. N.Y. Acad. Sci.* **287**, 86–96.

Ellingboe, A. H. (1975). Horizontal resistance: An artifact of experimental procedure? *Aust. Plant Pathol. Soc. Newsl.* **4**, 44–46.

Flor, H. H. (1946). Genetics of pathogenicity in *Melampsora lini. J. Agric. Res.* **73**, 335–357.

Flor, H. H. (1947). Inheritance of reaction to rust in flax. *J. Agric. Res.* **74**, 241–262.

Frankel, O. H. (1977). Genetic resources. *Ann. N.Y. Acad. Sci.* **287**, 332–844.

Frey, K. J., Browning, J. A., and Simons, M. D. (1977). Management systems for host genes to control disease loss. *Ann. N.Y. Acad. Sci.* **287**, 255–274.

Groth, J. V., and Person, C. O. (1977). Genetic interdependence of host and parasite in epidemics. *Ann. N.Y. Acad. Sci.* **287**, 97–106.

Harlan, J. R. (1976). Diseases as a factor in plant evolution. *Annu. Rev. Plant Pathol.* **14**, 31–51.

Horsfall, J. G., ed. (1972). "Genetic Vulnerability of Major Crops." *Natl. Acad. Sci.*, Washington, D.C.

Jaynes, R. A., Anagnostakis, S. L., and Van Alfen, N. K. (1976). Chestnut research and biological control of the chestnut blight fungus. *In* "Perspectives in Forest Entomology" (J. Anderson and H. Kaya, eds.), pp. 61–70. Academic Press, New York.

Jennings, P. R. (1974). Rice breeding and world food production. *Science* **186**, 1085–1088.

Johnson, R., and Taylor, A. J. (1976). Spore yields of pathogens in investigations of the race specificity of host resistance. *Annu. Rev. Plant Pathol.* **14**, 97–119.

Khush, G. S. (1977). Breeding for resistance in rice. *Ann. N.Y. Acad. Sci.* **287**, 296–308.

Krivchenko, V. I. (1977). Territorial utilization of resistance sources from the world collection of the Vavilov Institute for Wheat and Barley Breeding. *Ann. N.Y. Acad. Sci.* **287**, 29–34.

Lawrence, G. J. (1977). Genetics of pathogenicity in flax rust. Ph.D. Thesis, University of Adelaide.

Leijerstam, B. (1972). Studies in powdery mildew on wheat in Sweden. III. Variability of virulence in *Erysiphe graminis* f. sp. *tritici* due to gene recombination and mutation. *Natl. Swed. Inst. Plant Prot., Contrib.* **15**, 231–248.

Leonard, K. J. (1977). Selection pressures and plant pathogens. *Ann. N.Y. Acad. Sci.* **287**, 207–222.

Levings, C. S., III, and Pring, D. R. (1976). Restriction endonuclease analysis of mitochondrial DNA from normal and Texas cytoplasmic male-sterile maize. *Science* **193**, 158–160.

Loegering, W. Q. (1966). The relationship between host and pathogen in stem rust of wheat. *Hereditas, Suppl.* **2**, 167–177.

Mangelsdorf, P. C. (1966). Genetic potentials for increasing yields of food crops and animals. *Proc. Natl. Acad. Sci. U.S.A.* **56**, 370–375.

Marshall, D. R. (1977). The advantages and hazards of genetic homogeneity. *Ann. N.Y. Acad. Sci.* **287**, 1–20.

Martin, T. J., and Ellingboe, A. H. (1976). Differences between compatible parasite/host genotypes involving the *Pm4* locus of wheat and corresponding genes in *Erysiphe graminis* f. sp. *tritici*. *Phytopathology* **66**, 1435–1438.

Maynard Smith, J. (1976). A comment on the Red Queen. *Am. Nat.* **110**, 325–330.

Mercado, A. C., Jr., and Lantican, R. M. (1961). The susceptibility of cytoplasmic male sterile lines of corn to *Helminthosporium maydis*. *Nish. and Miy. Philipp. Agric.* **45**, 235–243.

Metcalf, H. (1908). Immunity of the Japanese chestnut to the bark disease. *U.S., Dep. Agric., Bur. Plant Ind. Bull.* **121**, 1–4.

Moore, F. J. (1977). Disease epidemics and host genetic resistance in British crops. *Ann. N.Y. Acad. Sci.* **287**, 21–28.

Murrill, W. A. (1906). A serious chestnut disease. *N.Y. Bot. Gard.* **7**, 143–153.

Osoro, M. O., and Green, G. J. (1976). Stabilizing selection in *Puccinia graminis tritici* in Canada. *Can. J. Bot.* **54**, 2204–2214.

Ou, S. H. (1977). Genetic defense of rice against disease. *Ann. N.Y. Acad. Sci.* **287**, 275–286.

Person, C., and Sidhu, G. (1971). Genetics of host-parasite interrelationships. *In* "Mutation Breeding for Disease Resistance," pp. 31–38. IAEA, Vienna.

Person, C., Groth, J. V., and Mylyk, O. M. (1976). Genetic change in host-parasite populations. *Annu. Rev. Phytopathol.* **14**, 177–188.

Robinson, R. A. (1976). "Plant Pathosystems." Springer-Verlag, Berlin and New York.

Robinson, R. A. (1977). The Food and Agriculture Organization International Program on Horizontal Resistance (IPHR). *Ann. N.Y. Acad. Sci.* **287**, 327–331.

Safeeulla, K. M. (1977). Genetic vulnerability, the basis of recent epidemics in India. *Ann. N.Y. Acad. Sci.* **287**, 72–85.

Trenbath, B. R. (1977). Interactions among diverse hosts and diverse parasites. *Ann. N.Y. Acad. Sci.* **287**, 124–150.

van der Plank, J. E. (1963). "Plant Diseases: Epidemics and Control." Academic Press, New York.

van der Plank, J. E. (1968). "Disease Resistance in Plants." Academic Press, New York.

Wheeler, H. (1975). "Plant Pathogenesis." Springer-Verlag, Berlin and New York.

Wolfe, M. S., and Minchin, P. N. (1976). Quantitative assessment of variation in field populations of *Erysiphe graminis* f. sp. *hordei* using mobile nurseries. *Trans. Br. Mycol. Soc.* **66,** 332–334.

Wolfe, M. S., and Schwarzbach, E. (1975). The use of virulence analysis in cereal mildews. *Phytopathol. Z.* **82,** 297–307.

Wolfe, M. S., Barrett, J. A., Shattock, R. C., Shaw, D. S., and Whitbread, R. (1976). Phenotype-phenotype analysis: field application of the gene-for-gene hypothesis in host-pathogen relations. *Ann. Appl. Biol.* **82,** 369–374.

Chapter 14

Diseases in Forest Ecosystems: The Importance of Functional Diversity

ROBERT A. SCHMIDT

Diversity is the bedfellow of Mother Nature

I. INTRODUCTION

A. Objectives, Scope, and Definitions

This chapter aims at an epidemiological discussion of tree diseases in natural forest ecosystems. Our topic concerns the relative amount of disease through time and space and the factors which condition this

287

disease incidence. A general notion pervades plant pathology that within natural forest ecosystems diseases are maintained at a low incidence and that epidemics are absent or at least infrequent. We will examine this premise in some detail.

This chapter emphasizes those factors which mitigate epidemics, although predisposing factors which favor epidemics must also be considered. Are there homeostatic mechanisms in natural forest ecosystems which prevent frequent or long-duration perturbations by forest tree pathogens? If so, what are they? Identification of such mechanisms might provide insight to disease management in forests or agroecosystems. To this end, this chapter seeks to interpret some existing knowledge of forest pathology in the context of contemporary epidemiology. Hopefully, this discussion will set the stage for the more difficult task of developing holistic yet synthetic quantitative theories of disease incidence in forest ecosystems. There is a great need and utility for such theories.

Our domain is that of natural forest ecosystems, primarily the deciduous and conifer forests of the eastern United States. I choose to emphasize these forests because they contain a wealth of examples, but also because of my experience with them. The specific details will differ among geographically and ecologically distinct forest ecosystems, but surely the general concepts will be applicable overall. An understanding of disease in natural forest ecosystems is facilitated by a comparison with disease in "unnatural ecosystems." Thus this discussion will, in part, be one of comparative epidemiology among natural forests, intensively managed forests or plantations, and even the proverbial "tree that grows in Brooklyn." Indigenous pathogens (primarily fungi, the most abundant and important forest tree pathogens) and diseases will be featured, but reference must be made to introduced pathogens so important to forests.

During the last two decades epidemiology has emerged as a more inclusive and integrating science in plant pathology (Schmidt, 1973; Zadoks, 1974). Originally epidemiology was defined as the study of secondary factors in disease development; primary factors, namely, the pathogen, being the bailiwick of etiology. Epidemiology considered the effects of environmental factors on disease development and as such was the ecology of disease. Early studies were primarily autecological, but with the appearance of disease progress curves (Large, 1945) a synecological or population approach to the study of plant disease was begun. van der Plank (1963) championed this approach and added greatly to the quantification of disease and disease control in populations. He emphasized the analysis of disease progress curves which showed the amount of disease as a function of time. Subsequently, it became fashionable to define epidemiology as the study of all factors—including host and pathogen genetics and disease control—as they influence disease progress curves. Although such curves are of paramount importance and utility, they do not encompass all that is pertinent to epidemiology. In addition there are important spacial concepts of epidemiology, e.g., disease gradient curves which show disease incidence as a function of distance. In fact, a holistic concept of epidemiology includes (1) disease assessment; (2) changes in the amount of disease in a population of plants in time and space as conditioned by any or all host, pathogen, and environmental factors and cultural practices; (3) yield loss relative to disease development; and (4) integrated disease management. This chapter is concerned primarily with the second and, to some extent, the fourth of these subjects.

To avoid ambiguity of terms within this chapter I have used the following definitions. Disease incidence is synonymous with amount of disease and includes both disease frequency and severity. Disease increase refers to the time dimension and disease spread to the space dimension and these are envisioned as disease progress and gradient curves, respectively. Indigenous species are native in contrast to introduced or exotic species. In my opinion an epidemic is best characterized as a disease gradient or disease progress curve regardless of the amount of disease or the magnitude and sign (+ or −) of the rate coefficients. Endemic means indigenous but also suggests limited disease incidence or more specifically limited disease severity. Unfortunately, the science of comparative epidemiology (Kranz, 1974) is only now emerging and the terminology to contrast and compare epidemics is ambiguous and lacking. This problem is especially acute in forest pathology because of its variable and expanded time dimension. Average infection rates have utility in forest pathology, especially for segments of the disease progress curve (Berger, 1977), but do not adequately address the problem. Appropriate terminology must be forthcoming, but in the meantime, for this chapter and notwithstanding my own concept of the epidemic, I will misuse the terms in the traditional sense. Thus, epidemic refers to a relatively large amount of disease and implies a relatively fast rate coefficient and endemic refers to a relatively small amount of disease and implies a relatively slow rate coefficient.

B. Historical Perspectives

Peace (1962) put the subject in perspective when he wrote ". . . but any effort to impose broad generalizations on a subject so imperfectly understood as forest pathology is to court disaster." This is good advice and so it is without "Peace of mind" that I proceed armed only with caution and the notion that the development of a synthetic perspective is desirable. In his original text and elsewhere Boyce (1948, 1954) considered disease incidence and control relative to natural and planted forests. He notes (1954) that in natural forest ecosystems pathogenic fungi and insects are invariably present and epidemics occur, appear alarming, and usually subside without having caused disastrous damage. Further, Boyce warns against the potential hazards of planted monocultures. He cites poor site selection, artificial regeneration, seed source, stand composition and management, native or introduced organisms, and climate change as factors which condition disease incidence. Peace (1957, 1961), on the other hand, defends the practicality of more intensively managed plantations and warns against the dangerous concept and limitations of wholly embracing the natural forest. In doing so, he recognizes the potential disease threat to planted monocultures, but stresses their silvicultural benefits and notes that unnatural is not necessarily the same as unhealthy. Baxter (1937, 1967) wrote on the development and succession of forest fungi and diseases in forest plantations and indicates that these are biological communities established under more or less unnatural conditions which may lead to attack by various fungi and result in a succession of diseases. In the latter treatise (1967) Baxter coins a most appropriate phrase indicating that forest tree diseases are ". . . the thief of time." More recently, in a symposium on disease in natural plant ecosystems, Edmonds and Sollins (1974) considered the impact of diseases on succession in coniferous forest ecosystems and Dinus (1974) discussed the utility of knowledge about natural ecosystems to disease control in a managed forest.

II. DIVERSITY IN FOREST ECOSYSTEMS

A. The "Natural" Forest and the Concept of Functional Diversity

If we consider the word "natural" to mean devoid of man's influence, then contemporary forests represents a continuum. At the one end, there is the truly natural forest as yet untouched by man, while at the other end there is the intensively managed forest plantation. Man's influences on the latter may involve site preparation, including mechanical destruction or burning of harvesting residue, cultivation, and bedding; artificial regeneration with seedlings or seeds from selected and improved sources; fertilization; burning to reduce competition; and precommercial thinnings and clear-cutting. The "natural" forests considered in this chapter are, at best, quasinatural forests having been managed or mismanaged to some extent by man. In general, management involves only harvesting of mature trees; regeneration occurs naturally. Some of these forests are comprised of a single tree species, others of many tree species of diverse phylogeny but, in any case, indigenous species. These natural forests may be even- or uneven-aged, regenerated by seed or by sprouts, and selectively or clear-cut.

Theoretically, a plant community designed to discourage widespread epidemics should include a variety of mechanisms which impede the initiation and rate of disease increase and spread. It is the thesis of this chapter that natural forests abound with such mechanisms which I will refer to collectively as functional diversity, implying functional against potential epidemics. These factors, which theoretically dampen perturbations caused by disease, also might be thought of collectively as ecosystem disease resistance. Others have suggested population resistance (van der Plank, 1975) and dilatory resistance (Browning et al., 1977). Whatever the terminology the general concept is certainly not new, having been stated succinctly in 1767 by Tozzetti (Zadoks and Koster, 1976) who wrote that "It is not so easy to render a reason why, Wheat growing seeded with Rye, or with Vetch, was not damaged by the rust, while a field of Wheat alone, standing between one of Rye, and Vetch, yielded scarcely any seed and that the most miserable."

Each forest has its own specific type and extent of functional diversity. What follows is an attempt to characterize and exemplify this diversity in eastern forest types. In passing it should be noted that annual crop pathologists have become interested in one aspect of functional diversity; namely, genetic diversity, especially as it relates to a more stable and perhaps more natural control of epidemics (Browning, 1974; van der Plank, 1963, 1968). As we shall see, this singular genetic diversity is inadequate to characterize forest ecosystems.

B. The Host Population

The natural forests of the eastern United States contain at least eight major forest types (forest stands of similar character as regards species composition and development due to given ecological factors) including more than 150 tree species (Society of American Foresters, 1954; U.S. Department of Agriculture, 1973). One such type; namely, the Appalachian Mixed Hardwoods, located in the central and southern Appalachian Mountains and adjacent Allegheny and Cumberland Plateaus exemplifies species diversity. Within this type at least 20 commercial tree species of diverse phylogeny—fifteen genera belonging to eleven plant families, including deciduous and conifer species—occur (Braun, 1950). In addition, these species occur in various associations (a community unique in its floristic composition) conditioned by environment, especially edaphic factors. Thus, a second order of species diversity exists. For example, red oak may occur in association with basswood and white ash on one site, but with other oaks and yellow poplar on another site; in still other nearby sites red oak may be absent (Society of American Foresters, 1954). This Appalachian Mixed Hardwood Type is bound on the north by the Northern Hardwoods and Cherry-Maple Types and on the south by the Oak-Pine Type. These types are not discrete entities—transition zones exist where types converge and where local environmental conditions favor one or the other type, e.g., at high elevations within types. Although not extensively documented except for a few species and traits, abundant intraspecific variation exists in these natural tree populations. Adding to this already rich species diversity are abundant shrub and annual plant species.

Uneven age distribution is characteristic of these mixed hardwood forests and varies with mode of regeneration, light tolerance of species, etc. Trees vary from 1 year to more than 100 years of age within the same forest stand. Furthermore, there exist within individual trees organs and tissues of varying physiological age and stages of development.

Mixed, uneven-age forests have a diverse horizontal and vertical structure. Horizontally, discontinuities of various distances exist among plants of a single species. Vertically, trees of the same or different species may occupy habitats which vary from a few centimeters to a 100 m or more above the earth's surface. As discussed more fully below this diversity in horizontal and vertical structure has major influences on the microclimate and incidence of forest tree diseases.

Unlike intensively managed annual or perennial plant ecosystems, natural forests embody to a large degree the concepts of competition, adaptation, succession, and climax. The climax forest is a plant community at the culmination of natural succession. This climax vegetation is determined by the climatic, edaphic, and biotic environment, including plant disease, and persists (Dansereau, 1957) unless visited by some unusual natural or man-induced disturbance. Until the climax vegetation is established, forest communities exhibit natural succession as one species and/or community replaces another. Natural forest ecosystems may be thought of as little disturbed by man but should not be considered as undisturbed or static. Even climax forests experience significant perturbations caused by drought, fire, insects, diseases, etc. Natural forests are dynamic communities characterized by change rather than the repeated anthropogenic cycles typical of annual crop ecosystems or, to some extent, those of intensively managed perennial crops, including forest plantations.

There is, in general, agreement among ecologists that natural succession leads to diversity and that ecosystem diversity is associated with species population stability (Pimentel, 1961; MacArthur, 1955). However, studies of insect predation (Watt, 1968) indicate that diversity in the predator population has not led to stability in the prey population. There appears to be no general agreement that diversity is either a singular or partial determinant of stability (May, 1973).

C. The Forest Environment

There exists among and within natural forest ecosystems a diverse array of climatic, edaphic, and biotic environments, many of which significantly affect plant diseases. Macroenvironments in themselves are quite diverse, depending on the geography of slope, aspect, and elevation. Additional environments arise from the interaction of vegetation and macroenvironment. In fact, the concepts of microenvironment and habitats come to life in the forest where a complex mosaic of unique habitats and biotopes can exist within relatively small areas.

The mixed hardwood forests of our concern transcend four of Köppen's climate types (Trewartha, 1954). In the north there are humid macrothermal climates with either warm or cool summers and no distinct dry season; in the south there are humid microthermal climates with either warm or cool summers and also no distinct dry season. The principal differences are minimum winter and maximum summer temperatures. These macroclimates are influenced locally by elevation, slope, and aspect so that adjacent stands with similar vegetation can have quite different climates. Changes in macroclimate, especially those which influence soil temperature, are associated with changes in disease incidence (Hepting, 1964).

The forest tree canopy is a meteorologically active surface, in addition to the earth's surface, and plays an important role in determining stand (crop) climate (Geiger, 1966). The canopy absorbs and thereby reduces shortwave radiation by about 80% and is essentially a black body for longwave radiation, absorbing and emitting most wavelengths. The canopy may reduce monthly maximum summer air temperatures about 5°C, rainfall 15–30%, and wind velocity 20–60% (Reifsnyder and Lull, 1965).

Because of this amelioration of the macroclimate by geography and vegetation and because of their complex horizontal and especially vertical structure, mixed, uneven-aged forests contain a multitude of microclimates. These microclimates condition the habitats of forest tree pathogens and, as indicated by Schmidt and Wood (1969, 1972), one pathogen may routinely experience several unique microclimates during various phases of one disease cycle. Van Arsdel (1961) has shown the importance of microclimate, relative to temperature and dew, for the development of white pine blister rust in the Lake States.

Forest soils, largely those not suitable for annual crop production, are notoriously heterogeneous in origin and composition. Within the area of the Appalachian Mixed Hardwoods Type there is a multitude of soil types (Soil Survey Staff, 1975). These differ widely with regard to chemical, physical, and biotic features including important parameters such as nutrients, organic matter, cation exchange capacity, pH, moisture holding capacity, and temperature. Significantly different soils can occur abruptly and within distances of a few meters. Soil moisture, which can change rapidly in both time and space, has a marked effect on vegetation, perhaps on disease resistant mechanisms, and on microflora, including beneficial symbionts and soil-borne pathogens.

The microflora and fauna populations of forest ecosystems are extremely large and heterogeneous. These organisms, many of which are microscopic, abound in the atmosphere and soil and on the surfaces of and within the leaves, stems, and roots of trees. In fact, were it possible to remove only the trees, the forest would remain, outlined by the organisms which live within or upon the trees. Bacteria, Actinomycetes, algae, fungi, protozoa, nematodes, worms, insect larvae, arthropods, and vertebrates comprise a significant portion of forest soils and their numbers and types are greatly influenced by litter, soil moisture and temperature, aeration, acidity, and organic and inorganic matter (Steubing, 1970; Edwards et al., 1970). The saprophytic organisms involved in nutrient cycling, beneficial symbionts such as nitrogen-fixing and mycorrhizal organisms, and bacteria and fungi antagonistic to plant pathogens vary greatly and affect disease development. Insect vectors and insects which attack trees weakened by disease also play a significant role in disease development. There is a natural succession of these organisms (Shigo, 1967; Giese et al., 1964) and diversity is further enhanced as they vary in numbers and species relative to substrate succession.

D. The Pathogen Population

The diversity of suscepts and environments in the forest conditions a similar diversity in the populations of pathogens. Curiously, however, most forest tree pathogens are fungi. The life and disease cycles of these pathogens are among the most unique and fascinating in plant pathology. They include abundant obligate and facultative saprophytes and obligate and facultative parasites. Most are naturally adapted components of the ecosystem and play functional roles in ecosystem regulation (Edmonds and Sollins, 1974) that are analogous to those suggested for insects by Mattson and Addy (1975). With the exception of decomposition these regulatory phenomena have been little considered by forest pathologists. Renewed interest in biological control prompts us to examine more closely in the future this regulatory aspect.

Often a multitude of pathogens attacks a single tree species. For example, one plant disease host index (U.S. Department of Agriculture, 1960) lists more than 75 genera of microflora that attack northern red oak (*Quercus rubra*). The total complement of microflora species on red oak might be double this number. Although some are obligate saprophytes and may not cause disease, still they compete for red oak substrate and potentially affect disease development. Since many tree species with different complements of microflora co-inhabit sites, the total number of microflora species is large and diverse. A simple example is provided in Table I which is a hypothetical life table of some fungal diseases on slash pine. In addition to a diverse microflora which includes fungi, bacteria, and seed plants, tree species also support diverse populations of insects and other microfauna.

Forest tree pathogens are notoriously ubiquitous. In fact, the very concept of opportunistic pathogens, basic to forest pathology, implies ubiquity. Damping-off fungi and other root pathogens such as *Armillaria mellea* (Shaw and Roth, 1976) and *Clitocybe tabescens* are common and persistent soil inhabitants. Ubiquity might also be claimed for *Fomes annosus* and numerous stem decay fungi. There are many forest tree pathogens for which it is safe to assume that inoculum is close at hand.

Forest tree pathogens are well adapted and can survive periods of unfavorable environment or dormant periods of the host as mycelium in host tissues, as complex vegetative structures such as sclerotia or rhizomorphs, and as spores in well-adapted, perfect, and imperfect reproductive structures. In the absence of living hosts, many tree pathogens have a saprophytic stage. Significantly, some facultative saprophytes are relatively unsuccessful as saprophytes and this greatly limits their survival and incidence.

Fruiting structures for asexual and sexual sporulation are varied and typical of the Phycomycetes, Ascomycetes, Basidiomycetes, and Deuteromycetes. Unlike agroecosystems, forests have an abundance of Basidiomycetes, especially Hymenomycetes with complex basidiocarps. Complex ascocarps are also very common. Many forest tree pathogens

TABLE I

Hypothetical Life Table of Some Fungus Diseases of Slash Pine

Age interval when disease is prevalent and damaging [a]	Disease	Pathogen(s)
0–3 Months	Damping-off	*Fusarium oxysporum, Pythium ultimum, Rhizoctonia solani, Sclerotium bataticola, Phytophthora* spp.
3 Months–3 years	Seedling root rot	*Sclerotium bataticola* and other damping-off fungi
1–5 Years	Needle blights	*Coleosporium* spp., *Hypoderma* spp., *Scirrhia acicola*
1–10 Years	Fusiform rust	*Cronartium fusiforme*
10–20 Years	Pitch canker	*Fusarium lateritium f. pini* (F. *moniliforme* var. *subglutanins*)
Thinning–rotation	Annosus root rot	*Fomes annosus*
15 Years–rotation	Cone rust	*Cronartium strobilinum*
20 Years–rotation	Root rots	*Polyporus schweinitzii, Clitocybe tabescens*
40 Years–rotation	Heartwood decay	*Fomes pini*
Harvest–utilization	Blue stain of logs and lumber	*Ceratocystis* spp.
Postharvest	Decay of wood in use	*Lenzites saepiaria, Poria incrassata,* many others

[a] Rotation age is 20–30 years for pulpwood and 60–80 years for lumber. Thinning age is 10–15 years.

produce prodigious amounts of inoculum. Notable examples include *Fomes applanatus* (White, 1919), *Cronartium fusiforme, Scirrhia acicola,* and *Endothia parasitica.*

Release of spores is both active and passive and both wet and dry spore types are abundant. Wind dissemination is most common but many wet spore types have insect associates which aid release and dissemination. Examples are the Nitidulidae beetle associated with the oak wilt fungus (*Ceratocystis fagacearum*) and the bark beetles associated with blue stain fungi (*Ceratocystis* spp.).

The energy available to spores for infection is probably not different from that of fungal pathogens of annual crops. However, certain Basidiomycetes produce rhizomorphs which greatly increase the energy available for infection. Basidiospores of many decay fungi germinate poorly or not at all in artificial culture (Merrill, 1970) but adequate germination occurs on natural substrates, especially those preconditioned by bacteria, fungi, or yeasts (Paine, 1968; Brown and Merrill, 1973). As such, inoculum potential is affected by a succession of organisms in forest ecosystems.

Although the extent of pathogenic variability—another component of inoculum potential—awaits characterization, the heterogeneity of the host population suggests that pathogenic variability is quite large. For example, the pathogenic variability of aeciospores of *Cronartium fusiforme* was as great among spores collected from one gall as among spores collected from many galls in different geographic locations (Snow *et al.*, 1975, 1976). Variation in temperature–growth relationships for monobasidiospore isolates of *Fomes annosus* from a single sporocarp was as great as that among 46 isolates from infected roots having worldwide distribution (Cowling and Kelman, 1964). In addition pathogenic variability is described in *Cronartium ribicola* (McDonald and Hoff, 1971), *Hypoxylon pruinatum* (Bagga and Smalley, 1974), *Armillaria mellea* (Raabe, 1967), and other forest tree pathogens.

The length of the latent and infectious periods and role of secondary sporulation cycles differ greatly among forest tree pathogens. While many foliage pathogens of trees are similar to their counterparts on annual plant foliage, fungi which cause vascular wilts, cankers, root rots, decays, rusts, and even some foliage pathogens of perennial plants may lack secondary cycles. In fact, life cycles of one or more years' duration are common. For example, rusts on conifers may require 2–4 years to complete their life cycle and some stem decay fungi even longer. There is little specific information on latent and infectious periods but some estimates are provided in Table II. Generally, forest tree pathogens have life cycles and latent and infectious periods of relatively long duration. Many tree disease cycles are interrupted annually by unfavorable environmental conditions and by changes in host physiology (either dormancy or a change in susceptibility).

Forest tree pathogens display the usual modes of penetration, entering the susceptible host (1) by direct penetration of intact, usually succulent tissue; (2) through natural plant openings; (3) through wounds; and (4) via insect vectors which create or visit wounds. Trees have evolved elaborate functional, preformed, prepenetration defense mechanisms and wounds which expose susceptible tissues of roots, stems, and branches play an important role in disease incidence. Wounds made by insects, nematodes, climatic factors, fire, and man are closely associated with the incidence of many forest tree diseases.

Modes of pathogenesis subsequent to penetration are typical of most pathogens; enzymes, toxins, and growth regulators are common. Lignification of xylem and phloem tissues provides a potential barrier to coloni-

TABLE II

Estimated Duration of Latent and Infectious Periods and Disease Cycles for Selected Diseases of Forest Trees in Eastern United States

Disease	Pathogen	Inoculum produced	Estimated duration [a]		
			Latent period	Infectious period	Primary disease cycle
Brown spot needle blight of longleaf pine	Scirrhia acicola	Conidia	2–6 weeks	1 month	1 year or less
Oak wilt of red oak	Ceratocystis fagacearum	Ascospores and conidia on mats	5–9 months	Several weeks	1 year
Nectria canker of hardwoods	Nectria galligena	Ascospores	Several months	1 month	1 year or less
Fusiform rust of pine	Cronartium fusiforme	Aeciospores	2 years	6 weeks, perennial	2–3 years
		Urediospores	2 weeks	Several weeks	
		Teliospores	2 weeks	6 weeks	
Southern cone rust of pine	Cronartium strobilinum	Aeciospores	Several months	6 weeks	1 year
		Urediospores	2 weeks	Several weeks	
		Teliospores	Several weeks	Several months	
Annosus root rot	Fomes annosus	Basidiospores	2–many months	2–6 months	1 year
Chestnut blight	Endothia parasitica	Conidia	4–6 weeks	Several months	Less than 1 year
Stem decay	Fomes spp.	Basidiospores	1–many years	1–many months, perennial	Less than 1 year to many years

[a] Durations are approximations intended only to exemplify the long duration of the latent and infectious periods and disease cycles of forest tree pathogens relative to annual crop pathogens.

zation by many organisms that cannot degrade lignin. Some wood-destroying fungi can degrade cellulose in the presence of lignin; others are unable to do so and these are limited to secondary roles in the succession of organisms which colonize woody tissues. The presence of heartwood decay fungi is also age dependent as suggested in Table I.

E. The Lack and Failure of Diversity

So far in this chapter we have noted the diversity within the natural forest ecosystem. We have suggested that functional diversity plays a key role in determining the incidence and distribution of diseases in the forest. The specific nature of this functional diversity is the topic of subsequent sections, but before proceeding further it is appropriate to make several points. First, not all forest ecosystems contain the same amount of diversity. By comparison with the mixed hardwood forests of the Appalachian Mountains, the natural pine forests of the southern United States contain must less diversity. For example, in certain areas in the Coastal Plains, the Slash Pine Type occurs in pure, even-aged stands. These forests are subclimax, having been maintained by wildfires which retard the invasion of oak and other hardwood climax species and which maintain even-age distribution in stands of the intermediately shade-tolerant and relatively fire-resistant slash pine. Species, age, and structural diversity are minimal, although not completely lacking. A variety of woody shrubs and annual plants is in evidence and age differences occur among adjacent stands naturally regenerated at different times. Even within a stand not all regeneration occurs in the same year. A similar diminution of biotic components, including pathogens, accompany the decrease in host species. Furthermore, these pine forests occur primarily within a relatively uniform mesothermic, warm, summer climate. Changes in climate associated with elevation, slope, and aspect are minimal. Although forest soils in the Coastal Plains are heterogeneous by agronomic standards, because of the similarity of parent materials these forest soils are much less variable than those in the Appalachian and Piedmont areas to the north. Despite this relative lack of diversity these natural pine forests contain considerable diversity when compared with planted stands of slash pine, fruit orchards, or annual crops.

Another important point is that diversity in itself is no safeguard against pathogens, especially introduced pathogens. The chestnut blight epidemic to be discussed later will illustrate this point very well. If diversity is to mollify epidemics, it must be of a kind that is functional against pathogens. In fact, some types of diversity will aid epidemics. Such is the case with heteroecious rusts where two species are required by the pathogen, and also with dwarf mistletoe, where uneven-age distribution may enhance the spread of the pathogen (Hawksworth, 1961).

III. SOME FOREST TREE DISEASES WHICH EXEMPLIFY FUNCTIONAL DIVERSITY

A. Brown Spot Needle Blight of Longleaf Pine

This local lesion foliage disease, caused by an ascomycete, *Scirrhia acicola*, is common on longleaf pine (*Pinus palustris*) in the southern United States and illustrates several important points. First, epidemics caused by indigenous pathogens are not rare in natural forest ecosystems. Local lesion foliar pathogens with primary and secondary cycles are often epidemic, e.g., brown spot needle blight, anthracnose of sycamore and oak, powdery mildews of various hardwood species. Nor are epidemics restricted to foliage diseases: wilt diseases, dieback-decline diseases, and even root and stem decay diseases become epidemic under certain conditions. Nevertheless, epidemics caused by indigenous pathogens in natural forest ecosystems are usually limited in time and space and ordinarily do not become pandemic.

Another point to be made is that forest trees are very tolerant of some diseases, e.g., foliage disease of deciduous trees. This tolerance is due in part to the large food reserves stored in the stems and roots and the ability of some angiosperms to produce a new flush of leaves subsequent to defoliation. In any case, a failure to distinguish between disease incidence and disease loss has contributed to the myth that epidemics are rare in natural forests. Conifers are not tolerant of foliage diseases which result in death of needles. For example, epidemics of brown spot needle blight of young longleaf pine result in significant growth loss and mortality.

Brown spot on longleaf pine also provides an example where age and structure of pine communities condition disease incidence. The fungus is favored by warm, moist conditions necessary for sporulation, germination, and penetration, and *S. acicola* produces abundant conidia and secondary cycles in such conditions. Indirect evidence that climatic factors can limit disease incidence is suggested, in that in nature epidemics are common, but attempts to duplicate epidemics artificially often met with failure until recently (Kais, 1975). Favorable climatic conditions (abundant dew and rain) occur and epidemics develop when longleaf pine is in the seedling or so-called grass stage of development. As trees become older needles may be less succulent and, thereby, resistant, but more important the trees grow out of the grass stage and thus out of the favorable environment for disease development. Similarly, seedlings beneath a closed canopy of older trees may escape infection. Grass-stage seedlings are fire-resistant so that controlled burning is useful to destroy inoculum in diseased foliage on the ground and on the seedlings. This is an example of the successful control of a high-infection-rate disease via sanitation. Breeding for disease resistance and for early height growth (i.e., reducing the duration of the grass stage and the period during which trees remain in a climate favorable for rapid inoculum increase) and use of uneven-age management to limit moisture on understory seedlings provide opportunities for silvicultural control of brown spot needle blight.

B. Southern Fusiform Rust

This disease, a rarity in natural stands before 1900, is now epidemic, even pandemic, in slash and loblolly pine plantations throughout much of the southern pine forests (Squillace 1976; Griggs and Schmidt, 1977). The incidence and distribution of fusiform rust has increased dramatically over the last 30 years, coincident with intensive forest management of southern pine (Dinus, 1974). While these planted pine forests are far from natural, both the hosts and the pathogen (*Cronartium fusiforme*) are indigenous species. Thus fusiform rust provides an opportunity to consider pandemics aside from those caused by introduced pathogens.

As discussed earlier, slash pine grows as a natural monoculture in some wet, poorly drained, coastal flatwoods or is mixed with a few other hardwood and conifer species on other sites. Slash pine is an intermediately shade-tolerant, fire-climax species which is replaced by oak and other hardwood species in the absence of fire on better drained sites. Monoculture in itself cannot be singularly blamed for the present pandemic, because slash pine grows naturally in pure stands and the pathogen, which requires oak for the uredial and telial stage, does not spread from pine to pine. What then is responsible for the existing and steadily worsening pandemic?

The pathogen is an obligate parasite, can penetrate intact, succulent tissues, and has tremendous inoculum potential, both in numbers of spores and in pathogenic variability (Snow *et al.*, 1975, 1976). The pathogen has specific climatic requirements and is favored by high relative humidity and abundant moisture on leaf surfaces. However, neither climate nor pine genotypes, both of which appear favorable to disease development over much of the range of slash pine (Hollis and Schmidt, 1977; Goddard and Schmidt, 1975), are likely to have changed significantly in this respect over the preceding 30 years. What then is different in the contemporary forest ecosystem?

Several important aspects have changed along with intensive forest management and appear to be involved in the increase in rust incidence: (1) Better growth of pine conditioned by intensive site preparation, fertilization, and genetic selection for rapid growth has provided more succulent and susceptible tissue; (2) The age distribution was changed from one that was predominantly old trees to one that is now predominantly young trees (I would estimate that in 1900 age frequency was normally distributed with an average of 35 years; now the distribution is severely skewed in favor of young plantations and average age is perhaps 8–10 years); the latter provides a more favorable microclimate and abundant rapidly growing succulent tissue; (3) In the natural monoculture there were age differences within stands and thereby structural variation which created less uniformly susceptible tissue and a less favorable microclimate for disease development (Rowan et al., 1975); (4) The range of slash pine has been extended by planting to drier sites where oaks abound or soon appear by natural succession; (5) Dense plantings on wet sites have significantly reduced soil moisture which favors oak colonization; (6) Both wildfire and control burnings are less in evidence and oaks reinvade pine sites; (7) Diseased pine were transplanted from nurseries onto sites where oaks were present or subsequently invaded.

The diversity of age, vertical structure, and microclimate are minimized in the South's Third Forest (the planted forest). Ironically, increased species diversity, i.e., increase in oak abundance also has favored the fusiform rust pandemic.

C. Oak Wilt

This vascular wilt disease caused by *Ceratocystis fagacearum* was first reported in 1942. At first the disease increased and spread rapidly (True *et al.*, 1960) and there was concern that oak wilt would become another chestnut blight. Although the geographic range of the disease continues to expand and local epidemics have occurred, the potential threat has not materialized and no widespread epidemic has developed. In addition to the influence of control measures against local (root graft) and long-distance (sporulating mat production) spread (Merrill, 1967), several natural conditions appear to have limited disease development. (1) Spread by root grafts between oaks is limited in the eastern forests by the diversity of species which grow adjacent to oak wilt infested trees. (2) Long-distance spread is limited by several factors: (a) fresh wounds into the xylem are required as infection courts; wounds made in the spring of the year when large springwood vessels are close to the surface and active are more susceptible than wounds made late in the summer; (b) neither asexual or sexual spores are wind disseminated; both spores are transported by insect vectors; primarily sap feeding beetles and bark beetles, both of which are apparently neither numerous nor efficient; (c) these insects feed on species other than oak which serves to further dilute inoculum; (d) the fungus is self-sterile and sexual sporulation results from cross-fertilization of the two mating types by insects; (e) sporulation for long-distance spread is limited by wood moisture content such that mats which produce asexual and sexual spores and rupture the bark via a unique pressure pad are not produced on all infected trees; in fact, infected white oaks do not produce sporulating mats; and (f) there are no secondary cycles except for trees infected very early in the spring; there is only one cycle per year for the majority of infected trees; the fungus has a rather long latent period and a short infectious period since mats produce spores for only a short period of time. All of the above factors tend to limit the inoculum potential of the fungus despite the fact that single spores can induce wilt and rapidly kill large oaks.

The oak wilt fungus is not competitive as a saprophyte. In fact, its inability to compete saprophytically is a limiting factor for the spread of *C. fagacearum* to the southern United States where susceptible oaks are abundant. Here a competitive saprophyte, *Hypoxylon punctulatum,* with a relatively high temperature optimum colonizes diseased trees and prevents sporulation of the oak wilt fungus by rapidly reducing substrate moisture critical for mat formation (Tainter and Gubler, 1973). This is one of several excellent examples in the forest ecosystem where biological diversity impedes a pathogen and disease development.

D. Dieback-Decline Diseases

Most, if not all, tree species are susceptible to dieback-decline diseases. These diseases have a diverse but similar symptomology and etiology. Symptoms reflect a gradual deterioration over a period of years and the etiology, which differs with each disease, illustrates the concepts of predisposition, opportunism and succession of pest organisms. Dieback-decline diseases such as birch dieback, oak mortality, ash dieback, sweetgum blight, and maple blight have been reported as epidemic in local (although not necessarily small) areas. The etiology is such that trees subject to a stress factor, often transient, are predisposed to attack by ubiquitous but normally innocuous facultative parasites or other secondary organisms, e.g., insects. The transient stress factor can be climatic, edaphic, or biotic. If this stress is alleviated before secondary organisms become established, the trees can return to a healthy condition; otherwise mortality results.

Maple blight, a typical dieback-decline disease, occurred in uneven-aged, mixed hardwood stands in the Lake States about 1950. The following sequence of events was associated with decline of residual maple subsequent to selective tree harvesting (Giese *et al.*, 1964): (1) large trees were harvested opening the stand to increased solar radiation; (2) higher soil temperatures and lower soil moisture resulted, but most important was an increase in light; (3) increased light favored several defoliating insects and their populations increased significantly; (4) maples were repeatedly defoliated for several years and decline began; (5) trees thus weakened were attacked and killed by the ubiquitous, opportunistic root pathogen *Armillaria mellea*. This general process is so natural as to appear a part of natural succession and suggests a functional role for forest pests.

E. Pitch Canker of Slash Pine

While pitch canker caused by *Fusarium lateritium* f. *pini* (recently identified as *F. moniliforme* var. *subglutanins*) is not a dieback-decline disease, similarities exist. Pitch canker was reported in 38 counties in Florida in 1953 (Hepting and Roth, 1953) and since then has been of sporadic occurrence in slash pine within the state. At the time of this writing, however, a sudden and severe epidemic has developed over several thousand acres of natural and planted slash pine in Central Florida. One hypothesis (Schmidt *et al.*, 1976) suggests that the epidemic, which appears worse in planted forests, is the result of dense, overstocked stands being subjected to drought at a critical stage in stand development; that is, when roots fully occupy the site and water requirements of the stand are maximum. Trees weakened by drought conditioned an increase in the population of *Pissodes* weevils coincident with the pitch canker epidemic. Although the relationship and sequence of causal agents is yet to be determined, trees so weakened succumbed to pitch canker, a disease which heretofore caused limited mortality in slash pine (Schmidt and Underhill, 1974).

Perturbations are to be expected in forest ecosystems and in this case drought, not necessarily atypical during one rotation interval, coincided with increased demand for water by pole-sized trees and thereby predisposed the trees to disease. One is reminded of the words of admonishment (concerning the neglect of timely thinnings in plantations) penned several decades ago by Boyce (1954) who wrote that "Vigorous early growth is no assurance of satisfactory long time development. The critical period for most stands, varying with species, is from approximately 20 to 40 years of age, that is, pole stage when stands make the greatest demand on the site." In this regard the carrying capacity of sites planted to trees is a subject about which we have much to learn. In this case it would have been safer to base planting densities on years with deficit rainfall rather than on 20 or 30-year averages or else to resort to shorter rotations.

F. Chestnut Blight

This canker disease of American chestnut (*Castanea dentata*), caused by an introduced pathogen (*Endothia parasitica*), virtually eliminated chestnut from the forests of the eastern United States. It is not my intent to dwell on introduced pathogens, but chestnut blight provides an opportunity to emphasize an important point; namely, that diversity, unless it is functional, is useless against epidemics.

It is difficult to imagine a more diverse ecosystem than that in which the chestnut occurred. This mixed-species forest was uneven-aged, naturally regenerated, and "selectively" cut. Climatic, edaphic, and biotic conditions were extremely diverse as previously described. In spite of this diverstiy the pandemic ran its course in about 20 to 30 years until virtually every American chestnut was attacked and killed at least to the root collar. The pathogen does not kill the root system and the species has survived primarily by remnant sprouts which, of course, are susceptible genotypes.

There was no resistance to stem infection by *E. parasitica* in the American chestnut. Neither was the chestnut tolerant and a single basal stem canker rapidly girdled and killed even the largest and most vigorous trees. Notwithstanding this extreme vulnerability, which led to the wholesale destruction of the chestnut, the reason for the rapid increase and spread of the pathogen was not the lack of genetic disease resistance to an introduced pathogen only. Climatic conditions were favorable for the pathogen and, curiously, infection sites (wounds) were not limiting. Paramount, however, was the tremendous inoculum potential of the fungus, which produced both sexual and asexual spores in abundance, most any time of year, on both living and dead trees (Boyce, 1948). The latent period was relatively short and the infectious period relatively long. Dispersal of ascospores was by wind and dispersal of pycnidiospores was by rain, insects, and other animals. Further, *E. parasitica* colonized the chestnut without significant competition from native saprophytes to which the chestnut was quite resistant.

The attempts to control the disease by sanitation were entirely futile as Stewart correctly said at the time (Hepting, 1964). The rate of disease increase and distance of spread were so great that effective sanitation was both theoretically and practically impossible. Sanitation was made more impractical because the pathogen attacked other species within these forests. The lesson is clear. Only functional diversity can mitigate epidemics in forest ecosystems.

G. Other Examples

A few other examples of limiting and predisposing factors are worthy of mention. Here it is important to note again that many native forest tree pathogens need wounds (some need fresh wounds) to gain entrance into and successfully colonize their host. For example, epidemics caused by some canker fungi and many sapwood and heartwood decay fungi are in part limited by the availability of infection courts. Some pathogens only sporulate on dead trees; others do not produce secondary cycles or the occurrence or role of imperfect spores is unknown. In the case of *Fomes annosus,* antagonistic (Nissen, 1956) and competitive saprophytes such as *Peniophora gigantea* (Rishbeth, 1952) can provide effective control of stump top colonization. Climatic limitations associated with stand structure inhibit development of blister rust on understory white pine (Van Arsdel, 1962). Bier and Rowat (1962) suggested that bark saprophytes limit facultative parasites that cause cankers. Mycorrhizae protect seedlings from damping-off fungi and older trees from feeder root pathogens (Marx and Davey, 1969).

IV. FUNCTIONAL DIVERSITY AND DISEASE MANAGEMENT IN FOREST ECOSYSTEMS

A. Components of Functional Diversity

The previous sections have attempted to show that forest ecosystems are multidimensional mosaics of diversity, some of which contribute to an innate suppression of epidemics. A more complete tabulation of this diversity and its possible effect on disease development is presented in Table III, and a listing follows.

Host factors which condition ecosystem resistance include: (1) genetic disease resistance such as immunity (nonhost), and specific and general disease resistance; (2) limited infection sites as many tree pathogens require wounds; (3) numbers of tree species and associations or mixtures of species; (4) physiological age; (5) horizontal and vertical community structure; and (6) natural succession of species.

Environmental factors which affect ecosystem resistance include: (1) microclimate resulting from the interaction of horizontal and vertical community structure, aspect, slope, and elevation with the macroclimate;

TABLE III

Components of Functional Diversity Which Influence Disease
in Natural Forest Ecosystems

Components of functional diversity	Effect on disease development
Host components	
Immunity	Inoculum has zero potential on nonhost species
Specific genetic resistance	Some genotypes of the pathogen have zero potential on resistant genotypes of a host species
General genetic resistance	May not limit initial amount of disease, but may limit colonization or production of inoculum and therein subsequent amount of disease
Tolerance	Does not limit amount of disease, but when amount of disease is not directly related to the amount of inoculum then a tolerant species can conceivably reduce amount of inoculum, i.e., tolerant species may also possess general disease resistance
Limited infection sites	Limits the amount of susceptible tissue similar to the correction factor $(1 - x)$ when x is near 1; however, this limitation is present throughout the epidemic
Physiological age	Limits amount and distribution of susceptible tissue and subsequent colonization and disease development
Numbers of species	Reduces the probability that inoculum will contact a susceptible host
Numbers of host associations	Reduces the probability that inoculum will contact a susceptible host
Horizontal dispersion of the host plants	Reduces the probability that inoculum will contact a susceptible host
Vertical dispersion of the host plants	Reduces the probability that inoculum will contact a susceptible host
Natural succession	Restricts the increase of pathogen populations by changing the host species
Environmental components	
Climatic factors such as temperature, moisture, etc.	May limit the amount of infection and conditions latent and infectious periods
Edaphic factors such as nutrients, moisture, etc.	May limit the amount of infection and subsequent rate of disease development
Biotic factors including Mycorrhizae	Protects susceptible species by limiting the amount of susceptible tissue
Competitive saprophytes, antagonistic saprophytes, hyperparasites, and succession of organisms	Limits amount of susceptible tissue and restricts vegetative and reproductive growth of pathogens

TABLE III (continued)

Components of functional diversity	Effect on disease development
Pathogen components	
Latent period	Long latent periods limit amount of inoculum
Infectious period	Short infectious periods limit amount of inoculum (usually not a limiting factor)
Genetic potential (for infection)	Limits number of successful infections as indicated for specific genetic resistance of the host
Requirement for a wound as an infection site	Limits amount of susceptible tissue
Dependence on insects for production, release, dissemination, or deposition of inoculum	Limits amount of susceptible tissue and probability that inoculum will reach the infection site if those insects are few or inefficient

(2) physical and chemical soil properties, especially soil moisture; and (3) mycorrhizae, competitive saprophytes, antagonistic organisms, hyperparasites, competitive insects, and the natural succession of organisms.

Characteristics of forest tree pathogens which condition ecosystem resistance include: (1) amount of inoculum including latent and infectious periods; (2) energy and genetic potential of inoculum; (3) requirement for wounds as infection sites; and (4) dependency on insects for spermatization, release, dissemination, and inoculation.

B. Aspects of Disease Management

1. Some Epidemiological Concepts

As pointed out previously it can be folly to generalize too much about forest tree diseases. Nevertheless, the following generalizations, far from complete, seem valid in many instances and are critical to the understanding of disease incidence relative to indigenous pathogens in natural forest ecosystems.

1. Epidemics are not infrequent in natural forest ecosystems but are limited in time and space.

2. Epidemics in natural forest ecosystems are limited by functional diversity of host, pathogen, and climatic, edaphic, and biotic environmental parameters.

3. Forest tree pathogens are ubiquitous, natural components of the forest and some play a significant regulatory role in the forest ecosystem.

4. Both host and pathogen populations interact within a dynamic natural succession that transcends time and space.

5. A delicate balance often exists between synthetic (growth) and destruction (disease) processes in natural forest ecosystems. These systems are subject to natural and man-induced perturbations which can shift the balance in favor of growth or disease.

2. Disease Management

Prevention of disease loss by silvicultural manipulation is the cornerstone of forest protection. With the notable exceptions of nurseries and seed orchards, forest managers are precluded by economic (and, more recently, societal) constraints against the use of chemicals and other direct methods of disease control.

Avoidance of disease is perhaps the most important to the forest manager and is utilized by (1) planting nonhost species; (2) preventing predisposing factors, especially wounds; (3) avoiding high-disease-hazard sites (forecasting); (4) stand manipulation to avoid microclimatic conditions favorable for the pathogen and disease development; (5) encouraging competitive saprophytes and mycorrhizal fungi; and (6) regulating rotation age to avoid disease of mature or overmature trees.

Reduction of inoculum by sanitation practices including control burning, cutting or removing diseased trees, and eradication of alternate hosts also provides common silvicultural controls. But, as van der Plank (1963) has suggested, sanitation is least helpful when most needed, i.e., against diseases with high infection rates. Nevertheless, sanitation has been useful against some forest tree diseases, e.g., larch canker, oak wilt, and Dutch elm disease. In fact, forest tree diseases capable of rapid increase can be controlled in some instances by sanitation either because the sanitation ratio (proportion of inoculum removed) is high or because the infection rate is low on some sites due to environmental conditions which are unfavorable for the rapid increase of disease, e.g., brown spot of long-leaf pine and white pine blister rust.

Finally, genetic disease resistance is the great hope of forest managers —if it is used wisely. If genetic diversity is discarded and a "supertree mentality" prevails, the results of selecting and breeding for disease resistance could be disastrous. If, as many plant pathologists believe, genetic homogeneity for disease resistance is hazardous for annual crops it has little to offer forest tree crops which require 25 to more than 100 years to mature. Used wisely so that genetic diversity is maintained and employed, resistance is a valuable tool for disease management in the forest. If, on the other hand, genes for disease resistance are misused, the results are likely to be disastrous. For example, the half-sib seed orchards and seed production areas of southern pines offer abundant opportunities for the maintenance of diverse resistant genotypes as a buffer against directed selection for pathogen virulence. But mass clonal production of a few resistant genotypes offers no such buffer against pathogenic variability. Theoretically a solution would be to couple many resistant genotypes with other kinds of functional diversity.

3. Comparative Epidemiology

However risky, I cannot forgo this opportunity to consider a few aspects of the comparative epidemiology of disease in forest ecosystems. From the start the problem of over-generalizing "rears its ugly head," but again goes unheeded. Such theorizing, right or wrong, has utility in the search for realistic models of increase and spread of forest tree diseases. To begin with, there are two levels of disease development to consider. The first is that within a focus which is relatively homogeneous, i.e., small areas within a natural forest or planted stand which lack the influences of functional diversity. The second level is the natural forest or even planted stands in which disease increase and spread are influenced by functional diversity.

In the first instance we can make use of and comparisons with familiar differential equations (van der Plank, 1963) for diseases of annual crops. Consider, for example, the model

$$\frac{dx_t}{dt} = R_c(x_{t-p} - x_{t-i-p})(1 - x_t)$$

where R_c is the basic infection rate of disease (x) increase with time (t) corrected for removals; $(x_{t-p} - x_{t-i-p})$ is the amount of infectious tissue; p is the latent period; i is the infectious period; and $(1 - x_t)$ is the amount of healthy tissue or the correction factor. Surely, within limited foci, the increase of local lesion foliar diseases of trees would be adequately described by this equation. For other types of forest tree diseases, those in which inoculum is nearly continuously available or those which have limited infection sites unrelated to plant growth, e.g., wounds, the role of p, i, and $1 - x$ would appear to be of less importance than is indicated in the above model. These types of forest tree diseases increase in proportion to the available infection sites, but it is as though a correction factor near zero was in operation throughout the epidemic; this in spite of continuously available inoculum. Of course, fluctuations in available infection sites, which, if unrelated to plant growth, are often discontinuous functions (in opposition to existing models), condition disease increase. Other forest tree diseases offer similar interesting and disparate comparisons.

If we consider the second level of interaction, where diversity plays a larger role, comparisons of disease spread are most important. For example, existing models of inoculum and disease gradients (Schrodter, 1960; Gregory, 1968) are primarily of unrestricted spread and are not meant to apply to forest ecosystems. In spite of abundant and sometimes continuous inoculum the probability of a spore reaching an infection site in the forest is diminished, not only by distance but also by intervention of non-host species, dispersion of host species, and limited infection sites. Fortunately some information on the spread of specific forest tree disease is available, e.g., white pine blister rust, Dutch elm disease, oak wilt, annosus root rot, dwarf mistletoe, etc.

It is of interest here to consider the threshold theorem (van der Plank, 1963) where $iR_c = I$. The quantity iR_c is the average number of daughter or progeny lesions produced by one parent lesion when there are unlimited infection sites. According to this theorem, I must be greater than 1 for disease incidence to increase; the larger I the greater the increase. Endemic forest tree diseases have I values around 1, sometimes greater and sometimes less than 1. Van der Plank (1975) used this theorem to indicate that for a population of plants to have a limited amount of disease it is not necessary for individual plants to be resistant but only that removals occur rapidly, i.e., a short infectious period (i). Thus, if i is of very limited duration, I is small regardless of the magnitude of R_c. This argument is really the essence of the role of functional diversity in the forest ecosystem. That is, in order that forests have limited disease it is not necessary for individual trees to be free of disease, but rather that these infected trees not result in the infection of many others. So, in concert the community functions to limit disease incidence. The other side of this coin is that forests stripped of population resistance must depend on the resistance of the individual.

In spite of the fact that models which relate yield loss to disease incidence were pioneered by forest pathologists studying stem decay, there is a paucity of such models for most forest tree diseases. Because tolerance of disease is a common trait of trees, models of disease incidence will be of limited value for disease management until companion yield-loss models are available. Unfortunately, loss data are difficult and costly to obtain in forest ecosystems. Nevertheless, the concept of tolerance index (Calpouzos et al., 1976) appropriately applied could aid forest managers who require disease loss models.

4. Summary

Natural forest ecosystems abound with diversity, some of which functions to stabilize the incidence of forest tree diseases and thereby mitigates epidemics. Forests without functional diversity are in the long run bad risks and increased disease incidence and loss are only a matter of time. If existing functional diversity is discarded for short-term gains other stabilizing factors must be substituted, or new management strategies devised to prevent epidemics and undue disease losses.

In the long term, disease losses, at present and especially in the future, in the intensively managed forests may depend, in part, on how well forest management understands predisposing factors and capitalizes on functional diversity. In the sense that silvicultural practices remain the prime tools for prevention and control of disease losses in the forest, it will be "business as usual" for the forest manager. But if the concept of functional diversity can be applied in the context of an integrated pest management program (Waters and Cowling, 1976), then this is surely a new dimension and much needs to be done by research and management alike. The challenge is not to avoid forest management, but to determine if and how diversity can be used in concert with known entities to control disease losses in managed forests. Our natural forests are "banks" of diversity just as they are gene banks or gene pools. As such, these forests offer unique opportunities to study the role and utility of functional diversity in disease management. Opportunities for knowledge disappear as natural forest ecosystems diminish.

References

Bagga, D. K., and Smalley, E. B. (1974). Variation in *Hypoxylon pruinatum* in cultural morphology and virulence. *Phytopathology* 64, 663–667.

Baxter, D. V. (1937). Development and succession of forest fungi and diseases in plantations. *Univ. Mich. Sch. For. Conserv., Circ.* 1, 1–45.

Baxter, D. V. (1967). Disease in forest plantations: The thief of time. *Cranbrook Inst. Sci., Bull.* 51, 1–251.

Berger, R. D. (1977). Application of epidemiology principles to achieve plant disease control. *Ann. Rev. Pathol.* 15,165–183.

Bier, J. E., and Rowat, M. H. (1962). The relation of bark moisture to the development of canker diseases caused by native, facultative parasites. VIII. Ascospore infection of *Hypoxylon priunatum* (Klotzsche). Cke. *Can. J. Bot.* 40, 897–901.

Boyce, J. S. (1948). "Forest Pathology." McGraw-Hill, New York.

Boyce, J. S. (1954). Forest plantation protection against diseases and insect pests. *FAO For. Dev. Pap.* 3, 1–41.

Baun, E. L. (1950). "Deciduous Forests of Eastern North America." McGraw-Hill (Blakiston), New York.

Brown, T. S., Jr., and Merrill, W. (1973). Germination of basidiospores of *Fomes applanatus*. *Phytopathology* 63, 547–550.

Browning, J. A. (1974). Relevance of knowledge about natural ecosystems to development of pest management programs for agro-ecosystems. *Proc. Am. Phytopathol. Soc.* 1, 191–199.

Browning, J. A., Simons, M. D., and E. Torres. (1977). Managing host genes: Epidemiologic and genetic concepts. *In* Plant Disease: An Advanced Treatise (J. G. Horsfall and E. B. Cowling, eds.), Vol. 1, pp. 191–212. Academic Press, New York.

Calpouzos, L., Roelfs, A. P., Madson, M. E., Martin, F. B., Welsh, J. R., and Wilcoxson, R. D. (1976). A new model to measure yield losses caused by stem rust in spring wheat. *Univ. Minn., Agric. Exp. Stn., Tech. Bull.* **307**, 1–23.

Cowling, E. B., and Kelman, A. (1964). Influence of temperature on growth of *Fomes annosus* isolates. *Phytopathology* **54**, 373–378.

Dansereau, P. (1957). "Biogeography: An Ecological Perspective." Ronald Press, New York.

Dinus, R. J. (1974). Knowledge about natural ecosystems as a guide to disease control in managed forests. *Proc. Am. Phytopathol. Soc.* **1**, 184–190.

Edmonds, R. L., and Sollins, P. (1974). The impact of forest diseases on energy and nutrient cycling and succession in coniferous forest ecosystems. *Proc. Am. Phytopathol. Soc.* **1**, 175–180.

Edwards, C. A., Reichle, D. E., and Crossley, D. A., Jr. (1970). The role of soil invertebrates in turnover of organic matter and nutrients. *Ecol. Stud.* **1**, 147–172.

Geiger, R. (1966). "The Climate Near the Ground" (transl. from the 4th ed.). Harvard Univ. Press, Cambridge, Massachusetts.

Giese, R. L., Houston, D. R., Benjamin, D. M., Kuntz, J. E., Kapler, J. E., and Skilling, D. D. (1964). Studies of maple blight. *Univ. Wis., Res. Bull.* **250**, 1–128.

Goddard, R. E., and Schmidt, R. A. (1975). Effect of differential selection pressure on fusiform rust resistance in phenotypic selections of slash pine. *Phytopathology* **65**, 336–338.

Gregory, P. H. (1968). Interpreting plant disease dispersal gradients. *Annu. Rev. Phytopathol.* **6**, 189–212.

Griggs, M. M., and Schmidt, R. A. (1977). Increase and spread of fusiform rust. *In* Management of fusiform rust in southern pines. (R. J. Dinus and R. A. Schmidt, eds.), 32–38. Symp. Proc. Univ. Fla., Gainesville.

Hawksworth, F. G. (1961). Dwarfmistletoe of ponderosa pine in the Southwest. *U.S., Dep. Agric., For. Serv., Tech. Bull.* **1246**, 1–112.

Hepting, G. H. (1964). Climate and forest diseases. *Annu. Rev. Phytopathol.* **1**, 31–50.

Hepting, G. H. (1974). Death of the American chestnut. *J. Forest History* **18**,61–67.

Hepting, G. H., and Roth, E. R. (1953). Host relations and spread of the pine pitch canker disease. *Phytopathology* **43**, 475 (abstr.).

Hollis, Č. A., and Schmidt, R. A. (1977). Site factors related to fusiform rust incidence in North Florida slash pine plantations. *For. Sci.* **23**,69–77.

Kais, A. G. (1975). Environmental factors affecting brownspot infection on longleaf pine. *Phytopathology* **65**, 1389–1392.

Kranz, J. (1974). Comparison of epidemics. *Ann. Rev. Phytopathol.* **12**,355–374.

Large, E. C. (1945). Field trials of copper fungicides for the control of potato blight. I. Foliage protection and yield. *Ann. Appl. Biol.* **32**, 319–329.

MacArthur, R. H. (1955). Fluctuations of animal populations, and a measure of community stability. *Ecology* **36**, 533–536.

McDonald, G. I., and Hoff, R. J. (1971). Resistance to *Cronartium ribicola* in *Pinus monticola*: Genetic control of needle-spots only resistance factor. *Can. J. For. Res.* **1**, 197–202.

Marx, D. H., and Davey, C. B. (1969). The influence of ectotrophic mycorrhizal fungi on the resistance of pine roots to pathogenic infections. IV. Resistance of

314   ROBERT A. SCHMIDT

naturally occurring mycorrhizae to infection by *Phytophthora cinnamomi*. *Phytopathology* **59**, 559–565.

Mattson, W. J., and Addy, N. D. (1975). Phytophagous insects as regulators of forest primary production. *Science* **190**, 515–522.

May, R. M. (1973). "Stability and Complexity in Model Ecosystems," *Monogr. Popul. Biol. No. 6*. Princeton Univ. Press, Princeton, New Jersey.

Merrill, W. (1967). The oak wilt epidemics in Pennsylvania and West Virginia: An analysis. *Phytopathology* **57**, 1206–1210.

Merrill, W. (1970). Spore germination and host penetration by heart-rotting Hymenomycetes. *Annu. Rev. Phytopathol.* **7**, 281–300.

Nissen, T. V. (1956). Soil Actinomycetes antagonistic to *Polyporus annosus* Fr. *Friesia* **5**, 332–339.

Paine, R. L. (1968). Germination of *Polyporus betulinus* basidiospores on non-host species. *Phytopathology* **58**, 1062–1063 (abstr.).

Peace, T. R. (1957). Approach and perspective in forest pathology. *Forestry* **30**, 47–56.

Peace, T. R. (1961). The dangerous concept of the natural forest. *Emp. For. Rev.* **40**, 320–328.

Peace, T. R. (1962). "Pathology of Trees and Shrubs." Oxford Univ. Press (Clarendon), London and New York.

Pimentel, D. (1961). Species diversity and insect population outbreaks. *Ann. Entomol. Soc. Am.* **54**, 76–86.

Raabe, R. D. (1967). Variation in pathogenicity and virulence in *Armillaria mellea*. *Phytopathology* **57**, 73–75.

Reifsnyder, W. E., and Lull, H. W. (1965). Radiant energy in relation to forests. *U.S., Dep. Agric., For. Serv., Tech. Bull.* **1344**, 1–111.

Rishbeth, J. (1952). Control of *Fomes annosus* Fr. *Forestry* **25**, 41–50.

Rowan, S. J., McNab, W. H., and Brender, E. V. (1975). Pine overstory reduces fusiform rust incidence in underplanted loblolly pine. *U.S., For. Serv., Res. Note SE* **SE-212**, 1–6.

Schmidt, R. A. (1973). The role of epidemiology in graduate plant pathology. *Abstr. Pap., Int. Congr. Plant Pathol., 2nd, 1973* Abstract No. 989.

Schmidt, R. A., and Underhill, E. M. (1974). Incidence and impact of pitch canker in slash pine plantations in Florida. *Plant Dis. Rep.* **58**, 451–454.

Schmidt, R. A., and Wood, F. A. (1969). Temperature and relative humidity regimes in the pine stump habitat of *Fomes annosus*. *Can. J. Bot.* **47**, 141–154.

Schmidt, R. A., and Wood, F. A. (1972). Interpretation of microclimate data in relation to basidiospore release by *Fomes annosus*. *Phytopathology* **62**, 319–321.

Schmidt, R. A., Wilkinson, R. C., Moses, C. S., and Broerman, F. S. (1976). Drought and weevils associated with severe incidence of pitch canker in Volusia County, Florida. *Univ. Fla., Inst. Food Agric. Sci., Prog. Rep.* **76-2**, 1–4.

Schrodter, H. (1960). Dispersal by air and water—the flight and landing. *In* "Plant Pathology: An Advanced Treatise" (J. G. Horsfall and A. E. Dimond, eds.), Vol. 3, pp. 169–227 Academic Press, New York.

Shaw, C. G., III, and Roth, L. F. (1976). Persistence and distribution of a clone of *Armillaria mellea* in a ponderosa pine forest. *Phytopathology* **66**, 1210–1213.

Shigo, A. L. (1967). Succession of organisms in discoloration and decay of wood. *Int. Rev. For. Res.* **2**, 237–299.

Snow, G. A., Dinus, R. J., and Kais, A. G. (1975). Variation in pathogenicity of diverse sources of *Cronartium fusiforme* on selected slash pine families. *Phytopathology* **65**, 170–175.

Snow, G. A., Dinus, R. J., and Walkinshaw, C. H. (1976). Increase in virulence of *Cronartium fusiforme* on resistant slash pine. *Phytopathology* **66**, 511–513.

Society of American Foresters. (1954). "Forest Cover Types of North America (Exclusive of Mexico)." *Soc. Am. For.*, Washington, D.C.

Soil Survey Staff. (1975). Soil taxonomy. *U.S., Dep. Agric., Agric. Handb.* **436**, 1–754.

Squillace, A. E. (1976). Geographic patterns of fusiform rust infection in loblolly and slash pine plantations. *U.S. Dept. Agric. Forest Serv. Res. Note* **SE-232**, 1–4.

Steubing, L. (1970). Soil flora: Studies of the number and activity of microorganisms in woodland soils. *Ecol. Stud.* **1**, 131–146.

Tainter, F. H., and Gubler, W. D. (1973). Natural biological control of oak wilt in Arkansas. *Phytopathology* **63**, 1027–1034.

Trewartha, G. T. (1954). "An Introduction to Climate." McGraw-Hill, New York.

True, R. P., Barnett, H. L., Dorsey, C. K., and Leach, J. G. (1960). Oak wilt in West Virginia. *W. Va., Agric. Exp. Stn., Bull.* **448T**, 1–119.

U.S. Department of Agriculture. (1960). Plant pests of importance to North American agriculture. Index of plant diseases in the United States. *U.S., Dep. Agric., Agric. Handb.* **165**, 1–531.

U.S. Department of Agriculture. (1973). Silvicultural systems for major forest types of the United States. *U.S., Dep. Agric., For. Serv. Agric. Handb.* **445**, 1–114.

Van Arsdel, E. P. (1961). Growing white pine in the Lake States to avoid blister rust. *U.S., Dep. Agric., For. Serv. Lake States For. Exp. Stn., Pap.* **92**, 1–11.

Van Arsdel, E. P. (1962). Some forest overstory effects on microclimate and related white pine blister rust spread. *U.S., Dept. Agric. Lake States For. Exp. Stn., Tech. Note* **627**, 1–2.

van der Plank, J. E. (1963). "Plant Diseases: Epidemics and Control." Academic Press, New York.

van der Plank, J. E. (1968). "Disease Resistance in Plants." Academic Press, New York.

van der Plank, J. E. (1975). "Principles of Plant Infection." Academic Press, New York.

Waters, W. E., and Cowling, E. B. (1976). Integrated forest pest management: A silvicultural necessity. *In* "Integrated Pest Management" (J. L. Apple and R. F. Smith, eds.), pp. 149–177. Plenum, New York.

Watt, K. E. F. (1968). "Ecology and Resource Management." McGraw-Hill, New York.

White, J. H. (1919). On the biology of *Fomes applanatus. Trans. R. Can. Inst.* **14**, 133–174.

Zadoks, J. C. (1974). The role of epidemiology in modern phytopathology. *Phytopathology* **64**, 918–923.

Zadoks, J. C., and Koster, L. M. (1976). A historical survey of botanical epidemiology: A sketch of the development of ideas in ecological phytopathology. *Meded. Landbouwhogeschool Wageningen* 76-12, 1–56.

Chapter 15

Climatic and Weather Influences on Epidemics

JOSEPH ROTEM

I. INTRODUCTION

"Epidemic" denotes a population of diseased individuals. Atmospheric factors affect an epidemic through their effects on the various phases of the pathogen's life cycle as they interact with specific responses of individual plants. In discussing atmospheric effects on epidemics, we shall summarize, integrate, and interpret the extent of atmospheric effects on individual plants and pathogens and thereby derive conclusions describing the epidemic as a unit.

Both climate and weather are considered in this chapter. "Climate" describes the average patterns of atmospheric conditions (temperature, precipitation, etc.) typical for a given location; "weather" represents the actual atmospheric conditions prevailing at a given site and time. Heptig (1963) has described the effects of long-term climatic change on diseases of perennial plants.

Soil-borne pathogens spread relatively slowly in a habitat which is relatively well buffered against sudden changes in weather. Fortunately for man, epidemics caused by soil-borne pathogens are relatively rare

317

events in the sense of an explosive increase of their population that causes devastating losses. In the soil environment, an epidemic results from a relatively slow and prolonged buildup of the inoculum and the final presence of a large pathogen population. These phenomena often take years and are therefore little influenced by short-term weather fluctuations (H. R. Wallace, personal communication). In such cases, climate, rather than weather, may determine which areas are more subject than others to epidemics caused by soil-borne pathogens. This subject is discussed more fully in Chapter 16, this volume.

By contrast, the habitat of pathogens, which invade the aerial parts of plants, is immediately and profoundly influenced by weather. These pathogens usually reproduce abundantly and with the onset of favorable conditions spread rapidly from a minimum amount of initial inoculum. The resulting epidemics are often sporadic and of relatively short duration, but devastating. Their frequency is determined by the occurrence of suitable combinations of weather rather than by climatic factors.

Atmospheric parameters usually are expressed as daily, monthly, or seasonal means. However, the mean values are only a mathematical fiction and disease development actually depends upon the frequency of occurrence of favorable and unfavorable conditions (Schröder, 1963). The analysis and ranking of these conditions is often difficult. Substantial differences often occur in the limits of favorable and unfavorable conditions for the recurrent and often overlapping phenomena of sporulation, dispersal, infection, etc. In fact, epidemics can and do develop under a wide range of weather parameters until one of these exceeds a submarginal level, as discussed in Section V.

There are other complicating effects. Disease is affected by the microclimatic rather than macroclimatic values measured in a standard meteorological station. In formal meteorological terms, conditions present up to 2 m above ground level are the microclimatic conditions. Phytopathologists, however, refer to the microclimate as conditions within the plant canopy, whether composed of grass or tall trees.

Only a few out of thousands of relevant publications are cited in this chapter. The references provide us with specific examples, but scientists often do not claim that the described phenomenon occurs beyond the specific cases studied. Based on these references, my conclusions and hypotheses may be too speculative. The intention is to show that an epidemic results from a combination of many factors interacting with each other, compensating for limitations in each other, and attaining a different level of importance in different locations, seasons, and climatic zones.

II. INFLUENCE OF WEATHER FACTORS ON AIRBORNE PATHOGENS

A. The Effects of All Factors on Survival

The ability of a pathogen to survive during periods of adverse conditions enables it to carry over from one season to another (overwintering and/or oversummering). This ability influences the transfer of a pathogen from one area to another (dispersal) and from one part of a given field to another. Survival is also important over the relatively short time of sporulation until the time of infection.

Spores or resting structures of fungi survive better at low temperature. Spores of many fungi survive longer at low relative humidity. There are some notable exceptions to this generalization, among some (but not all) *Phycomycetes*, powdery mildews and the basidiospores of rust fungi and some other species which are dispersed mainly at night (Cohen and Rotem, 1971; Hsieh and Buddenhagen, 1975; Rotem, 1968). The ability to survive is adversely affected by solar radiation (Hsieh and Buddenhagen, 1975; Visser *et al.*, 1961). Spores still attached to lesions appear to be more resistant than dispersed spores (Hunter and Kunimoto, 1974; Rotem and Cohen, 1974). However, survivability is basically a genetic feature of any given species and ranges from several minutes, as with *Phytophthora palmivora* (Hunter and Kunimoto, 1974), to years, as with *Alternaria porri* f. sp. *solani* (Rotem, 1968).

From the epidemiological point of view, survival is most important when adverse conditions exist between two growing seasons or between the time of sporulation and of infection. Thus, survival is more important in hot than in cool areas or seasons. A study of the role of survivability in late blight epidemics (sporangia sensitive to heat and desiccation) in different seasons in Israel revealed that (1) in the hot season, survivability of dispersed sporangia is a key factor conditioning the extent of epidemics in both the humid and the arid parts of the country; (2) in the cool season, survivability remains a major factor in the arid zone, but plays a minor role in epidemics developing in the humid region (Rotem *et al.*, 1970).

Under all conditions, the factor of survivability is relatively unimportant in epidemics involving species with resistant spores. (See Section II,C for a discussion of the critical role of survival in the ability of pathogens to benefit from intermittent wet periods and dewfall, and in the adaptation of pathogens to areas with adverse climatic conditions.)

B. The Effects of Temperature, Radiation, and Light of Epidemic Development

Although critical, temperature is less often a limiting factor than mois-ture, because the effective range of the former is wider. In addition, the minimum and maximum temperature for the development of the patho-gen usually does not cause its death. Consequently, unfavorable tempera-tures in the field often inhibit an epidemic temporarily, but do not eradicate the pathogen, unless they last for prolonged periods. In temperate zones or in cool seasons, inhibition results from the low temperatures which sometimes prevail at night and reduce the rate of infection and sporulation processes. If the low night temperatures ex-tend over many nights, the rate of the epidemic's development is reduced via a decrease in the number of new infection sites and lack of inoculum, rather than by an effect on the pathogen already present in the infected tissue; the latter continues to develop slowly during the usually warmer daytime hours. This slow development in lesions preserves the sporulat-ing potential of the pathogen, which is thus able to produce large quan-tities of inoculum when temperature conditions improve.

In hot zones, inhibition of epidemics is due to the high temperatures which prevail during the daytime. Under these conditions, infection and sporulation are not affected, because the lower night temperatures are often in the optimum range for these processes. High day temperatures endanger viability of dispersed spores, speed up lesion development or necrotic reactions of the host (the optimum temperature for which is usually higher than that for colonization), and thus decrease the sporu-lating potential of obligate parasites. High temperatures may also shorten the sporulating period of facultative parasites even though these are not affected adversely by necrosis. Since evapotranspiration is increased in general, and in infected leaves in particular, high temperatures can speed up the destruction of infected leaves in spite of inhibited development of the pathogen. In extreme cases high temperatures can eradicate patho-gens present in leaves exposed to sun since, because of the effects of radiation, the temperatures are higher there than in the air. Indeed, intense solar radiation tends to inhibit epidemics, as found with late blight on potatoes (Rotem *et al.*, 1970), and with *Exobasidium vexans* in tea (Visser *et al.*, 1961).

The situations described above are extreme. In many cases the adverse effects of high temperatures are restricted to only several hours per day. My own observations on the effects of hot spells (midday maximum of up to 44°C) on potato late blight and tomato *Stemphylium* blight showed that, although temporarily inhibited, the disease was not eradicated and its activity was renewed when the weather returned to "normal." In addition to the relatively short duration of the extreme temperatures each day, the noneradication of the disease was also due to the much lower temperatures prevailing inside the canopy of dense stands and crops.

The indirect effect of light intensity and photoperiod (acting via the host's susceptibility) has been proved on several occasions (Colhoun, 1973) but needs further clarification. Some experiments were carried out in growth chambers, where the light intensities were lower than in the field. By contrast, in field studies researchers were not able to separate the effects of the varying conditions of light from many associated factors, such as temperature and humidity.

A direct influence of light on infection is mentioned in the literature, but with some exceptions it seems to be a rather minor factor. By contrast, through its effects on photosynthesis by the host, light stimulates sporulation of all obligate parasites (Cohen and Rotem, 1970), and, through a phenomenon of induction, that of some facultative parasites (Bashi and Rotem, 1976). Under most environmental conditions, however, light intensity in the field is not the factor limiting sporulation. More important is the fact that in most species the final stage in sporogenesis is inhibited by light in conjunction with high (but not low) temperature (Bashi and Rotem, 1975b; Lukens, 1966). For this reason spores are usually formed at night in warm localities. The fact that light in conjunction with low temperature does not inhibit sporulation explains why epidemics of some pathogens develop at high latitudes despite the short summer nights.

C. The Comparative Effects of Moisture Sources on Epidemic Development

Except for powdery mildews and some "wound" pathogens, a film of free moisture on the surface of leaves is essential for infection by most pathogens. Some pathogens, like *Phytophthora palmivora* on papaya, penetrate after a wetting period as brief as 15 min (Hunter and Kunimoto, 1974), and some, like *Stemphylium botryosum* f. sp. *lycopersici* on tomatoes, need to have free water present for 24 hr or more. The latter fungus also can cause infection after several short wet periods interrupted by dry intervals (Bashi and Rotem, 1974; Bashi et al., 1973). Although sporulation of some species occurs at a relative humidity close to 100% (with the exception of powdery mildews, rusts, and some minor pathogens), the most abundant sporulation usually is induced by free water on the leaf surface. In general, sporulation requires longer wet periods than infection, but some fungi sporulate after several short wet periods interrupted by dry intervals (Bashi and Rotem, 1975a; Nelson and Tung, 1973; Rotem et al., 1976).

The three major sources of moisture for plant disease epidemics are rain, dew, and overhead irrigation (sprinkling). Epidemics of some pathogens have been attributed to high relative humidity. But it is not clear whether high humidity affected these epidemics directly or only by indicating conditions of condensation (such as dew formation) which lead to the development of an epidemic. Fog is an important factor at sites where it occurs frequently.

Rain is a macroclimatic factor which, in contrast to sprinkling, affects large areas and occurs, in contrast to dew, also by day. The literature mentions epidemics affected by hurricanes and epidemics associated with the total amount of rainfall. Very few published reports indicate the effects of intensity or seasonal distribution of rain on plant disease epidemics. The apparent preference of certain pathogens for different patterns of rainfall seems to be connected with the specific characteristics of a given pathogen and climatic region. For instance, in dewless areas, foliage epidemics are expected to be affected by frequent and evenly distributed rains rather than by occasional heavy but short periods of rain. Under different conditions, however, heavy and short rains may be more effective than prolonged drizzles because the former are more efficient in washing off any applied fungicides and in splash dispersal of the spores. Pathogens not adapted to dispersal by wind may benefit more from heavy showers than those which are easily dispersed by wind. Pathogens which require prolonged moist periods for infection, or those whose spores are very sensitive to dryness, will benefit from drizzles. The following examples illustrate preferences for a specific rain regime due to specific characteristics of the pathogen or host. *Spongospora subterranea* needs a prolonged moisture period to become established in potatoes. Therefore, rainfall evenly distributed over the whole season in England has little effect on potato scab, but several days of continuous rain in an otherwise dry season can lead to heavy damage (Hims and Preece, 1975). In the case of sunflower downy mildew, rain is critical during the first fortnight of growth, because only then are the seedlings susceptible to systemic infection (Zimmer, 1975). Due to a requirement for prolonged periods of moisture, morning or evening rains which extend the periods of leaf wetness caused by dew are critical for epidemics of *Alternaria longipes* on tobacco in Malawi; the sensitive germ tubes suffer if hot and dry conditions follow a shower in the daytime (Norse, 1973). It should be stressed that the most effective types of rain for the above-mentioned diseases may be different in other climatic regions.

In contrast with natural rainfall, the amount of water applied by sprinklers and the frequency and duration of the irrigations can sometimes be manipulated for optimum development of the host and minimum benefit to its pathogens (Rotem and Palti, 1969).

Dew is a microclimatic factor. The duration of dew periods is more important than the amount of water deposited. Several dew recorders have been developed by phytopathologists, but different instruments exposed to similar conditions recorded different durations of surface wetness (Lomas and Shashoua, 1970). In addition, an instrument records the duration of dewfall on its sensor, which does not necessarily react in the same way as the leaf surface it is designed to simulate. Also, outer leaves in the foliage canopy may be subjected to dewfall for longer periods than inner leaves (see Section III,C) and, therefore, interpretations drawn from the "average" dew duration (as shown by an instrument) can be misleading. In spite of these limitations, dew records and visual observations have helped to establish a causal relationship between dewfall and epidemics (Rotem and Reichert, 1964; Ullrich, 1962; Wallin, 1967).

The following comparisons can be made between the effects on epidemics of rain, sprinkling, and dew. These three sources of moisture differ in their influence on the various phases of a pathogen's life cycle. Thus, rain and sprinkling cause splash dispersal of spores, but dew does not. All three sources of moisture can facilitate germination and penetration which, in most cases, do not require long periods of moisture and are only slightly, if at all, affected by light. On the other hand, sprinkling, usually performed by day for a limited number of hours, is likely to have less effect than rain or dew on sporulation, which requires longer wet periods and often is inhibited by light. Dew, which largely coincides with darkness, will usually promote sporulation, while the effect of rain on sporulation will depend on whether it falls during the day or at night, and on whether the day temperature is low enough to repress the inhibiting effect of light. This is apparently the case in cool zones or seasons. Nocturnal rainfall in cool zones or seasons may be accompanied by temperatures too low for the pathogen to multiply fast enough to cause an epidemic. In hot areas a decrease in leaf temperature brought about by rainfall or sprinkling may favor the development of an epidemic.

Although the importance of dew in epidemics has been recognized for temperate zones (Ullrich, 1962; Wallin, 1967), its effects may be even more pronounced in hot and rainless areas or seasons (Rotem et al., 1970; Rotem and Reichert, 1964; Stover, 1970). The importance of dew in such areas appears to stem from the scarcity of rain in some of these areas, but most particularly from the relatively high night temperatures which accompany dewfall. Thus, in Israel, dew promotes epidemics more markedly in summer (night temperatures above 20°C) than in winter (night temperatures below 10°C).

The influence of rain, sprinkling, and dew continues until the drops or film of water evaporate from the leaf surface. The duration of surface wetness depends upon the prevailing weather conditions and also upon the density of the foliage canopy. The hotter, drier, sunnier, and windier the weather, the quicker the drying process.

For epidemics to occur, some fungi require wet periods longer than those provided by a single source of surface moisture. For example, the combined moisture periods resulting from sprinkling and dew have been associated with epidemics of *Stemphylium* blight in tomatoes (Rotem and Cohen, 1966), and those from rain and dew with epidemics of *Alternaria longipes* in tobacco (Norse, 1973).

The ability of a pathogen to benefit from dewfall for infection is conditioned by the ability of its deposited spores to withstand the adverse daytime conditions. Thus *Alternaria porri* f. sp. *solani*, having spores extremely resistant to heat and desiccation, is able to utilize dew for infection after daytime exposure of the dispersed spores to conditions which would kill the more sensitive spores of many other pathogens (Rotem and Palti, 1969). By contrast, *Phytophthora infestans* with its sensitive sporangia, can utilize dew only when daytime weather conditions permit survival of dispersed sporangia (Rotem and Palti, 1969; Rotem *et al.*, 1970). It may be expected that all pathogens with robust spores will enter the infection phase as soon as dew wets the foliage. If night temperatures are sufficiently high for infection to occur during the time of dewfall, then rain or sprinkling (in many areas less frequent than dew) at other times will be of little additional benefit. Under marginal night temperatures, however, rain or sprinkling during the warmer daytime will facilitate development of an epidemic. Other pathogens, the dispersed spores of which are sensitive to hazardous daytime conditions, will not be able to benefit from dewfall and will be able to infect only during periods of rainfall or sprinkling. Under less adverse conditions, sensitive spores will be able to survive the day and to benefit from dew, provided the night temperatures are within a suitable range.

III. HABITAT OF AIRBORNE DISEASES

A. Exchange of Energy: The Key Process

All the "bricks" in the complicated structure of epidemics of foliage diseases (the phases of infection, sporulation, dispersal, etc.), are directly affected by microclimatic conditions in the phylosphere. The microclimate of a plant tissue is affected by the ambient climate through an exchange of energy, radiation flux, and air temperature. Wind speed, relative humidity, and transpiration all influence the final balance. Although attempts to calculate the crop's microclimate according to macroclimatic data have been made (Goudriaan and Waggoner, 1972; Waggoner, 1965), establishment of a quantitative relationship between the two is difficult in the dynamic environment.

Conditions interfering with energy exchange increase the difference between the air and the leaf temperature. In hot and dry areas, with very strong radiation and limited transpiration by the plant, the temperature of leaves exposed to the sun occasionally may be as much as 20°C higher than that of the ambient air; differences of 4° to 8°C are common. When abundant soil moisture facilitates efficient transpiration, however, evaporative cooling often is sufficient to keep the leaf temperature below the air temperature even on a hot day (Lomas et al., 1972). Free moisture on leaf surfaces cools the leaves even more.

Because of the escape of long-wave radiation at night, leaves are then cooler than the air. In fact, water vapor may condense on the cool leaves and dew may be formed. According to different combinations of relative humidity, temperature, wind, and topography, dew may form almost every night in some areas and very rarely in others.

All the microclimatic phenomena mentioned in this section are influenced by a number of topographic and agricultural factors as discussed below.

B. The Effects of Topography and Direction of Exposure

Because the aspect of exposure affects the intensity and duration of solar radiation, slopes facing the sun have higher soil and leaf temperatures, lower relative humidity, and less (or no) dewfall than those not so exposed. On slopes facing away from the sun, however, dewfall also is limited due to the downward movement of cool air during the night. The cool air sinks to the valley bottoms, where precipitation of moisture may occur as dew or fog. Observations on duration of dew and the frequency of Peronospora tabacina epidemics on tobacco planted on a north-facing slope and at its bottom, showed a 3-month average dew period of 4 hr per night and only traces of disease on the upper part of the slope, compared to a 9-hr dew period per night and complete destruction of plants at the valley bottom (J. Rotem, unpublished).

The direction of planting also can affect the persistence of dew and the rate of development of epidemics in row crops grown on level fields. Thus in trellised tomatoes aligned in different directions, the incidence of lesions caused by *Stemphylium* sp. and *Alternaria porri* f. sp. *solani* and *Xanthomonas vesicatoria* was, for the first three months, 20 to 30% lower on the south or southeast side of each row, than on the north or northwest side where more shade and longer periods of dewfall were recorded. These differences disappeared toward the end of the growing season because, even though there were fewer lesions present on the south-facing side of each row, leaf damage was accelerated there by higher leaf temperature so that the final impact of diseases was similar on both sides of the row (J. Rotem, unpublished). Shaded sites created by trees, windbreaks, buildings, etc., are often foci of primary infection. These phenomena result from the changed conditions of temperature and radiation that affect different phases in the pathogen's life cycle.

C. The Effect of Plant Density

Plant density plays an important role in determining the degree to which the microclimate within a plant canopy differs from that of the surrounding air.

In a young field crop, direct sunlight reaches most parts of the plant and the leaf temperature in the daytime may be higher than that of the air. The field is well aerated because the sparse stand does not interfere greatly with wind movement. The good aeration results in more rapid drying of previously deposited rain or dew drops and limits the number of ecological niches helpful for the initial establishment of the invading pathogen. Radiation energy also penetrates the ground and raises the temperature of the upper layer of soil (Geiger, 1957). This results in relatively unfavorable conditions for foliage pathogens, more so in a hot and/or continental climate with generally stronger radiation than in a maritime and/or temperate zone. These conditions often coincide with the beginning of the growing season when the amount of inoculum is low. Observations on the relation between age (i.e., density of foliage) of potatoes and the amount of *Phytophthora infestans* inoculum needed to cause initial infection have shown that more inoculum or better macroclimatic conditions (rain, optimum temperatures, etc.) were needed for the establishment of the disease in a young rather than in an older (and more dense) field (J. Rotem, unpublished). Unless especially favorable conditions occur, the late blight appears later, i.e., when a more dense stand of potatoes interferes with aeration and creates numerous ecological niches (provided by the outer leaves shading the inner ones). Hirst and Stedman (1960), who measured the weather parameters inside and outside the canopy of potatoes, pointed out the "self-destructiveness" of the growing crop. They found that the development of the crop progressively modified the microclimate within the canopy and made it more favorable for an epidemic of *Phytophthora infestans*. When the pathogen finally defoliated the crop, the environmental conditions again became less favorable for disease development and resembled those present in a young crop.

In the dense stand of an older crop, radiation acts primarily on the outer leaves of the canopy, resulting in higher temperatures in this section. Because of shading by the outer leaves, the inner part of the plant canopy is less exposed to the effects of direct solar radiation; it remains cooler and is surrounded by a more humid atmosphere. At night, the outer foliage of a dense crop cools off and becomes covered by dew. Because of protection by the outer leaves, the escape of long-wave radiation from the inner leaves is partly inhibited. The inner leaves are then warmer than the outer leaves and less covered by dew. In spite of receiving less dewfall, but because of the protection from sunlight, the inner leaves often form favorable sites for development of many pathogens, and especially of those species which are sensitive to the radiation, temperature, and humidity conditions prevailing on the outer leaves during the daytime.

Whether the outer or the inner part of a plant canopy is the most suitable site for disease development depends on the macroclimate of a given area and season and on the characteristics of the pathogen involved. Thus, in a hot and rainless climate where the outer leaves are daily exposed to strong radiation, abundant dew on the outer leaves will not favor pathogens sensitive to dryness and high temperature. This is due to the elevated leaf temperature and decreased humidity in the outer leaves. Much more favorable conditions for development of these pathogens during hot seasons will then be on the inner leaves in spite of the less frequent dewfall that occurs there. However, some pathogens may prefer the outer leaves when the external conditions do not endanger the sensitive spores, or when spores of a given species are robust. For instance, during the mild autumn in Israel, the sensitive sporangia of *Phytophthora infestans* frequently infect the outer leaves of potatoes; but during the hotter and drier spring, the inner leaves and stems are the favored sites for infection. By contrast, the robust spores of *Alternaria porri* f. sp. *solani* on the same host infect the outer leaves in all seasons.

The density of the plant canopy also affects the patterns of spore dispersal. Among all the factors affecting dispersal (for details, see Chapter 8, this volume), wind speed is the most important. Dense stands reduce the wind speed and may decrease it to a force at which the detachment of spores is diminished. Temperature, and especially relative humidity, also affect the detachment of spores. Some fungi disperse their spores earlier and terminate the dissemination more abruptly under hot and dry conditions than under cool and humid conditions (Rotem and Cohen, 1974). It is therefore expected (although not proven) that the spores present inside a dense stand and under cooler, more humid, and less windy conditions will be dispersed later in the day than spores produced in a sparse stand or on the outer leaves of the canopy. Since the spores inside a dense stand are better and longer protected from the macroclimatic hazards, they may be a valuable source for infections. More studies are needed to ascertain how true and common these phenomena are and how they affect the epidemic process.

IV. HABITAT OF SOIL–BORNE DISEASES

Fungal soil-borne pathogens have, in comparison with the foliage pathogens, a simpler life cycle in which the sporulation and dispersal of spores play a minor role, if any. Growth (often saprophytic) in the soil system, and the processes of infection and survival of the population between the seasons or plantings of susceptible crops, are less affected by atmospheric conditions than similar phenomena in the airborne pathogens. For instance, survival of many airborne pathogens often extends for a few days or hours, while that of most soil inhabitants is usually measured in months or years. Such an insensitive response results partly from the protection afforded by the soil from some lethal (e.g., ultraviolet rays) or extreme factors (e.g., high temperature).

Although the physical factors in the soil are less subject to the rapid changes which frequently occur in the phylosphere, their interactions with biotic factors apparently have a greater effect on soil inhabitants than do similar interactions on airborne pathogens. In soil, these interactions result in an extremely complex environment in which a change in one factor (temperature, moisture, oxygen, carbon dioxide, biotic factors) leads to a change in the others (Griffin, 1969; Raney, 1965; Sewell, 1965). The levels of these factors are also affected by the characteristics of the plant cover, which depletes the environment of moisture, reduces the temperature by shading, adds nutrients to the root zone, and influences the development of antagonistic or competing microflora. Similar changes also are created by agricultural practices (Sewell, 1965) which directly or indirectly improve or worsen the conditions for epidemics. This complex situation is so difficult to study in the field that the habitat of soil-borne pathogens is much less well understood than that of the airborne pathogens.

Soil temperatures are stable at depths of 30–40 cm or more. Diurnal fluctuations increase toward the soil surface. The soil surface is cooler than the air by night but hotter in the daytime, especially in a continental climate and/or a light soil exposed to solar radiation. In a sparse stand, sun scald may be induced by these conditions and lead to epidemics of collar rots or to other diseases, like that caused by *Sclerotium bataticola* in potato tubers, which are predisposed to invasion by high soil temperatures.

Soil moisture is an important regulator of temperature because more heat is needed to raise the temperature of wet soil, which conducts the heat to deeper layers. Although the development of most soil-borne pathogens is enhanced by abundant soil moisture (but below saturation), development of some pathogens (e.g., *Fusarium roseum* on wheat) is facilitated by relatively drier conditions (Cook, 1973). In the complex soil habitat, however, it is not sufficient to define the effect of moisture without accounting for that of other factors acting simultaneously and/or conditioned by soil moisture (Griffin, 1969). Aeration, which is an inverse function of the moisture content of soil (the more water the less space for air), has profound influences on disease induced by many soil-borne pathogens. Thus although some *Pythium* sp. develop well in wet and poorly aerated soil, damage to the peanut crop increases when the wet soil is also well aerated (Frank, 1967). The role of the host in these interactions is illustrated by *Phytophthora citrophthora* in citrus, in which excessive wetness or experimental limitations of the availability of oxygen, inhibits growth and regeneration of roots thus predisposing them to more decay (Stolzy *et al.*, 1965).

In the absence of comparative studies, we can only speculate about the direct and indirect effects of temperature and moisture in different types of soil and in various climatic zones. It is obvious that a similar amount of water (from rain or irrigation) will result in longer periods of abundant moisture and deficient aeration in heavy soils than in light, well-drained soils with limited water-holding capacity. This increases the probability of epidemics in heavy soils, provided they are not saturated with water for long periods and the low oxygen and high carbon dioxide concentrations do not inhibit disease development. On the other hand, it is not the deficient aeration but rather the lack of moisture that is expected to limit most diseases in light soils.

A similar amount of water in the same type of soil may be expected to increase the soil's moisture content for a longer period in cool than in hot regions. This is due to enhanced evapotranspiration which rapidly depletes the soil of moisture. Consequently, less rainfall (or irrigation) will be needed in cool than in hot regions (1) to provide the abundant moisture needed for epidemics of most soil-borne pathogens but also

(2) to prolong the saturation conditions detrimental to some of these diseases. On the other hand, high temperatures are expected to intensify the detrimental effect of temporary saturation by increasing the biotic activities which lead to a deficiency of oxygen and the production of toxic materials. Under moist (but not saturated) conditions, high temperatures are also expected to increase the epidemic potential by prolonging the growing season and, consequently, the pathogen's activity. If the last assumption is true, then the relative importance of soil-borne pathogens in the development of epidemics will be greater in hot and humid than in cool and humid areas. Pertinent statistical data in support of this claim are lacking.

The many assumptions made in relation to soil-borne epidemics point again to the profound underdevelopment of our knowledge of soil-borne pathogens in comparison to airborne ones. A more aggressive approach to these problems is needed. It is technically possible to conduct field trials in which epidemic development is tested against different temperature and moisture regimes, and in which most of the resulting physical and chemical changes in the root zone are monitored. The author is not aware that such experiments are being carried out.

V. THE HYPOTHESES OF COMPENSATION

Because of limitations of climate, the epidemic development of some pathogens is restricted to specific geographical zones (Reichert and Palti, 1967; Weltzien, 1972). However, the occurrence of other pathogens seems to be limited mainly by the distribution of their hosts. For instance, inspection of the Commonwealth Mycological Institute Distribution Maps of Plant Diseases reveals that *Phytophthora infestans,* restricted for many years to temperate and humid zones, has spread to a number of tropical and semidesert localities where potatoes or tomatoes have been introduced. There is little evidence to justify the belief that the population of a pathogen must contain environmental races if it has become established in an area from which its former absence was attributed to climatic limitation.

A partial explanation for these "new arrivals" is provided by the influence of a change in microclimate brought about by changes in agricultural or forest practices (see Chapter 2, this volume). Nevertheless, there are definite limitations of the extent to which a change in microclimate can facilitate the development of epidemics under continually adverse macroclimatic conditions.

In the following theses, advanced to explain the worldwide occurrence of epidemics of some pathogens, it is postulated that more than one eco-logical pathway leads to (1) epidemics of the same disease in a variety of climatic regions; and (2) epidemics, in the same climatic region, of disease induced by pathogens which differ in their environmental demands.

According to these theses, the ability of a pathogen to succeed in several differing climatic regions stems from its ability to complete its life cycle regularly in spite of the marginal state of one physical or biological factor, provided that optimal states of other factors concurrently compensate for the limiting effect of the marginal one. For example, a pathogen may be able to penetrate its host, or to sporulate, when one factor (e.g., temperature) is near a limit of the range tolerable for this phase of the life cycle provided other factors (e.g., abundance and duration of surface wetness) remain nearly optimal throughout the duration of the phase. Similarly, epidemics may develop when some phases in the life cycle of the pathogen are temporarily arrested, provided other phases meet with favorable conditions.

Situations rarely occur in which all phases of the life cycle proceed in an orderly succession under optimal conditions (see also Chapter 3, this volume). Analysis of the patterns of meteorological factors which occurred during several epidemics of late blight of potato crops in Israel have shown that, even under apparently favorable overall conditions, some phases of the pathogen's life cycle at times encountered highly marginal conditions. However, it was demonstrated that the optimal and marginal states of some physical and biological parameters cannot be represented by fixed values. Their effects on the pathogen are relative rather than absolute, the contribution of each being conditioned by the states of the others (Bashi et al., 1973; Rotem et al., 1971, 1976). These observations form the basis for the "hypotheses of compensation."

The first hypothesis is: "A highly favorable state of one factor essential for development of a given phase in the life cycle of a pathogen can compensate for the limitations imposed by the simultaneously unfavorable state of another factor."

An example is in the infection phase of the life cycle of Phytophthora infestans on potatoes, where the requirements for temperature, duration of wetting, or spore supply are relative and vary in accordance with the status of the two other variables. This explains why similar levels of disease occurred after potato plants had been exposed to different combinations of these three factors (Rotem et al., 1971).

Sometimes the importance of one factor dominates that of the others, as is found in the case of Stemphylium botryosum f. sp. lycopersici on tomatoes. A long period of wetting compensated for suboptimal temperatures and sparse inoculum in leading to infection, but short wet periods were not compensated for by highly favorable temperatures and an abundant supply of spores (Bashi et al., 1973).

For development of *Peronospora tabacina* on tobacco, the factors determining the required period of wetting are temperature, age of leaves, susceptibility of cultivar, and abundance of spores (Kröber, 1967). Relative effects of temperature and/or moisture (interacting in specific cases with biotic factors) were also described for the sporulation process of *Rhynchosporium secalis* in barley (Rotem et al., 1976), and for the infection processes of *Ophiobolus graminis* in wheat (Henry, 1932), *Fusarium* wilt in peas (Schroeder and Walker, 1942), and *Helminthosporium sativum* in wheat (McKinney, 1923).

According to the first compensation hypothesis, interactions among several physical and biotic factors in the microenvironment of the pathogens may result in the development of similar amounts of disease in dissimilar habitats or seasons. For example, a similar "quantum" of sporulation, germination, or some other phase in the life cycle of a pathogen could result from either a short period of dewfall (e.g., 8 hr) at a relatively high temperature (e.g., 20°C) in the midst of a hot zone or season, or a longer period of dewfall (e.g., 12 hr) coinciding with a lower temperature (e.g., 10°C) in a cool zone or season.

The second hypothesis is: "A specific 'weakness' in a pathogen can be compensated for by a specific 'strength'."

This concept may be illustrated by the following examples. If a pathogen requires a long period to complete one of the phases in the infection process, this represents a "weakness" likely to restrict its development in a semiarid area. For example, *Stemphylium botryosum* f. sp. *lycopersici* and *Alternaria porri* f. sp. *solani* need a long period for penetration and/or sporulation. In semiarid areas, the occurrence of periods when the foliage remains continuously wet for sufficient time to allow completion of these processes is a rare event. Nevertheless, these pathogens are able to thrive because of a "strong" feature: their conidia, germ tubes, and conidiophores will tolerate intermittent desiccation during germination, penetration, and/or sporulation. This feature enables them to infect and/or sporulate, under a regime of several short, wet periods interrupted by intervals when the foliage is dry.

Phytophthora infestans demonstrates a different kind of "strength." This pathogen has sporangia able to germinate and send a germ tube into the host tissue during a very short wet period; this compensates for a "weakness" in the extreme sensitivity of its sporangia to desiccation during germination. Thus, in semiarid regions *P. infestans* is able to cause potato blight epidemics as severe as those which occur in the cool, wet regions that have been the traditional habitat of this pathogen (Bashi and Rotem, 1974, 1975a; Bashi *et al.*, 1973).

These examples of compensation by a "strength" for a "weakness" help us to understand better why EPIDEM, a simulation program for early blight developed by Waggoner and Horsfall (1969), indicated that both temperate and semiarid conditions can favor epidemic development of the pathogen *Alternaria solani.*

The third hypothesis is: "A high frequency of occurrence of one phase in the life cycle of a pathogen can compensate for a low frequency of occurrence of another phase in the life cycle of the same pathogen."

This hypothesis is illustrated by cases in which an abundant supply of viable spores (resulting in nature from frequent and prolific sporulation and/or from a high rate of survival of dispersed spores) results in frequent infection periods despite generally unfavorable temperature and/or moisture conditions (Kröber, 1967; Rotem *et al.*, 1971). With the soil-borne pathogens *Plasmodiophora brassicae* in *Brassica* spp., *Fusarium culmorum* in wheat, and *Phoma chrysanthemicola* in chrysanthemum, the presence of abundant inoculum in soils where physical and chemical factors favored the pathogen and compensated for suboptimal soil moisture and/or temperature (Colhoun, 1953; Colhoun *et al.*, 1968; Peerally and Colhoun, 1969). In England, epidemics of *Diplocarpon rosae* begin when there is frequent rain but, once a critical amount of inoculum is present, the epidemic continues to progress even though rainfall may become less frequent (Saunders, 1966).

More direct evidence of the compensation of one phase by another has been obtained in potato late blight epidemics induced in growth chambers. The most severe development of disease was favored by continuous leaf wetness throughout the day and night, simulating a prolonged period of drizzle which sometimes occurs in cool temperate regions. In other experiments where we simulated epidemics in the hot summer season of a semiarid region, allowing some hours of wetting at night only, more disease developed under a regime of low (50%) than of high (80%) daytime humidity. This relationship ended only when the plants were subjected to day temperatures high enough to reduce the survival of the dispersed sporangia. According to the daytime regime of temperature and relative humidity to which the plants were exposed, either the dispersal phase (enhanced by low relative humidity and suppressed by high relative humidity) or the survival phase (unaffected by low temperatures but depressed by high temperatures) was dominant in determining the rate of progress of the epidemic. Thus, the favored phase compensated for the suppressed phase in allowing the epidemic to proceed.

Despite numerous careful experiments, no compensation phenomena have so far been detected in epidemics of powdery mildew (*Sphaerotheca fuliginea*) on squash. Although certain phases in the life cycle of this pathogen are favored by dry weather and others by more humid conditions, it appears to have a sufficiently broad tolerance in all of its phases for epidemics to occur under a wide variety of weather conditions (Reuveni and Rotem, 1974). Such pathogens, which have a wider tolerance of environmental conditions in all phases of their life cycle, may be expected to succeed in a broad range of climatic zones without the need for compensation phenomena.

Obviously there are limits beyond which no degree of compensation can overcome the rigors of climate which preclude the development of epidemics of some pathogens in certain regions of the world. We hope, however, that the hypotheses set forth here may stimulate additional observations of epidemics in the field and of simulated epidemics in controlled environment chambers. From this may develop a more rational understanding of why some pathogens thrive in habitats traditionally regarded as unsuitable territory.

Acknowledgments

The critical reading of the manuscript by Drs. M. V. Carter, Y. Cohen, J. H. Haas, J. Kranz, and J. Palti is very much appreciated.

References

Bashi, E., and Rotem, J. (1974). Adaptation of four pathogens to semi-arid habitats as conditioned by penetration rate and germinating spore survival. *Phytopathology* **64**, 1035–1039.

Bashi, E., and Rotem, J. (1975a). Sporulation of *Stemphylium botryosum* f. sp. *lycopersici* in tomatoes and of *Alternaria porri* f. sp. *solani* in potatoes under alternating wet-dry regimes. *Phytopathology* **65**, 532–535.

Bashi, E., and Rotem, J. (1975b). Effect of light on sporulation of *Alternaria porri* f. sp. *solani* and of *Stemphylium botryosum* f. sp. *lycopersici in vivo*. *Phytoparasitica* **3**, 63–67.

Bashi, E., and Rotem, J. (1976). Induction of sporulation of *Alternaria porri* f. sp. *solani in vivo*. *Physiol. Plant Pathol.* **8**, 83–90.

Bashi, E., Rotem, J., and Putter, J. (1973). Effect of wetting duration, and of other environmental factors, on the development of *Stemphylium botryosum* f. sp. *lycopersici* in tomatoes. *Phytoparasitica* **1**, 77–94.

Cohen, Y., and Rotem, J. (1970). The relationship of sporulation to photosynthesis in some obligatory and facultative parasites. *Phytopathology* **60**, 1600–1604.

Cohen, Y., and Rotem, J. (1971). Dispersal and viability of sporangia of *Pseudoperonospora cubensis*. *Trans. Br. Mycol. Soc.* **57**, 67–74.

Colhoun, J. (1953). A study of the epidemiology of club-root disease of Brassicae. *Ann. Appl. Biol.* **40**, 262–283.

Colhoun, J. (1973). Effects of environmental factors on plant disease. *Annu. Rev. Phytopathol.* **11**, 343–364.

Colhoun, J., Taylor, G. S., and Tomlinson, R. (1968). Fusarium diseases of cereals. II. Infection of seedlings by *F. culmorum* and *F. avenaceum* in relation to environmental factors. *Trans. Br. Mycol. Soc.* **51**, 397–404.

Cook, R. J. (1973). Influence of low plant and soil water potential on diseases caused by soilborne fungi. *Phytopathology* **63**, 451–458.

Frank, Z. R. (1967). Effect of irrigation procedure on *Pythium* rot of groundnut pods. *Plant Dis. Rep.* **51**, 414–416.

Geiger, R. (1957). "The Climate near the Ground." Harvard Univ. Press, Cambridge, Massachusetts.

Goudriaan, J., and Waggoner, P. E. (1972). Simulating both aerial microclimate and soil temperature from observations above the foliar canopy. *Neth. J. Agric. Sci.* **20**, 104–124.

Griffin, D. M. (1969). Soil water in the ecology of fungi. *Annu. Rev. Phytopathol.* **7**, 289–310.

Henry, A. W. (1932). Influence of soil temperature and soil sterilization on the reaction of wheat seedlings to *Ophiobolus graminis* Sacc. *Can. J. Res.* **7**, 198–203.

Heptig, G. H. (1963). Climate and forest diseases. *Annu. Rev. Phytopathol.* **1**, 31–50.

Hims, M. J., and Preece, T. F. (1975). *Spongospora subterranea* f. sp. *subterranea. In* "CMI Descriptions of Pathogenic Fungi and Bacteria," No. 477.

Hirst, J. M., and Stedman, O. J. (1960). The epidemiology of *Phytophthora infestans.* I. Climate, ecoclimate and the phenology of disease outbreaks. *Ann. Appl. Biol.* **48**, 471–488.

Hsieh, S. P. Y., and Buddenhagen, I. W. (1975). Survival of tropical *Xanthomonas oryzae* in relation to substrate, temperature and humidity. *Phytopathology* **65**, 513–519.

Hunter, J. E., and Kunimoto, R. K. (1974). Dispersal of *Phytophthora palmivora* sporangia by wind-blown rain. *Phytopathology* **64**, 202–206.

Kröber, H. (1967). Der Einflus der Benetzungsdauer auf die Entstehung der Blauschimmelkrankheit an Tabak. *Phytopathol. Z.* **58**, 46–52.

Lomas, J., and Shashoua, Y. (1970). The performance of three types of leaf-wetness recorders. *Agric. Meteorol.* **7**, 159–166.

Lomas, J., Schlesinger, E., Zilka, M., and Israeli, A. (1972). The relationship of potato leaf temperatures to air temperatures as affected by overhead irrigation, soil moisture and weather. *J. Appl. Ecol.* **9**, 107–119.

Lukens, R. J. (1966). Interference of low temperature with the control of tomato early blight through use of nocturnal illumination. *Phytopathology* **56**, 1430–1431.

McKinney, H. H. (1923). Influence of soil temperature and moisture on infection of wheat seedlings by *Helminthosporium sativum. J. Agric. Res.* **26**, 195–217.

Nelson, R. R., and Tung, G. (1973). Influence of some climatic factors on sporulation by an isolate of race T of *Helminthosporium maydis* on a susceptible male-sterile corn hybrid. *Plant Dis. Rep.* **57**, 304–307.

Norse, D. (1973). Some factors influencing spore germination and penetration of *Alternaria longipes. Ann. Appl. Biol.* **74**, 297–306.

Peerally, M. A., and Colhoun, J. (1969). The epidemiology of root rot of Chrysanthemums caused by *Phoma* sp. *Trans. Br. Mycol. Soc.* **52**, 115–123.

Raney, W. A. (1965). Physical factors of the soil as they affect microorganisms. In "Ecology of Soil-Borne Plant Pathogens" (K. F. Baker and W. C. Snyder, eds.), pp. 115–119. Univ. of California Press, Berkeley.

Reichert, I., and Palti, J. (1967). Prediction of plant disease occurrence: A phytogeographical approach. Mycopathol. Mycol. Appl. 32, 337–355.

Reuveni, R., and Rotem, J. (1974). Effect of humidity on epidemiological patterns of the powdery mildew (Sphaerotheca fuliginea) on squash. Phytoparasitica 2, 25–53.

Rotem, J. (1968). Thermoxerophytic properties of Alternaria porri f. sp. solani. Phytopathology 58, 1284–1287.

Rotem, J., and Cohen, Y. (1966). The relationship between mode of irrigation and severity of tomato foliage diseases in Israel. Plant Dis. Rep. 50, 635–639.

Rotem, J., and Cohen, Y. (1974). Epidemiological patterns of Phytopththora infestans under semi-arid conditions. Phytopathology 64, 711–714.

Rotem, J., and Palti, J. (1969). Irrigation and plant diseases. Annu. Rev. Phytopathol. 7, 267–288.

Rotem, J., and Reichert, I. (1964). Dew—a principal moisture factor enabling early blight epidemics in semi-arid region of Israel. Plant Dis. Rep. 48, 211–215.

Rotem, J., Palti, J., and Lomas, J. (1970). Effects of sprinkler irrigation at various times of the day on development of potato late blight. Phytopathology 60, 839–843.

Rotem, J., Cohen, Y., and Putter, J. (1971). Relativity of limiting and optimum inoculum loads, wetting duration and temperatures for infection by Phytophthora infestans. Phytopathology 61, 275–278.

Rotem, J., Clare, B. G., and Carter, M. V. (1976). Effects of temperature, leaf wetness, leaf bacteria and leaf bacterial diffusates on production and lysis of Rhynchosporium secalis spores. Physiol. Plant Pathol. 8, 297–305.

Saunders, P. J. W. (1966). Epidemiological aspects of blackspot disease of roses caused by Diplocarpon rosae Wolf. Ann. Appl. Biol. 58, 115–122.

Schröder, H. (1963). The utilization of mean values and of frequency values of meteorological factors in analyzing the temperature response of fungi. NATO Adv. Study Inst. Epidemiol. Biometeorol. Fungal Dis. Plants, pp. 249–250.

Schroeder, W. T., and Walker, J. C. (1942). Influence of controlled environment and nutrition on the resistance of garden pea to Fusarium wilt. J. Agric. Res. 65, 221–248.

Sewell, G. W. F. (1965). The effect of altered physical condition of soil on biological control. In "Ecology of Soil-Borne Plant Pathogens" (K. F. Baker and W. C. Snyder, eds.), pp. 479–494. Univ. of California Press, Berkeley.

Stolzy, L. H., Letey, J., Klotz, L. J., and Labanauskas, C. K. (1965). Water and aeration as factors in root decay of Citrus sinensis. Phytopathology 55, 270–275.

Stover, R. H. (1970). Leaf spot of bananas caused by Mycosphaerella musicola: role of conidia in epidemiology. Phytopathology 60, 856–860.

Ullrich, J. (1962). Beobachtungen über die Infektionsbedingung waehrend der Ausbreitung von Phytophthora infestans in Kartoffelfeld. Nachrichtenbl. Dtsch. Pflanzenschutzdienst (Berlin) 14, 149–152.

Visser, T., Shanmuganathan, N., and Sabanayagam, J. V. (1961). The influence of sunshine and rain on tea blister blight, Exobasidium vexans Massee, in Ceylon. Ann. Appl. Biol. 49, 306–315.

Waggoner, P. E. (1965). Microclimate and plant disease. Annu. Rev. Phytopathol. 3, 103–126.

Waggoner, P. E., and Horsfall, J. G. (1969). Epidem. *Conn., Agric. Exp. Stn., New Haven Bull.* 698.

Wallin, J. R. (1967). Agrometeorological aspects of dew. *Agric. Meteorol.* 4, 85–102.

Weltzien, H. C. (1972). Geophytopathology. *Annu. Rev. Phytopathol.* 10, 277–298.

Zimmer, D. E. (1975). Some biotic and climatic factors influencing sporadic occurrence of sunflower downy mildew. *Phytopathology* 65, 751–754.

Chapter 16

Geophytopathology

HEINRICH C. WELTZIEN

I. INTRODUCTION

A. Definitions

Phytopathology as a life science frequently deals with distribution patterns of hosts, diseases, and their abiotic or biotic causes. As such it becomes part of biogeography, the science describing the distribution of organisms on the earth. At the same time, phytopathology is understood as an experimental, analytical science, dealing with the causal analysis of disease phenomena in plants. In addition, its medical aspects are clearly visible in all actions to achieve plant health. Thus, the prefix "geo-" to phytopathology is added to indicate that we deal with distribution patterns of plant diseases, the causal understanding of these patterns, and the geographic aspects of disease control.

339

B. Historical Background

The concept of geobotany goes back to A. von Humboldt's famous journey through Latin America from 1799 to 1803 (von Humboldt and Bonpland, 1807). It has since developed into a well-established branch of botany, presenting an impressively complete picture of the geographic distribution of plants (Walter, 1965ff). The same principles were used by Reichert (1953, 1958) and by Reichert and Palti (1966, 1967) for their concept of pathogeography of plant diseases. The similar concept of geophytopathology was discussed by Weltzien (1967, 1972), who closely connected it with the concept of geomedicine as developed for human medicine by Zeiss (1942) and Rodenwaldt (1952a,b, 1961, 1966). Thus Weltzien (1973) arrived at the term "geophytomedicine," to stress the medical aspects in view of the diseased or damaged plant as the patient, to be protected or treated.

II. THE VALUE OF MAPS

A. General Background

Maps are the main instruments for all types of geographic studies. They are also the main source of information for geophytopathological studies. Thus, some introduction to the use and structure of maps must be discussed here. Maps are used as either pictures, documents, or tools (Wilhelmy, 1972). The picture map is widely used as a visual aide in extension, teaching, and advertising. Its abstractions stimulate the viewer's imagination and improve his recognition and understanding of disease. Picture maps can be neglected for research purposes.

A document map is almost unsurpassed as a source of data. All information can be seen in logical relationships. A document map carries an immense data load and is easy to read with very little training and without expensive equipment. Its potential as a working tool is very great, too, although not at all fully explored in our science. As the concept of geophytopathology has gained acceptance, the use of maps has increased. A survey of the world literature of plant pathology up to 1971 (Weltzien, 1972) yielded only 805 maps. Five years later when the same sources were surveyed again for this publication, another 311 maps were registered. This was the addition of more than one-third as many maps in 5 years as in all previous years combined. This increased interest in maps may also be an indicator of more involvement of plant pathologists in planning for disease management, pesticide management, control strategy, and environmental protection.

B. Types of Maps

All maps used in plant pathology can be classified as thematic maps. They are usually analytical in nature, singling out selected facts (Wilhelmy, 1972). But as various phenomena are combined in one map, new, previously unknown facts may become visible in what is called a synthetic map. The map by Shukla and Schmelzer (1974) is a good example (Fig. 1). It demonstrates a large complex of information: (1) frequency of viruses in the Brassicaceae within the territory of the German Democratic Republic, (2) the proportion of aphid- and nematode-transmitted forms of viruses, and (3) the major soil types and areas with low risks for aphid-transmitted viruses. From this the close relation of soil type and nematode transmission of viruses becomes clearly visible and adds new information.

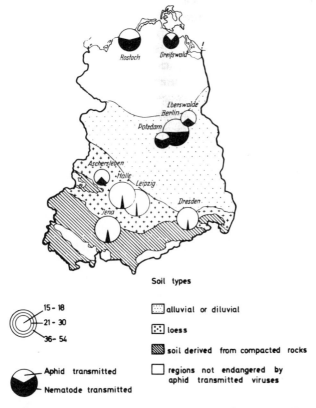

Fig. 1. Frequency of aphid- and nematode-transmitted viruses on Brassicaceae in relation to soil types within the territory of the German Democratic Republic (after Shukla and Schmelzer, 1974).

Fig. 2. *Puccinia* path of North America and the three regions into which genes for vertical resistance to crown rust are deployed (from Frey *et al.*, 1973).

Usually our maps will be based on primary observed facts by which generalizations are made by an inductive method. In some cases transformed information is also used to develop deductive maps with a high degree of abstraction and generalization, such as the ones used by Stewart *et al.* (1970) or Frey *et al.* (1973), delineating ecological areas for rust races in the United States (Fig. 2). Other maps may deal with the more or less constant factors such as soil or climatic zones. Still others deal with disease phenomena that are highly variable. For example, King (1972) has mapped the varying severity over England of powdery mildew in spring barley for the years 1967–1970 (Fig. 3), thus showing the necessity for an annual and local forecasting. Diagram maps are also useful, although they indicate only approximate locations. Well-known examples are the climatogram maps of Walter and Lieth (1966). The map in Fig. 1 with the circle diagrams also belongs to this group.

The structure of a map is the result of its content. The most frequent type is the position or location map. Here, the facts are shown in their true geographic location, but, to a large extent, the scale controls the exactness. The location map remains the most widely used in plant

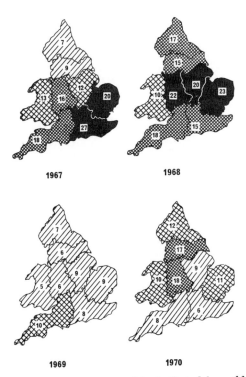

Fig. 3. Mean percentage area on second leaf affected by mildew in each of the administrative regions 1967–1970 of the Ministry of Agriculture, Forestry and Fishery (from King, 1972).

pathology. Out of 1114 collected maps 47% were classified under this category. The map by Leonard (1974) (Fig. 4) is a good example and many others could be mentioned here. It can be pointed out that these maps are the most direct, but least comprehensive, because they use only a very limited portion of the potential of the map as a data carrier.

Area maps are another type of great importance to our science. From the above-mentioned collection 31% were classified under this category. The disease areas are shown either in outline form or by shading. The map by Holdeman (1970) (Fig. 5) on walnut cankers in California may be given as a typical example for both. If many locations are registered in an area by single dots, an area map may also develop by some kind of a pointillistic method, a technique often used for mapping of agricultural crops.

The isoline map, quite familiar from weather maps, is frequently used to document the epidemic spread of plant diseases such as is seen in

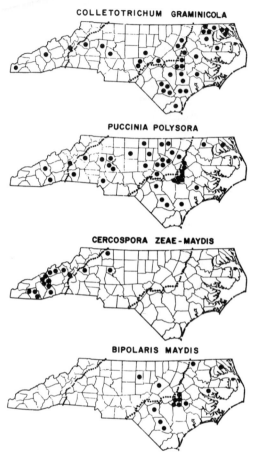

Fig. 4. Location map for four different leaf pathogens of corn in North Carolina in 1972 and 1973 (from Leonard, 1974).

the map in Fig. 6 (Hoffmann *et al.*, 1976). In these cases, so-called "isochrons" connect all points where the disease was observed at the same time.

Finally, dynamic processes are often shown through vector maps, where direction of movement is indicated by arrows. The quantity of moving entities may be indicated by arrow strength. A good example is the map by van Kraayenoord *et al.* (1974) (Fig. 7) showing the invasion of two poplar rusts from Australia into New Zealand in 1973. This map is also a good example of the large quantity of information that can be included on a map without a loss in readability. This map shows the

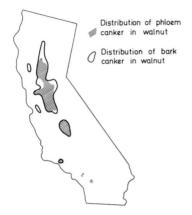

Fig. 5. Distribution of walnut cankers in California (after Holdeman, 1970).

invasion of two organisms, the true spots where they were located at various dates, the isochrons developed from these locations, and the main directions of movement. The map is also a true synthetic map. Here the many possibilities which a map offers were used as documents and tools of scientific significance.

Fig. 6. Spread of *Endothia* chestnut blight in the United States (from Hoffmann *et al.*, 1976).

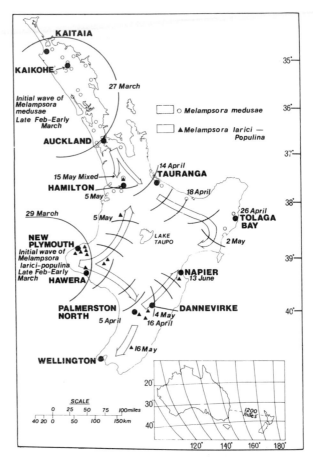

Fig. 7. Invasion of two poplar rusts, *Melampsora medusae* and *Melampsora larici-populina* into New Zealand from Australia (from van Kraayenoord *et al.*, 1974).

III. DISTRIBUTION PATTERNS OF PLANT DISEASES

A. Distribution Patterns by Area

It is widely recognized that plant diseases are not uniformly distributed over the range where their hosts are cultivated or growing. This fact inspired the development of the serial publication of "Distribution Maps of Plant Diseases" (Commonwealth Mycological Institute, 1942), a widely recognized standard reference in all studies of plant disease distribution. They allow a classification of pathogens according to their distribution patterns. "Cosmopolitic pathogens" that occur in most parts of the world are *Agrobacterium tumefaciens* (CMI map 137), *Phytoph-thora infestans* (map 109), or *Alternaria solani* (map 89). "Endemitic pathogens" with a rather limited distinct distribution are *Gymnosporan-gium juniperi-virginianae* (map 61), peach yellows virus (map 60), or *Venturia cerasi* on cherry (map 196). The continuity of distribution is another characteristic demonstrated by these maps. Pathogens which cover a rather "continuous area" are the rust of broad beans *Uromyces viciae fabae* (map 200) and *Taphrina deformans* (map 192). "Discon-tinuous" or "disjunct areas" are characteristic for *Phytophthora phaseoli* (map 201) or beet curly top virus (map 24). Of course, all these areas are not uniformly covered by the disease and their borders are fictive lines. They must, therefore, be understood as "pseudoareas" as defined by Imhoff (1972).

B. Difference in Disease Frequency

The worldwide occurrence of one family of pathogens, the Erysipha-ceae, was studied by Hirata (1966). Using his data, Weltzien (1977) developed a world map of powdery mildew frequency. This map (Fig. 8) is based on the number of reported hosts in different countries, states, or regions. Centers of host density are therefore interpreted as centers

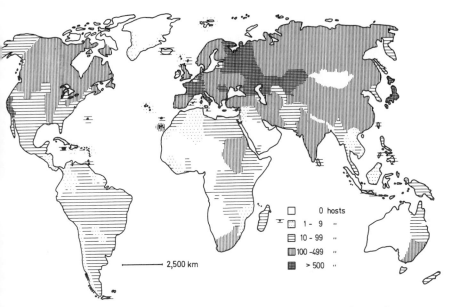

Fig. 8. World map of powdery mildew frequency with areas harboring between 0 and 500 recorded host plants (from Weltzien, 1977).

of higher powdery mildew frequency. The differences in powdery mildew frequency visible on this map are very striking. In large areas of Africa and Asia no powdery mildews have ever been reported or fewer than 10 hosts are known. On the other hand, there are clear centers for these diseases in central Europe, California, and Japan with more than 500 reported host plants. If areas with more than 100 hosts are considered, the Eurasian continent, parts of North America, and only smaller areas in Africa and Australia are prominent. Of course, it can not be excluded, that inefficient surveys may obscure the true distribution pattern in some cases. However, the differences in geographic distribution of powdery mildew frequency are clearly visible. Similarly the difference in disease frequency in a smaller national territory was demonstrated for sugar beet virus diseases in the German Democratic Republic by Hartleb (1975). Figure 9 shows the very irregular distribution pattern for beet yellows, beet mosaic, and beet leaf curl, at a 5 and 15% frequency level.

C. Distribution by Intensity

It is even more important to understand that within its area of distribution the intensity of disease is highly variable and should be mapped. Although disease assessment has recently drawn much attention and many standardized methods were suggested (FAO, 1971), they are mostly designed to allow assessment of disease frequency or intensity in single fields. Estimates of diseases in larger areas are more difficult to obtain and standard methods are lacking. All authors discussing the

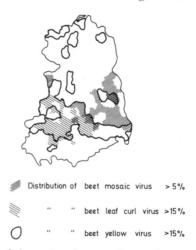

Fig. 9.   Frequency of three virus diseases of sugar beets in areas of major attack in the German Democratic Republic (after Hartleb, 1975).

TABLE I

Terminology Used by Various Authors to Describe Three Areas of Different Disease or Pathogen Significance within the Natural Distribution Pattern

Zone I	Zone II	Zone III	Reference
Normal occurrence	Occasional occurrence	Possible occurrence	Cook, 1925
Permanent distribution	Main distribution	Distribution	Bremer, 1929
Main damage	Damage	Distribution	Eidmann, 1934
Main damage	Marginal damage	Sporadic attack	Weltzien, 1967
Permanence area (area of main or permanent damage)	Gradation area (area of damage)	Latence area (no-damage area)	Schwerdtfeger, 1968

problem of different disease intensity within the area of distribution have therefore used only three classes of disease intensity (Table I). In theory they can be understood as three concentric rings with decreasing disease frequency from the center to the edge. For each typical area of distribution of a pathogen or disease there should be a center area with high disease frequency, regular epidemics, and heavy damage, respectively (Zone I). It is surrounded by an area of fluctuation and decreasing severity (Zone II). The total area of distribution is finally completed by Zone III, where the attack or occurrence is insignificant. The terminology used by various authors to define these three zones is summarized in Table I. Occurrence (Cook, 1925), distribution (Bremer, 1929), damage (Eidmann, 1934; Weltzien, 1967), and terms used in population dynamics (Schwerdtfeger, 1968) are the vocabulary presently used. They are necessary tools to describe and analyze a natural biogeographic phenomenon which obviously exists without being understood.

Figure 10, taken from Baxter (1952), shows as an example the zones for the three levels of intensity of *Cronartium fusiforme* on southern pines in the southeastern part of the United States.

IV. ROLE OF ENVIRONMENT IN DISTRIBUTION PATTERNS

A. A Static Model

The causes of the distribution patterns of disease can only be found in environmental factors. If the environment varies, then disease distribution can vary concomitantly and may be classified accordingly (Table

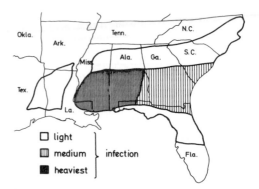

Fig. 10. Disease intensity of *Cronartium fusiforme,* the fusiform gall rust, on southern pines (after Baxter, 1952, and Lamb and Sleeth, 1940).

I). In principle, this is independent of the actual host distribution. Environmental factors are equally significant in areas of actual and potential disease occurrence.

Having established maps for areas where host and pathogen occur together, one can therefore give a prognosis for the likelihood of disease in areas where the host occurs but where a pathogen has not yet been introduced. The same holds true for areas where the host is imported as a new crop and the potential threat by pathogens or pests has to be estimated.

In Fig. 11 we show a graphic model for the distribution of plant disease. It is similar to the models of Schwerdtfeger (1968) and Franz and Krieg (1976) for the abundance of insects. Suppose we consider the two most common environmental factors that influence disease, namely, temperature and wetness. This gives three variables for Fig. 11: host range, optimum temperature range, and optimum wetness range.

In Fig. 11 the geographic distribution of the host is shown as a slightly tilted oval in the center. Outside that oval, there is no host. The oval leaning toward the left is the area of the optimum temperature for the pathogen. Likewise, the one leaning toward the right depicts the area of the optimum wetness. On this model map one can see the zones for the three classes of disease damage described in Table I. In that part of the host range, where both temperature and wetness are favorable, we deal with disease Zone I. Where only one of the two environmental factors is favorable, disease is medium (Zone II), and where neither is favorable, disease is low or practically absent (Zone III).

All three zones occur within the actual area of disease occurrence where the host is grown and the potential areas of disease occurrence if the host or pathogen is introduced. The prerequisite for such an

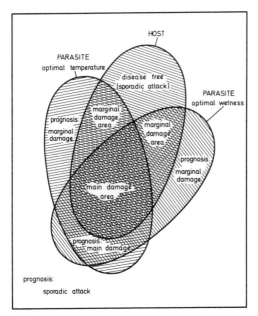

Fig. 11. Model to describe the areas of effective and potential occurrence of a plant pathogen, controlled by two predominant environmental factors (temperature and wetness). Zones I–III as in Table I. (For details see text.)

analysis and prognosis under practical conditions is obviously knowledge about the environmental conditions controlling disease occurrence and intensity and the possible compensation by which one optimal factor may compensate for deficiencies in another, suboptimal one.

The picture becomes more complicated, however, if more than two factors must be considered. Many of the sometimes confusing distribution patterns in nature are probably based on multifactorial dependencies. However, it seems reasonable to assume that in most cases rather few dominating factors control the distribution in general, with some minor factors occasionally complicating the picture.

B. Some Practical Examples

The influence of climate on epidemics is discussed by Rotem in the Chapter 15, this volume. However, some remarks may be added here to show the direct influence of environmental factors on plant disease distribution patterns. Wetness and temperature are the main factors controlling apple scab infection. Studt (1975) has analyzed a scab epidemic in Lebanon during 1971–1974, where the disease intensity varied greatly between different parts of the country. Measuring the wetness periods and temperatures in various locations, he could clearly correlate the local disease intensity with the frequency of infection periods according to the model of Mills and La Plante (1951).

The apple scab situation in the state of Washington was reported by Blodgett and Semler (1970). Figure 12 shows a map developed from their data. The climatograms indicate that the coastal region (1) is clearly a main damage area (Zone I, Table I). High monthly rainfall during spring and summer allows regular epidemics, and arid conditions occur only during short periods in summer. Area 2 must also be classified as a main damage area, though the conditions are clearly less favorable for the disease. But the rainfall data during spring months are still high enough to allow regular infections at favorable temperatures. Area 3, however, represented by the climatogram of Penticton, must be considered as an area with only occasional epidemics (Zone II, Table I). The rainfall is rather low throughout the year; arid conditions prevail from May to September and in many years chances of infection must be very limited. Finally, areas 4 and 5 are clearly nondamage areas (Zone

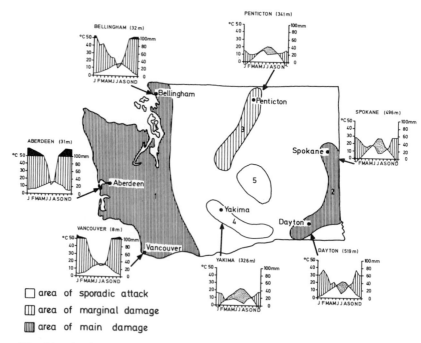

Fig. 12. Apple-growing areas in the state of Washington and their classification according to apple scab intensity (after Blodgett and Semler, 1970, and Walter and Lieth, 1966).

III, Table I) with only sporadic attack, if the only available climatogram for Yakima is representative. Rainfall becomes rare as soon as the temperature rises in spring and aridity prevails throughout the vegetation period. If overhead sprinkling is used, however, the infection may become more dangerous because it widens the favorable wetness area. The same may be true if in some years exceptionally late rains coincide with favorable temperature and the presence of ascospores.

Local deviations from these general classifications may occur, as the rainfall data for Tieton reported by Blodgett and Semler (1970) indicate. This shows the importance of local measurements, as microclimatic differences of epidemiological significance may occur even between nearby places.

Another example of environmentally controlled disease distribution is the different disease intensity of Cercospora beticola and Erysiphe betae in the sugar-beet-growing areas of the world. Figure 13 is based on the data by Bleiholder and Weltzien (1972) on Cercospora and by Drandarevski (1969) on Erysiphe. Summer aridity or humidity are the differentiating factors, since both diseases require high average temperatures of 18°C or more during the main part of the vegetation period for maximum development. In regions where the temperatures are generally too low, there may be only sporadic attacks of both diseases. Areas of medium damage result from intermediate temperatures and occasional aridity for Erysiphe or humidity for Cercospora. An analysis of the sugar-beet-growing areas of the world, with help of the climatic diagrams by Walter and Lieth (1966), resulted in the world map shown in Fig. 13. It represents both the effective and the potential areas of attack for the regions with sugar beet cultivation, but not for regions where the host is not yet cultivated. Thus it is also an attempt to develop a prognosis map for areas where the pathogen has not yet been introduced. The validity of such a prognosis has been elegantly proved. Drandarevski (1969) predicted a damage area for the southwestern United States (California, Arizona, New Mexico, and Utah). Five years later these states were actually hit by heavy epidemics.

Ruppel et al. (1975) mapped the sequence of disease occurrence for 1974 in a diagram map which showed that the affected areas are identical with the ones predicted. Figure 14 is based on their data. The isochrons show the spread from southern California in a northwesterly direction. Since crop damage is directly related to the date of infection in summer grown beets (Weltzien and Ahrens, 1977), one may consider the April, May, and June period as that of greatest damage (Zone I), the July and August period as that of medium damage (Zone II), and the September

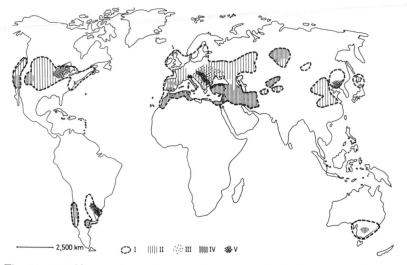

Fig. 13. World map of sugar-beet-growing areas with the three zones of disease. intensity for leaf spot (*Cercospora beticola*) and powdery mildew (*Erysiphe betae*). I. Areas of sporadic attack for powdery mildew and leaf spot (Zone III).   II. Areas of occasional powdery mildew epidemics (Zone II).   III. Areas of occasional leaf spot epidemics (Zone II).   IV. Areas of regular powdery mildew epidemics (Zone I).   V. Areas of regular leaf spot epidemics (Zone I).

period as that of no damage at all or only sporadic damage (Zone III). There is little doubt that the same principle of prediction can be used for any new project for sugar beet cultivation in the world.

The cultivation and selection of disease-free areas or areas with sporadic attack is another possible application of this type of disease distri-

Fig. 14. Isochrons for powdery mildew first occurrence on sugar beets in the western United States for 1974 (after Ruppel *et al.*, 1975).

bution study. In Fig. 13 one can easily identify the northern coastal areas of Europe as not threatened by either one of the two leaf diseases and the same is true for some zones in North and South America. However, the value of such an interpretation is limited as long as not all major pests and pathogens of a crop have been mapped in the same differentiated way. Such a crop-orientated pest and disease atlas could be a highly valuable source of information for teaching, extension, and crop protection planning.

C. The Epidemic Spread

The dynamics of epidemics are another target for cartographic documentation. Figure 7 showed the introduction of *Melampsora medusae* and *Melampsora larici-populina* into New Zealand. Other recent maps deal with the introduction of fire blight, *Erwinia amylovora*, into western Europe (Fliege, 1973) and the spread of coffee rust, *Hemileia vastatrix*, into Latin America (Kranz, 1970). Some very detailed studies were made by Joshi and Palmer (1973), Nagarajan and Singh (1975), and Nagarajan *et al.* (1976) on the cereal rust epidemics in India. Figure 15 shows the isochrons for the south–north movement of wheat stem rust (*Puccinia graminis tritici*) in three successive years, together with the parallel running isotherms for 14°C low monthly averages. Typical air trajectories transporting spores from south to north and east as reported by Nagarajan *et al.* (1976) are included in Fig. 15. This example shows that maps on epidemic spread may also be composed as synthetic maps, showing the causes of spread together with the isochrons and allowing the planning of crop protection strategies.

V. PLANNING OF CROP PROTECTION

All aspects discussed in the previous sections of this chapter are certainly important for the planning of crop protection activities. This is equally true for survey work to obtain data on disease occurrence, for the interdependent variables between plant diseases and environment, and for the results of studies on epidemic spread. However, there are other cases where a geographic approach may be an important base for further studies. This is the case in problems of biological or integrated control, where the occurrence of a hyperparasite is a prerequisite for success. A map by Smith *et al.* (1973), for example, gives the location for *Colletotrichum*-infected weeds in Arkansas rice fields. A series of maps has been developed by Hanf (1972), showing the use of pesticides by state for the main crops in Western Germany. The maps can

356   HEINRICH C. WELTZIEN

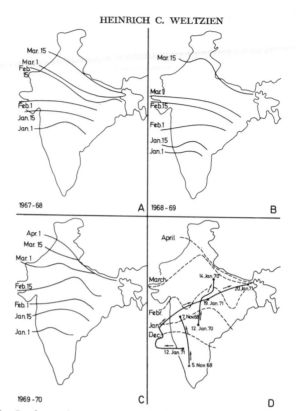

Fig. 15. Isochrons for appearance of wheat stem rust (*Puccinia graminis tritici*) in India 1967–1970 (A, B, C) and isotherms for low monthly means of 14°C plus air trajectories traced after spore catches in 1968, 1970, and 1971 (D) (after Joshi and Palmer, 1973, and Nagarajan *et al.*, 1976).

be used for planning of sales and extension work and may serve as a base for the development of modern legislation. The special problems of legislation on the trade and use of lindane is the theme of several continent maps by Blaquiere (1976). The residue problem in humans and environmental contamination as one of the most unwanted side effects of chemical plant protection was mapped by Davies (1975). His data for Europe (Fig. 16) indicates not only the average concentration of DDT-type compounds in human fat but also reflects on different legislative efforts and economic systems. For the United States he could show the higher residues in the fat of people in the states with warm climates and higher residues in blacks than in whites.

The most general, almost strategical approach, however, was tried by Grainger (1968). Assuming that high-value crops are more threatened by disease than low-value crops, he calculated one index figure for a country, state, or region by weighting the crop areas according

Fig. 16. Concentrations of DDT-type compounds (in ppm) in human adipose tissue for some European states (from Davies, 1975).

to their agricultural intensity. He based a second index on the crop protection intensity that a group of crops can carry with economic profit. The product of both figures, which may vary between 0 and 300, was called "G." It should represent the agricultural intensity and the potential for successful crop protection. The higher the numerical value of "G," the higher the potential of this area or territory for high agricultural output and for a successful, intensive crop protection. "G" was calculated and mapped from the FAO production data for 1959. Figure 17

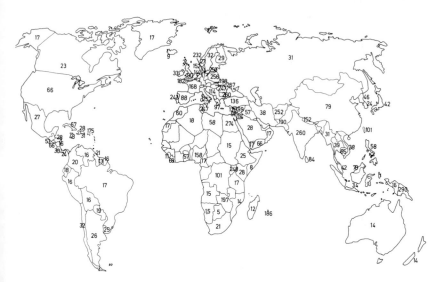

Fig. 17. "G" values as index for the potential intensity of agricultural production and of profitable crop protection (after Grainger, 1968). (For details see text.)

shows Grainger's data on a world map. Because they are of such a speculative nature, showing possibilities rather than facts, they stimulate the imagination and are quite informative. The rather high figures for some Mediterranean and tropical countries stand out, while others, e.g., in Africa and South America, are surprisingly low.

One may well agree that recognizing the potential for agricultural production and plant protection is a vital problem in our struggle to increase and maintain the world food supply. This may also be considered as one of the possible contributions of geographic approaches in plant pathology to agriculture, and all efforts toward reaching that goal are welcome.

References

Baxter, D. V. (1952). "Pathology in Forest Practice," 2nd ed. Wiley, New York.

Blaquiere, C. (1976). Legal regulations pertaining to trade and use of Lindane, 74–112. In Ulmann, E. "Lindane" 2, Suppl. K. Schillinger, Freiburg.

Bleiholder, H., and Weltzien, H. C. (1972). Beiträge zur Epidemiologie von Cercospora beticola Sacc. an Zuckerrübe. III. Geopathologische Untersuchungen. Phytopathol. Z. 73, 93–114.

Blodgett, E. C., and Semler, L. F. (1970). Apple scab in the Yakima valley of Washington. Plant Dis. Rep. 54, 63–65.

Bremer, H. (1929). Grundsätzliches über den Massenwechsel der Insekten. Z. Angew. Entomol. 14, 254–272.

Commonwealth Mycological Institute. (1942). "Distribution Maps of Plant Diseases." Commonw. Mycol. Inst., Kew.

Cook, W. C. (1925). The distribution of the Alfalfa Weevil (Phytonomus pesticus Gyll.). J. Agric. Res. 30, 479–491.

Davies, J. E. (1975). Current medical problems of pesticide management. Plant Prot. News Manila-Eschborn 4, 48–57.

Drandarevski, C. (1969). Untersuchungen über den echten Rübenmehltau Erysiphe betae (Vanha) Weltzien. III. Geophytopathologische Untersuchungen. Phytopathol. Z. 65, 201–218.

Eidmann, H. (1934). Zur Epidemiologie der Forleule. Mitt. Forstwirtsch. Forstwiss. 5, 13–27.

FAO (1971). "Crop Loss Assessment Methods," FAO Manual on the Evaluation and Prevention of Losses by Pests, Diseases and Weeds. Commonw. Agric. Bur., England.

Fliege, H. F. (1973). Feuerbrand. Erwerbsobstbau 15, 24–27.

Franz, J. M., and Krieg, A. (1976). "Biologische Schädlingsbekämpfung," 2nd ed. Parey, Berlin.

Frey, K. J., Browing, J. A., and Simons, M. D. (1973). Management of host resistance genes to control diseases. Z. Pflanzenkr. Pflanzenschutz 80, 160–180.

Grainger, J. (1968). Disease assessment and the prevention of crop loss. World Crops 20, 21–28.

Hanf, M. (1972). Pflanzenschutzentwicklung in Deutschland 1946–1971. BASF-Mitt. Landbau 3/72, 1–26. BASF, Limburger Hof.

Hartleb, H. (1975). Der Befall von Beta-Rüben durch Viruskrankheiten in der Deutschen Demokratischen Republik in den Jahren 1972 bis 1974. *Nachrichtenbl. Pflschutz D. D. R.* **29**, 45–50.

Hirata, K. (1966). "Host Range and Geographical Distribution of the Powdery Mildews." Niigata University, Niigata, Japan.

Hoffmann, G., Nienhaus, F., Schönbeck, F., Weltzien, H. C., and Wilbert, H. (1976). "Lehrbuch der Phytomedizin." Parey, Berlin.

Holdeman, Q. L. (1970). Varietal and geographic distribution of walnut phloem canker and bark canker in California. *Plant Dis. Rep.* **54**, 373–376.

Imhoff, E. (1972). "Lehrbuch der allgemeinen Geographie," Vol. 10: Thematische Kartographie. Walter de Gruyter, Berlin.

Joshi, L. M., and Palmer, L. T. (1973). Epidemiology of stem, leaf, and stripe rusts of wheat in northern India. *Plant Dis. Rep.* **57**, 8–12.

King, J. E. (1972). Surveys of foliar diseases of spring barley in England and Wales 1967–70. *Plant Pathol.* **21**, 23–35.

Kranz, J. (1970). Kaffeerost in Brasilien. *Bild Wiss.* 1133–1139.

Lamb, H., and Sleeth, B. (1940). Distribution and control for the southern pine fusiform rust. *U.S. South. For. Exp. Stn., Occas. Pap.* **91**, 1–5.

Leonard, K. J. (1974) Foliar pathogens of corn in North Carolina. *Plant Dis. Rep.* **58**, 532–534.

Mills, W. D., and La Plante, A. A. (1951). Diseases and insects in the orchard. *Cornell Ext. Bull.* **711**, 21–27.

Nagarajan, S., and Singh, H. (1975). The Indian stem rust rules—an epidemiological concept on the spread of wheat stem rust. *Plant Dis. Rep.* **59**, 133–136.

Nagarajan, S., Singh, H., Joshi, L. M., and Saari, E. E. (1976). Meteorological conditions associated with long-distance dissemination and deposition of **Puccinia graminis tritici** uredospores in India. *Phytopathology* **66**, 198–203.

Reichert, I. (1953). A biogeographical approach to phytopathology. *Proc. Int. Bot. Congr., 7th, 1950* pp. 730–31.

Reichert, I. (1958). Fungi and plant diseases in relation to biogeography. *Trans. N.Y. Acad. Sci. [2]* **20**, 333–339.

Reichert, I., and Palti, J. (1966). On the pathogeography of plant diseases in the Mediterranean region. *Proc. Congr. Mediterr. Phytopathol. Union, 1st, 1966* pp. 273–280.

Reichert, I., and Palti, J. (1967). Prediction of plant disease occurrence; a pathogeographical approach. *Mycopathol. Mycol. Appl.* **32**, 337–355.

Rodenwaldt, E. (1952a). Introduction to "World Atlas of Epidemic Diseases," Vol. 1, pp. 11–12. Verlag Falk, Hamburg.

Rodenwaldt, E. (1952b). "World Atlas of Epidemic Diseases," Vol. 1. Verlag Falk, Hamburg.

Rodenwaldt, E. (1961). "World Atlas of Epidemic Diseases," Vol. 2. Verlag Falk, Hamburg.

Rodenwaldt, E. (1966). "World Atlas of Epidemic Diseases," Vol. 3. Verlag Falk, Hamburg.

Ruppel, E. G., Hills, F. J., and Mumford, D. L. (1975). Epidemiological observations on the sugar beet powdery mildew epiphytotic in western U.S.A. in 1974. *Plant Dis. Rep.* **59**, 283–286.

Schwerdtfeger, F. (1968). "Demökologie." Parey, Berlin.

Shukla, D. D., and Schmelzer, K. (1974). Ergebnisse virologischer Untersuchungen an Öl- und Futterpflanzen sowie an Zier- und Wildpflanzen aus der Familie der Kohlgewächse in der Deutschen Demokratischen Republik. *Nachrichtenbl. Pflanzenschutz D. D. R.* **28**, 232–235.

Smith, R. J., Daniel, J. T., Fox, W. T., and Templeton, G. E. (1973). Distribution in Arkansas of a fungus disease used for biocontrol of northern jointvetch in rice. *Plant Dis. Rep.* **57**, 695–697.

Stewart, D. M., Romig, R. W., and Rothman, P. G. (1970). Distribution and prevalence of physiologic races of **Puccinia graminis** in the United States in 1968. *Plant Dis. Rep.* **54**, 256–260.

Studt, G. (1975). Apfelschorf **Venturia inaequalis** (Cooke) Winter-Untersuchungen zur Sporulation und zur Entwicklung von Bekämpfungsverfahren im mediterranen Klimabereich. Dissertation, Landwirtschaftliche Fakultät, Bonn.

van Kraayenoord, C. W. S., Laundon, G. F., and Spiers, A. G. (1974). Poplar rusts invade New Zealand. *Plant Dis. Rep.* **58**, 423–427.

von Humboldt, A., and Bonpland, A. (1807). "Ideen zu einer Geographie der Pflanzen." Cotta, Tübingen (New ed., Wiss. Buchges, Darmstadt, 1963).

Walter, H. (1965ff). "Vegetationsmonographien der einzelnen Grossräume," Vols. I–X. Fischer, Stuttgart.

Walter, H., and Lieth, H. (1966). "Klimadiagramm-Weltatlas." Fischer, Jena.

Weltzien, H. C. (1967). Geopathologie der Pflanzen. *Z. Pflanzenkr.* (*Pflanzenpathol.*) *Pflanzenschutz* **74**, 175–189.

Weltzien, H. C. (1972). Geophytopathology. *Annu. Rev. Phytopathol.* **10**, 277–298.

Weltzien, H. C. (1973). Geophytomedizin. *Geogr. Z., Beih., Fortschr. Geomed. Forsch.* pp. 110–114.

Weltzien, H. C. (1978). Geographical distribution of powdery mildews. *In* Spencer, D. M. "The Powdery Mildews." Academic Press, London. (in press).

Weltzien, H. C., and Ahrens, W. (1977). Ertragsrückgang durch den echten Mehltau der Zuckerrübe, **Erysiphe betae** (Van.) Weltzien. *Z. Zucker* **30**, 288–291.

Wilhelmy, H. (1972). "Kartographie in Stichworten," Vol. III. Hirt, Kiel.

Zeiss, H. (1942). "Seuchen-Atlas." 2 vols. Perthes, Gotha.

Chapter 17

Agricultural and Forest Practices that Favor Epidemics

ELLIS B. COWLING

I. INTRODUCTION

Our friends in farming and forestry often accuse plant pathologists of being more enthusiastic about sick plants than healthy ones. When we survey the devastation induced by one of our favorite pathogens, who among us has not exclaimed, a little too cheerfully for the farmer's liking, "What a splendid disease you have here!" My friends in forestry have chastised me about this: "Can't you pathologists tone down your enthusiasm over the sick ones?" Bedside manner aside, enthusiastic study of sick plants helps us learn how to keep plants healthy. Similarly, enthusiastic study of factors that favor epidemics can help us learn how to prevent future epidemics.

361

It seems paradoxical that certain management practices should favor epidemics. Usually the paradox derives from conflicts between desirable objectives—for example, market demands for uniformity versus the biological imperative to maintain genetic diversity, or the lower cost of continuous monoculture versus the greater safety of diverse crops in rotation. Sometimes the conflicts are between long- and short-term objectives, or between the interests of producers and consumers, or sometimes both.

Most agricultural and forest practices are designed to meet the demands of the marketplace for wholesome food and useful fiber, to increase yields, or to decrease labor or other costs of production. When a given practice permits sustained yields at reasonable prices and has few undesired side effects, we all can be proud. But when a given practice increases the risk of loss due to disease, then producers, consumers, and scientists alike should recognize the risk and seek to minimize it.

II. FACTORS THAT FAVOR EPIDEMICS

Five factors are critical to the development of every epidemic: (1) the population of host plants must be susceptible, preferably as uniformly so as possible; (2) the plants should be clustered or crowded together—the closer, the better: (3) a virulent pathogen(s) must be present and have the potential to increase in abundance—the faster, the better; (4) the weather and other factors of the environment should be favorable for dissemination of the pathogen and for development of the disease; and (5) the timing of favorable conditions should be sufficient to sustain the epidemic.

Any practice that makes one or more of these factors more favorable will increase the chances of an epidemic. We will discuss each, in turn, below.

A. The Host Susceptibility Factor

The susceptibility of a given population of host plants to disease is determined by two major features—the genetic characteristics of the population and predisposing environmental conditions that may increase susceptibility.

1. The Role of Genes

Man has used about 3000 species of plants as a source of food during his history (Mangelsdorf, 1966). At present, however, the great majority of the 4 billion people on this planet depend on 15 species of plants for most of their food. These 15 species include five cereals—rice, wheat, maize, sorghum, and barley; two sugar plants—sugar cane and sugar beets; three root crops—potatoes, sweet potatoes, and cassavas; three legumes—beans, soybeans, and peanuts; and two other crops—coconuts and bananas. The widespread use of these few species has greatly narrowed the genetic and functional diversity in the major crops of the world and thus greatly increased their vulnerability to epidemic disease. As Stevens (1939) has shown, corn and buckwheat are among the healthiest of the major grain crops. He attributes this to the fact that corn and buckwheat are out-crossing species and therefore relatively heterogeneous, in comparison with wheat and rice which are self-fertile and therefore more homogeneous.

Unfortunately, the pressures for uniformity and homogeneity do not stop with the species—they also extend to varieties within the species. The National Academy of Sciences (1972) and later Day (1973) presented data to show the extent to which specific varieties dominate in the United States. In 1969, four varieties of dry beans made up 93% of the land area planted in beans; five inbred lines occur in varieties of maize planted on 66% of the area; only two varieties of peas covered 96% of the area. The figures for Australia were similar—two varieties of peanuts on 99% of the land; three varieties of barley on 69% of the land (see also Chapter 13, this volume).

The more uniform man makes his crops, the more vulnerable they will be to disease.

2. The Role of Predisposition

Predisposition is an effective way to make a population of host plants more susceptible to disease (Schoeneweiss, 1975). Predisposition has been defined as "environmentally conditioned susceptibility . . . the tendency of nongenetic conditions, acting before infection, to affect the susceptibility of plants to disease" (Yarwood, 1959).

The results of many management practices can predispose plants to disease. For example, wounding plants opens the road for many pathogens to penetrate their hosts. Planting when the temperature and moisture are highly favorable for growth of the pathogen but unfavorable for the host makes infection more likely. Fertilizing some plants with nitrogen increases their susceptibility to certain pathogens (Huber and Watson, 1974). Essentially everything a farmer does has the potential of increasing or decreasing susceptibility to disease.

B. The Clustering Factor

Clustering plants together is convenient for the grower; it is also convenient for the pathogens that induce disease. Clustering can be achieved in both space and "time"; pathogens are favored by both.

Clustering plants in space can be achieved within a given region, locale, field, glasshouse, or storage bin.

Clustering fields and crowding plants in the same field shorten the distance that pathogens and their vectors must travel (Waggoner, 1962; van der Plank, 1963). Splash dispersal of inoculum will become more efficient. Wounding will increase during cultivation of the crop. When plants are clustered so close that their leaves and/or roots are in contact, abrasion of leaves and stem or fruits will occur. Under these conditions, sporulation by fungal pathogens may even become unnecessary because vegetative growth or passive transport of the pathogen from one plant to another will become possible. Packing harvested products tightly together also offers these same conveniences to postharvest pathogens.

The closer the plants or their parts are to each other, the more they will dominate the microenvironment within the canopy of foliage, the root zone, or the storage bin. The relative humidity and the duration of leaf wetness after dewfall and rain will increase. The temperature will become more uniform. The number of leaf and root contacts will increase.

Clustering plants together in "time" will also favor epidemics. Plants in fields that are seeded at the same time will be highly uniform in age and stage of development. Because susceptibility to disease often changes with time (see Chapter 12, this volume), uniform planting dates will ensure that most plants will be equally susceptible at the same time. What a convenience for the pathogen!

When crop rotations or fallow are used, a pathogen must have survival mechanisms strong enough to bridge the host-free period. When continuous and/or multiple cropping is practiced, the pathogen is spared the rigors of surviving long periods without a suitable host. When the same resistant variety of the host is used year after year, the pathogen is encouraged to develop virulent genotypes which can overcome the genes for resistance in the host. Successive crops planted and harvested in the same field in the same year favor the pathogen. When alternate or alternative hosts are planted nearby or permitted to grow in fence rows or idle land nearby pathogens are likely to increase.

C. The Virulence Factor

Pathogens induce epidemics among susceptible hosts in much the same way that sparks ignite fire in a dry forest. If there are no sparks, there will be no fire. If there are sparks but the forest is wet, there is little chance of fire. On the other hand, if the sparks are present, and especially if they are abundant when the forest is dry, the chances of fire will be high.

Similarly, the pathogen must be present to cause disease. If the weather does not favor an increase in inoculum, there will be little chance of an epidemic. If the pathogen is present and virulent, and especially if the weather favors an increase in inoculum over a long period of time, however, the chances of an epidemic disease will be high.

Numerous agricultural and forest practices determine the presence and relative virulence of pathogens. Some practices affect not only the amount of initial inoculum but the rate of its increase over time. Other practices can favor the development of more virulent races of a given pathogen.

D. The Weather Factor

As long as man has cultivated crops, farmers have been aware that certain weather conditions were more favorable than others for epidemics of plant disease. Only in recent times, however, have farmers recognized the extent to which their own choices about, how, when, and where to cultivate their crops can change the microclimate within a given field or forest and thus increase or decrease the chances of an epidemic. Even though the effects of management on microclimate are limited, certain practices can have marked influences on the development of epidemics. Growers can change the microclimate within their crops by their choice of planting date and site, use of enclosures and shading, method and timing of irrigation, and many other practices (see Chapter 7, Vol. I). Improper conditions for harvesting, transport, storage, and processing are well known to favor epidemics in perishable food and fiber products.

E. The Timing Factor

An epidemic of plant disease requires a series of specific events that make up the disease cycle. Each event must occur in its proper sequence. The initial inoculum must be available when the plants are susceptible. The pathogen must penetrate and then colonize the host. The pathogen must reproduce. The new propagules must be dispersed and survive long enough to penetrate a new host.

If the conditions of weather, host nutrition, and development, etc., are not favorable for any one of these sequential events, the epidemic will stop or develop very slowly. If the conditions are favorable for each event, in turn, the spread of disease in time and space will proceed quickly. Under optimal conditions, certain pathogens such as *Phytophthora infestans* and *Helminthosporium maydis* can complete their disease cycle in as few as four days (Day, 1973). If such conditions prevail for a long period of time, a catastrophic epidemic may be in the making.

Many management practices affect the timing of epidemics—time of planting, time of pruning, time of fertilization and irrigation, and time of harvesting and storage before processing.

III. SPECIFIC MANAGEMENT PRACTICES THAT FAVOR EPIDEMICS

In the discussion that follows, specific practices that favor epidemics will be used to illustrate the factors we have just discussed. Some practices illustrate only a single factor but most illustrate more than one.

A. Practices that Increase Genetic Vulnerability

Genetic vulnerability is a new term for an old problem in agriculture and forestry (Zadoks and Koster, 1976). The term became popular after the corn blight epidemic in the United States in 1970. Genetic vulnerability is an inverse function of the heritable variability and functional diversity in the crops man cultivates (see Chapters 13 and 14, this volume). Man wants uniformly high quality in his crops and he has developed both simple and sophisticated technological systems to achieve that uniformity. The trouble is that most pathogens like uniformity too.

Marshall Ward (1901) recognized the hazard of uniformity 75 years ago:

> It is clear from our study of the factors of an epidemic that one of the primary conditions which favors the spread of any disease is provided by growing any crop continuously in "pure culture" over large areas . . . The history of all great planting enterprises teaches us that he who undertakes to cultivate any plant continuously in open culture over large areas must run the risk of epidemics.

1. Breeding Practices

Selection and breeding are two of the least expensive ways by which farmers and foresters can increase yields, improve quality, and decrease the cost of crop production. Plant breeders and plant pathologists have worked together since the rediscovery of Mendel's laws about the turn of this century. The need for further cooperation is spelled out clearly in a recent paper by Peter Day (1973):

> Plant pathology and entomology were spawned out of the need to devise ways to protect vulnerable crops. Plant breeders introduced genetic resistance to restore the balance but only succeeded in making the crops still more uniform . . . plant breeders and agronomists, caught up in the success of modern methods, should be aware of their long-term consequences to avoid laying foundations for intractible problems in the years ahead.

The need for cooperation also has been emphasized in eloquent papers by Jack Harlan (1972) and Norman Borlaug (1968). The special needs to conserve germplasm resources are outlined in a series of persuasive case studies by David Timothy (1972). He wrote his paper as part of "The Careless Technology"—a case by case analysis of the meager role of ecological thinking in international development programs around the world (Farar and Milton, 1972).

The objective of crop selection and breeding are determined by the demands of the growers, processors, and consumers of crops. Producers want abundant yields of quality produce that will command high prices and thus increase their incomes. To achieve this goal, producers want uniformity in time of germination and maturation, adaptability to mechanization, efficient use of fertilizers, and resistance to lodging, wind, diseases, insects, drought, and other stress factors. Processors want uniformity in size and shape as well as ease of harvesting, transport, and storage. In food crops, consumers want attractive appearance, flavor, texture, and both cooking and nutritional quality. In fiber crops, consumers want strength and durability. Breeders want convenience in pollination, hybridization, and other genetic manipulations as well as adaptability to differences in day length, growing season, and other climatic and environmental factors. Processing firms and producers often establish marketing standards and seed certification programs which tend to increase uniformity. Sometimes these standards find their way into laws such as the single variety law governing the type of cotton that can be grown in the San Joaquin Valley of California (see Chapter 6, Vol. I). All these pressures tend to limit genetic diversity within species and varieties of crops, and thus to increase genetic vulnerability.

The corn blight epidemic of 1970 emphasized that, in addition to genetic uniformity, uniformity in heritable determinants contained in the cytoplasm can also influence susceptibility to disease (Ullstrup, 1972; National Academy of Sciences, 1972). It is helpful to trace the background of the breeding practices that led to this destructive pandemic.

In 1931, a corn plant was discovered in Texas that produced infertile pollen. Later, it was found that this trait was heritable and the determining factor could be transferred to other corn plants through the cytoplasm of female parents. After the discovery of the pollen "restorer" gene, Texas male sterile cytoplasm proved to be very useful in producing seed of hybrid corn. It was so convenient, in fact, that by 1970 most of the inbred lines of corn in the United States carried this source of male sterility. This cytoplasmic factor made obsolete the expensive process of hand detasseling corn to ensure against self-pollination. At a cost of $1 billion in diseased corn plants, it was discovered in 1970 that this same cytoplasmic factor also transfers high susceptibility to a virulent strain of *Helminthosporium maydis,* the causal agent of southern corn leaf blight.

The unforeseen consequence of this simple practice, which was intended to decrease the cost of hybrid seed corn, led to an epidemic which destroyed more food than any other epidemic man had yet known! As Ullstrup (1972) said at the end of his dramatic account of this epidemic:

> In terms of human suffering, it is particularly fortunate that this epidemic occurred in a developed nation with a highly diversified agriculture. . . . nations whose agriculture is dependent on only a few crops will be wise to learn . . . how important it is to diversify their agriculture and maintain adequate genetic (and cytoplasmic) diversity in their major crops . . . Never again should a major cultivated species be molded into such uniformity that it is so universally vulnerable to attack by a pathogen, an insect, or environmental stress.

In many parts of Asia, Africa, and both Central and South America the widespread adoption of high-yielding dwarf wheat and rice varieties has increased the abundance of these crops but it has also decreased their genetic diversity. As a result, it is possible that locally adapted cultivars, so long used in Asia, may be lost irrevocably unless precautions are taken promptly to conserve these genetic resources in some permanent way (Frankel, 1971, 1972; Harlan, 1972; Day, 1973; Timothy, 1972; see also Chapter 13, this volume).

2. Forest Practices

Most agricultural crops have long ago lost their ability to compete in natural habitats. They are thus utterly dependent on man for their conservation and perpetuation. Although forests are not as dependent on man, cutting practices that favor one particular species over another, and widespread planting of trees of superior form and rate of growth are diminishing the functional diversity in various types of forests (see Chapter 14, this volume). For example, the current epidemic of fusiform rust in the southern pines of the United States is due in part to widespread replacement of resistant longleaf pine with rust-susceptible loblolly and slash pines (Dinus and Schmidt, 1977).

Similarly, in the northern United States the constant cutting of commercially desirable birch and maple in mixed hardwood forests left behind a constantly increasing population of beech that was less readily utilizable because it was difficult to dry the timber properly. Recently, wood technologists have overcome the difficulties of drying beech timber, but, in the meantime, many of the residual trees became overmature and weak. Soon the weakened trees were attacked by a complex of scale insects and *Nectria* sp. which cause the beech bark disease. Now this disease is spreading to younger, more vigorous trees as well as residual ones. The end result is an ever-growing epidemic of this disease which now affects American beech in much of the northeastern United States and the Maritime Provinces of Canada (Shigo, 1964).

Another illustration is the large-scale replacement of genetically diverse native forests with plantations of exotic pines in New Zealand, Australia, east and west Africa, and Chile. This practice has drastically increased the genetic vulnerability of the forest resources of these nations and regions of the world. The current epidemics of *Dothistroma* blight of *Pinus radiata* in all of these regions provides persuasive evidence of the hazards inherent in this forest practice (Gibson, 1972).

3. Landscaping Practices

The problem of genetic vulnerability of species can also be viewed in microcosm by comparing the landscaping for two homes in Raleigh, North Carolina. One home is in a grove of pines which shade the house and has a lush, green lawn of Kentucky 31 Red Fescue. The foundation plantings are limited to azaleas. The other house stands on a lot wooded with trees that are the products of natural regeneration—24 species of angiosperms and three species of gymnosperms, all of various ages. The lawn consists of a mixture of grasses and the shrubbery includes a variety of species. The vegetation around the first house stands vulnerable to an epidemic of pine bark beetles now spreading across this city. By contrast, the diversity of species surrounding the second house makes it almost immune to epidemic disease.

The current epidemic of the so-called Dutch elm disease in city after city across the United States also illustrates the risk of wide-scale planting of a genetically narrow, uniformly susceptible segment of a given host population. Genetically homogeneous selections of elms were planted as street trees in cities and towns all across North America. In regions such as southern New England and New York State, where the disease has been active for nearly four decades, the most susceptible trees have been killed, but the more resistant portions of the native elm population persist. If only the city and town planners of earlier years had understood the value of genetic diversity in the selection of trees for trees and parks! If they had, New Haven, Connecticut would perhaps still deserve its former nickname—the Elm City.

B. International Trading Practices that Favor Epidemics

Numerous epidemics on every continent have resulted from international trade of propagating materials and harvested products. The classic cases are known to us all—chestnut blight, white pine blister rust, and the so-called Dutch elm disease were introduced from Europe to North America by international tradesmen. Similarly, the powdery and downy mildews of grape, the downy mildew of hops, and the blue mold of tobacco were carried by man from North America to Europe. International trade also brought citrus canker from Asia to North America, as well as fire blight from North America to Asia, New Zealand, and Europe. Man also brought the golden nematode from Europe to both North and South America and the soybean cyst nematode from Asia to Europe and then to North America.

In discussing his longstanding interest in the close ties between agricultural practices and plant disease, Ten Houten (1974) gives special attention to the jet-speed international exchange of plant pathogens. He used chrysanthemum cuttings as an example:

> At present, cuttings for the European market are produced in the USA, South Africa, Italy and Malta, the French Riviera, and most of the northwest European countries. They are flown from one country to another, and this incurs the risk that certain plants may carry a disease, which due to quick transportation and a long incubation time, will not show before the plants are grown commercially in the growers' greenhouses. Thus *Ascochyta* ray blight has been spread over the world. In a similar way *Puccinia horiana*, one of many chrysanthemum rusts endemic in Japan, spread globally from one of the few private cutting producers in South Africa.

Ten Houten also mentions that

> the preparation and distribution of potting soil from greenhouse crops is done also by only a few specialized farms in The Netherlands. This has resulted in the spread of Arabis mosaic in cucumber, a soil-borne virus transmitted by nematodes. . . . These are only a few out of several examples of the rapid spread of hitherto unknown diseases due to modern agricultural practices and rapid transportation.

C. Horticultural Practices that Favor Epidemics Below Ground

Apple varieties propagated on size-controlling rootstocks offer many advantages in the management of apple orchards. The "dwarfed" trees are precocious and make for easier pruning, spraying, and harvesting. Their compact form permits many more trees per hectare. Pomologists have enthusiastically promoted their use and the idea has spread around the world. But this practice, together with other common horticultural practices, has led to epidemics of collar rot caused by *Phytophthora cactorum*.

Although many different size-controlling rootstocks are known, many of the most widely used (Malling-Merton 104 and 106, for example) are highly susceptible to collar rot. Many orchardists purchase grafted trees from commercial nurseries that propagate the size-controlling rootstocks vegetatively. The rooting beds are commonly used for many years. Infestation of these beds by *P. cactorum* leads to infection of the rootstocks. Because collar rot develops very slowly, latent infections are difficult to detect, and much infected planting stock is certified as "disease-free" prior to shipment to orchards throughout North America. When the trees arrive for planting, tractor-powered augers are commonly used to plant the grafted trees. In heavy soils, these augers tend to pack the sides of the hole forming a "well." As the soil settles, a depression or "basin" is formed around the tree. Both "well" and "basin" collect water and drain poorly. The end result of selecting disease-susceptible rootstocks, propagating them in infected nursery beds, and then providing environmental conditions conducive to collar rot is an increase in epidemics of this serious orchard disease.

A similar increase in amount of *Verticillium* wilt was reported by Dimock (1951) when New York State nurserymen stopped grafting roses to imported rootstocks on their own lands and instead purchased plants that were already budded to rootstocks grown on the Pacific Coast. Some of the Pacific Coast rootstocks also were infected before shipment.

D. Epidemics Induced by Early Planting

Early planting of onion sets in Michigan has led to epidemics of onion smut caused by *Urocystis colchici*. When onion sets are planted, a "race" is started between this pathogen and the emergence of the cotyledonary leaves, at which time the sets become immune to infection. The fungus usually invades the succulent young tissues. When onion sets are planted in cold soil (below 24°C) they mature very slowly and the fungus often wins the "race." When planting is delayed until soil temperatures are warmer (above 24°C), the plants usually become immune before the pathogen becomes established (Stienstra and Lacy, 1972).

E. Epidemics Induced by Crowding

In the late 1950's, the United States was producing more food than it could consume or market abroad. This helped to stabilize world food supplies, but the resulting low prices upset the farmers. The government, therefore, developed a "soil bank program" that would pay farmers to take some of their land out of production and plant trees instead. This tripled the demand for tree seedlings. Forest nurserymen responded to the challenge by expanding the size of the nurseries and by increasing the density of seedlings in their nursery beds. The results of the latter were disastrous. Crowding the seedlings together induced epidemics of *Pythium, Fusarium, Rhizoctonia,* and *Cylindrocladium* foliage and root diseases. These diseases spread rapidly from seedling to seedling, wiping out entire beds and sometimes whole sections of nurseries.

F. Epidemics Induced by Handling and Wounding Plants

Agricultural and forest practices that cause wounds favor epidemics of many types of pathogens by providing open avenues for infection. The list could be endless, but a few examples can be cited.

Even the most gentle handling of tobacco and tomato plants by field personnel can produce epidemics of tobacco mosaic virus. Pruning of recently blighted branches with nonsterile tools can accelerate the spread of fire blight in pear and apple orchards. Use of controlled burning in hardwood forests can increase the amount of discoloration and decay in tree stems. Similarly, wounds resulting from selective logging practices have greatly increased damage by many different decay-causing fungi. Use of paraformaldehyde tablets to increase the duration of sap flow during maple sugaring operations increases the amount of decay (Walters and Shigo, 1977). Injections of benomyl and other chemicals may be useful in slowing epidemics of Dutch elm disease but when carelessly administered, cause greatly increased decay in treated elms (Shigo and Campana, 1977). In Brazil, pruning of plants without sterilizing machetes between cuts has caused epidemics of the bacterial wilts caused by *Pseudomonas solanacearum* in banana and *Xanthomonas manihotis* in cassava.

G. Epidemics Induced by Fertilization Practices

Many investigators have reported fertilization practices that increase the susceptibility of host plants to disease and the rate of spread of epidemics (van der Plank, 1963). Huber and Watson (1974) have enumerated 20 root or cortical rots, six vascular diseases, seven foliar diseases, three nematode, gall or other diseases, and three virus diseases in which applications of nitrate or ammonium nitrogen increased disease susceptibility. Similar effects have been reported for phosphorus and certain micronutrients including copper, boron, and zinc (Yarwood, 1959). By contrast, fertilization with potassium and calcium generally decreases disease susceptibility (Walker, 1957).

Since fertilization usually increases the vigor of plants, pathogens that prefer vigorous hosts may be favored by fertilization treatments. Fertilization treatments that result in nutrient imbalances generally increase disease susceptibility whereas treatments which correct a previously existing nutrient imbalance often decrease disease susceptibility. Fertilization of peanuts during the growing season can lead to epidemics of pod breakdown disease caused by *Pythium myriotylum* (Hallock, 1973). Apparently, midseason fertilization increases the amount of succulent root tissue which is readily attacked by this pathogen.

H. Epidemics Induced by Biocides

Use of biocides to control various plant pests is increasing enormously (Altman and Campbell, 1977). Most chemicals used to control plant pests have a range of biological activities. Some compounds, such as the antibiotics used to control bacteria, are not known to inhibit fungi, insects, or weeds. Other compounds are much less discriminating, however. Benomyl, for example, will inhibit many different types of fungi, including plant pathogens, beneficial mycorrhizal fungi, and soil saprophytes that are natural antagonists of plant pathogens (Backman *et al.*, 1975). In many cases, herbicides that are used to control weeds also have effects on crop plants, sometimes increasing their susceptibility to pathogens (Katan and Eshel, 1973; Altman and Campbell, 1977).

Despite these known, and frequently unknown, biological activities, it is a common agricultural practice to name compounds according to their *intended use* rather than their *actual activity*. For example, benomyl was first tested as a nematicide, then later as an insecticide, and finally was developed as a fungicide. Also, we commonly say "2,4-D is a herbicide" even though it has been reported to stimulate growth of certain pathogens such as *Helminthosporium sativum* and to favor development of disease caused by *Botrytis fabae* on bean and *Fusarium oxysporum* on tomato (Katan and Eshel, 1973).

Epidemics of various pathogens often are induced by farmers who are not sufficiently aware of these unintended side effects on "nontarget" organisms. For example, peanut farmers in the southeastern United States have often applied benomyl, Bravo, Difolatan, or Fungisperse to control *Cercospora* leaf spot only to find that these compounds induce epidemics of foliar mites. What happened? Apparently these compounds kill both *Cercospora* and certain parasitic fungi that regulate mite populations (Campbell and Batts, 1974). For this reason it is desirable to call these four compounds "broad-spectrum fungicides" instead of simply "fungicides."

Unexpected increases in disease sometimes result from changes from one pesticide to another or from interactions among different pesticides used on the same crop. For example, sooty blotch of apples became more prevalent when captan replaced the copper fungicides formerly used to control apple scab and when organic insecticides such as DDT and Guthion were substituted for lead arsenate for control of insects in apple orchards (Groves, 1953). Some unexplained interaction between captan and both DDT and Guthion apparently makes these combinations less effective than captan used together with lead arsenate (Daines, 1962).

Natti *et al.* (1956) discuss an intriguing interaction between an emulsifying agent, a solvent, a fungicide, and an insecticide when these materials were used together on broccoli. The solvent needed for the insecticide evidently dissolved part of the cuticle. As a result, leaves remained wet longer, thus providing conditions more favorable for downy mildew. These authors also describe varietal-selection practices which led to an epidemic of downy mildew on broccoli in New York State in 1954 and 1955.

The interactions described above among combinations of pesticides, "inert ingredients," and their differential effects on disease complexes are part of the reason that Cox (1971) so strongly advocates apprenticeship experiences and course work in entomology, weed sciences, plant nutrition, and plant pathology in the education of private consultants for agriculture (see also Chapter 21, Vol. I).

I. Epidemics Induced by Tillage Practices

Recently, increased interest in reduced tillage systems of crop culture has stimulated renewed interest in the effects of tillage practices on plant disease. Planting peanuts without previous plowing of the land has induced epidemics of stem rot caused by *Sclerotium rolfsii*. A similar effect has been noted with anthracnose of corn caused by *Colletotrichum graminicola*. Plant refuse left on the soil surface provides a food base for production of inoculum of these pathogens. Because of such repeated observations, so-called "no-till" systems of land preparation are generally considered hazardous for agronomic crops, at least in the southeastern United States.

J. Epidemics Induced by Irrigation Practices

Cook and Papendick (1972) have discussed the water potential of plants and soil and their substantial effects on pathogens of wheat and other field crops. By careful management of water potential it is possible to diminish the rate of spread of disease. But careless irrigation can provide conditions highly favorable for diseases. Cook and Papendick list 11 diseases favored by dry soil and eight diseases favored by wet soil.

Overhead irrigation of corn with water from certain surface ponds has been shown to increase the abundance of bacterial stalk rot whereas similar irrigation with water from wells had no such effect (D. L. Thompson, personal communication). The hypothesis is currently being tested that the stalk-rot bacteria drain into surface ponds from infected plants and then are redistributed over the crop in the irrigation water.

Other reports of increases in disease due to irrigation include those of Crossan and Lloyd (1956) on anthracnose and *Rhizoctonia* fruit rots of tomato; Curl and Weaver (1958) on leaf-spotting diseases of various forage crops; and Menzies (1954) on *Fusarium* foot rot of beans and rust of spearmint and beans. It stands to reason that overhead systems of irrigation will inevitably increase the hazard of splash dispersal of pathogens and increase periods of leaf wetness. These disease-favoring conditions can be avoided by use of subsoil, trickle, or furrow systems of irrigation (Cook and Papendick, 1972; Dinus and Schmidt, 1977).

K. Epidemics Induced by Machine Harvesting

Until the early 1960's the cherries in Door County, Wisconsin, were picked by hand. Normally, very little fruit was left on the trees. Brown rot of the fruit caused by *Monilinia fructicola* occurred periodically but usually was controlled by standard sprays. During the 1960's, tree shakers were developed to decrease the rising labor costs of harvesting, but the shakers left many more cherries on the trees. The unpicked fruit became infected with *Monilinia* and dried into the characteristic mummies which serve as overwintering sites for the brown-rot fungus. Following spring rains, conidia of the fungus were produced in such abundance that epidemics of brown rot became more and more difficult to control. The increasingly high cost of labor has made it impossible to return to hand picking. Thus, increased severity of brown rot induced by increases in initial inoculum caused by changes in mechanical harvesting has contributed to the decline of the cherry-growing industry in Door County.

Wounds induced by mechanical fruit-harvesting devices have also increased epidemics of *Ceratocystis* canker of stone fruit trees (DeVay *et al.*, 1960). This relationship was summarized well by Barnes (1964):

> *Ceratocystis fimbriata* . . . is commonly present in almond orchards but apparently becomes parasitic only when in contact with injured bark tissues. After penetration in the wounded area, the fungus invades adjacent bark, becomes perennial, enlarges year after year, and eventually girdles and kills the branches.

For many years, mallets with rubber padding have been used to strike large limbs to dislodge the almonds during harvest. Injuries caused by the mallets were the main avenue of entrance. The extensive use of mechanical shakers in recent years, however, has increased the extent of injuries during harvest and the prevalence of *Ceratocystis* cankers. Moreover, the extension of harvest procedures involving mechanical shakers to other stone fruits such as prunes, peaches, and apricots has resulted in rapid spread of *C. fimbriata* to these fruit trees, especially to prunes. In two French prune orchards in California, each approximately 50 acres in size, 87 and 74% of the trees have become severely diseased by use of mechanical tree shakers. In contrast to bruise-type injuries, pruning wounds have not been an important avenue of entry for *C. fimbriata* in stone fruits (DeVay *et al.*, 1962). Fruit decay of mechanically harvested peaches and apricots has been reported (Ogawa, Sandeno and Mathre, 1963). Such effects are not necessarily limited to fruits; the incidence of *Fusarium* dry rot of potatoes increases after mechanical grading (Eide, 1955).

L. Epidemics Induced by Harvesting and Storage Practices

Certain harvesting and storage practices can favor epidemics of storage diseases in many crops. Using a standard method for determining the decay potential of potato tubers, Lund and Kelman (1977) have shown that: (1) machine-dug tubers are much more susceptible to bacterial soft rot than hand-dug tubers; (2) tubers that retain a film of water after washing can become anaerobic very quickly and then are much more susceptible to decay than tubers that are dried thoroughly after washing; (3) even with thorough drying, washed tubers retain an inherently greater susceptibility to decay than nonwashed tubers; and (4) water films on tuber surfaces and use of certain chemical sprout inhibitors can inhibit healing of mechanical wounds and thus predispose tubers to decay.

IV. CONCLUSIONS

In this chapter we have discussed both general factors and specific management practices that favor epidemics of plant disease. Preparing this chapter has also led to certain conclusions about the relationship between agricultural and forest practices and epidemics of plant disease.

1. Most epidemic-favoring practices result from one or more of the following: (a) inadequate knowledge of disease processes; (b) inadequate extension of available knowledge; (c) unforeseen consequences of changing management practices; (d) poor communication or lack of coordination among crop-production and crop-protection specialists; and (e) administrative and institutional constraints within research, extension, or marketing organizations.

2. Numerous authoritative studies in recent years have warned

against the dangers of uniformity in our major crops and recommended against further dwindling of genetic and germ plasm resources. These warnings have come from Marshall Ward (1901) and Neil Stevens (1939, 1942), from blue-ribbon panels headed by James Horsfall for the National Academy of Sciences (1972), and by George Sprague for the United States Department of Agriculture (1973). Similar warnings and recommendations were issued from special meetings called by the Food and Agriculture Organization of the United Nations in 1961, 1967, and 1972; by the International Biological Program in 1967; by the Rockefeller Foundation in 1971; by the Consultative Group on International Agricultural Research in 1971 and 1972; by the Conservation Foundation and The Center for the Biology of Natural Systems, Washington University in 1971 (Farvar and Milton, 1972); by participants in various symposia at the International Congress of Plant Pathology in 1973; by the American Phytopathological Society in 1974 and 1975; and by the New York Academy of Sciences (Day, 1977). But all these reports with all of their constructive recommendations have not been heeded adequately. The major crops of the world are still less variable and, therefore, still more vulnerable than they were in 1901, 1939, or even as recently as 1970 or 1975. It appears that the forces in man's nature and his economic system that demand uniformity in crops are stronger than the combined persuasive force of argument and reason that can be generated by all of these agencies and some of their most persuasive spokesmen.

3. The criteria for advancement and recognition in plant pathology, particularly at the earliest stages of a scientific career, place a higher premium on discovery and publication of new knowledge than on vigilance in the protection of crops against disease. Status is achieved by being the first to discover and publish new knowledge. Thus, success is judged primarily by the quantity, quality, and priority of manuscripts published in respected scientific journals and only secondarily by the benefits society derives from the discovery.

Society rightfully expects a return from its investment in agricultural research and extension; scientists have a responsibility to ensure that this return is substantial. More effective packaging and delivery of information to users is essential. Innovative leadership and other incentives apparently are needed to draw plant pathologists, plant breeders, and other agricultural and forest scientists into well-coordinated, interdisciplinary teams with the primary mission to stabilize crop production by protecting crops from epidemic disease and other stresses. Ideally these teams should be national and preferably international in scope and participation. Thoughtful consideration and dedicated effort by every agricultural scientist will help ensure that such teams are created and that they function effectively.

4. Plant pathologists have not often concerned themselves with management practices that favor epidemics. Only Barnes (1964) and Ten Houten (1974) appear to have dealt with this subject systematically. This lack of concern is reflected in the conspicuous absence of words relating to agriculture, horticulture, agronomy, and silviculture in the indexes of the textbooks and the original, review, and abstracting journals of plant pathology. In the future we hope plant pathologists will deal with this subject more openly so that farmers, foresters, and pathologists alike can know "practices to avoid" as well as "practices to follow" in minimizing loss due to disease.

Two recent publications by the North Eastern Forest Experiment Station in Upper Darby, Pennsylvania, illustrate this approach. These blunt publications are entitled: "How to Kill Your Tree 'Let Me Count the Ways'" and "Your Trees' Trouble May Be You!"

5. The priorities that guide investment in research on internationally dangerous plant diseases are usually low until a crisis is upon us (Thurston, 1973; Gibson, 1972; United States Department of Agriculture, 1973. See also Chapter 6 in this volume.). Time after time, with disease after disease, the resources with which to pursue research needed to cope with an introduced pathogen have become available only after the disease has been introduced. Crisis management is an understandable mechanism by which to establish priorities for research, but it is terribly costly and, in many cases, contrary to a reasoned assessment of costs and benefits. Let us hope that reason will prevail more often in the future.

6. In a facetious "toast to changing technology" after a field excursion during the summer of 1968, the late Erik Björkman summed up an important implication of this chapter: "Every change in technology will bring forth new ecological conditions in which plant pathogens must live. These changes will bring on new disease problems for plant pathologists to study. Let us drink to the future changes in technology that will provide work for us all!"

Acknowledgments

Many colleagues have provided examples and made suggestions for this chapter. I am especially grateful to Jay Julius, Paulo de Souza, Carlyle Clayton, Turner Sutton, Arthur Kelman, Alex Shigo, Marvin Beute, Charles Main, Donald Thompson, and Arthur Verrall.

References

Altman, J., and Campbell, C. L. (1977). Effect of herbicides on plant diseases. *Annu. Rev. Phytopathol.* **15**, 361–385.

Backman, P. A., Rodriguez-Kabana, R., and Williams, J. C. (1975). The effect of peanut leafspot fungicides on the nontarget pathogen, *Sclerotium rolfsii. Phytopathology* 65, 773–776.

Barnes, E. H. (1964). Changing plant disease losses in a changing agriculture. *Phytopathology* 54, 1314–1319.

Borlaug, N. E. (1968). Wheat breeding and its impact on world food supply. *Proc. Third Internl. Wheat Genetics Symposium.* Australian Academy of Sciences, Canberra, Aust. pp. 1–36.

Campbell, W. V., and Batts, R. W. (1974). Effect of fungicides and insecticides on spider mite buildup and suppression on peanuts. *J. Am. Peanut Res. Educ. Assoc.* 6, 51.

Cook, R. J., and Papendick, R. I. (1972). Influence of water potential of soils and plants on root disease. *Annu. Rev. Phytopathol.* 10, 349–374.

Cox, R. S. (1971). "The Private Practitioner in Agriculture." Solo Publications, Lake Worth, Florida.

Crossan, D. F., and Lloyd, P. J. (1956). The influence of overhead irrigation on the incidence and control of certain tomato diseases. *Plant Dis. Rep.* 40, 314–317.

Curl, E. A., and Weaver, H. A. (1958). Diseases of forage crops under sprinkler irrigation in the southeast. *Plant Dis. Rep.* 42, 637–644.

Daines, R. H. (1962). Spray programs and the control of sooty blotch of apples. *Plant Dis. Rep.* 46, 513–515.

Day, P. R. (1973). Genetic variability of crops. *Annu. Rev. Phytopathol.* 11, 293–312.

Day, P. R. (ed.) (1977). "The Genetic Basis of Epidemics in Agriculture." N.Y. Acad. Sci., New York.

DeVay, J. E., English, H., Lukezic, F. L., and O'Reilly, H. J. (1960). Mallet wound canker of almond trees. *Calif. Agric.* 14(8), 8–9.

DeVay, J. E., Lukezic, F. L., English, H., Uriu, K., and Hansen, C. J. (1962). Ceratocystis canker. *Calif. Agric.* 16(1), 2–3.

Dimock, A. W. (1951). Bud transmission of *Verticillium* in roses. *Phytopathology* 41, 781–784.

Dinus, R. J., and Schmidt, R. A. (1977). "Management of Fusiform Rust in Southern Pines." *Symp. Proc. Univ. Fla.*, Gainesville, Florida.

Eide, C. J. (1955). Fungus infection of plants. *Annu. Rev. Microbiol.* 9, 297–318.

Farvar, M. T., and Milton, J. P., eds. (1972). "The Careless Technology: Ecology and International Development." Natural History Press, New York.

Frankel, O. H. (1971). Genetic dangers in the green revolution. *World Agric.* 19(3), 9–13.

Frankel, O. H. (1972). Genetic conservation—parable of the scientists' social responsibility. *Search* 3, 193–201.

Gibson, I. A. S. (1972). *Dothistroma* blight of *Pinus radiata. Annu. Rev. Phytopathol.* 10, 51–72.

Groves, A. B. (1953). Sooty blotch and fly speck. *In* "Plant Diseases." Yearbook of Agriculture, pp. 663–666. U.S. Dep. Agric., Washington, D.C.

Hallock, D. L. (1973). Soil fertility relationship in pod breakdown disease of peanuts. *J. Am. Peanut Res. Educ. Assoc.* 5, 152–159.

Harlan, J. R. (1972). Genetics of disaster. *J. Environ. Qual.* 1, 212–215.

Huber, D. M., and Watson, R. D. (1974). Nitrogen form and plant disease. *Annu. Rev. Phytopathol.* 12, 139–165.

Katan, J., and Eshel, Y. (1973). Interactions between herbicides and plant pathogens. *Residue Rev.* 45, 145–177.

Lund, B. M., and Kelman, A. (1977). Determination of the potential for development of bacterial soft rot of potatoes. *Am. Potato J.* **54**, 211–225.

Mangelsdorf, P. C. (1966). Genetic potentials for increasing yields of food crops and animals. *Proc. Natl. Acad. Sci. U.S.A.* **56**, 370–375.

Menzies, J. D. (1954). Plant disease observations in the irrigated areas of central Washington during 1953. *Plant Dis. Rep.* **38**, 314–315.

National Academy of Sciences. (1972). "Genetic Vulnerability of Major Crops." Nat. Acad. Sci., Washington, D.C.

Natti, J. J., Hervey, G. E. R., and Sayre, C. B. (1956). Factors contributing to the increase of downy mildew of broccoli in New York State and its control with fungicides and agrimycin. *Plant Dis. Rep.* **40**, 118–124.

Ogawa, J. M., Sandeno, J. L., and Mathre, J. H. (1963). Comparisons in development and chemical control of decay-causing organisms on mechanical- and hand-harvested stone fruits. *Plant Dis. Rep.* **47**, 129–133.

Schoeneweiss, D. F. (1975). Predisposition, stress, and plant disease. *Annu. Rev. Phytopathol.* **13**, 193–211.

Shigo, A. L. (1964). Organism interactions in the beech bark disease. *Phytopathology* **54**, 263–269.

Shigo, A. L., and Campana, R. (1977). Discolored and decayed wood associated with injection holes in American elm. *J. Arboricul.* **3**, 230–235.

Stienstra, W. C., and Lacy, M. L. (1972). Effect of inoculum density, planting depth, and soil temperature on *Urocystis colchici* infection of onion. *Phytopathology* **62**, 282–286.

Stevens, N. E. (1939). Disease, damage and pollination types in "grains." *Science* **89**, 339–340.

Stevens, N. E. (1942). How plant breeding programs complicate plant disease problems. *Science* **95**, 313–316.

Ten Houten, J. G. (1974). Plant pathology: Changing agricultural methods and human society. *Annu. Rev. Phytopathol.* **12**, 1–11.

Thurston, H. D. (1973). Threatening plant diseases. *Annu. Rev. Phytopathol.* **11**, 27–52.

Timothy, D. H. (1972). Plant germ plasm resources and utilization. *In* "The Careless Tehcnology," M. T. Farvar and J. P. Milton, eds., pp. 631–656. Natural History Press, New York.

Ullstrup, A. J. (1972). The impacts of the southern corn leaf blight epidemics of 1970–71. *Annu. Rev. Phytopathol.* **10**, 37–50.

U.S. Department of Agriculture. (1973). "Recommended Actions and Policies for Minimizing the Genetic Vulnerability of Our Major Crops," Numbered special report by an Ad Hoc Subcommittee of the Agricultural Research Policy Advisory Committee. U.S. Dep. Agric., Washington, D.C.

Van der Plank, J. E. (1963). "Plant diseases: Epidemics and Control." Academic Press, New York.

Waggoner, P. E. (1962). Weather, space, time, and chance of infection. *Phytopathology* **52**, 1100–1108.

Walker, J. C. (1957). "Plant Pathology." McGraw-Hill, New York.

Walters, R. S., and Shigo, A. L. (1977). Discoloration and decay associated with paraformaldehyde-treated tapholes in sugar maple (*Acer saccharum* Marsh.). *Can. J. For. Res.* (in press).

Ward, H. M. (1901). "Diseases in Plants." MacMillan, New York.

Yarwood, C. E. (1959). Predisposition. *In* "Plant Pathology: An Advanced Treatise"

(J. G. Horsfall and A. E. Dimond, eds.), Vol. 1, pp. 521–562. Academic Press, New York.

Zadoks, J. C., and Koster, L. M. (1976). "A Historical Survey of Botanical Epidemiology; A Sketch of the Development of Ideas in Ecological Phytopathology." Mededelingen Landbouwhogeschool, Wageningen, The Netherlands, 76, 1–56.

Chapter 18

People-Placed Pathogens: The Emigrant Pests

RUSSELL C. McGREGOR

I. INTRODUCTION

Man is a mobile animal. He moves himself, his family, his domesticated crops, and his livestock. If he moves, his pests move too. If he emigrates, they emigrate too; he places them down wherever he settles. These may be called "people-placed pathogens" or "the emigrant pests."

In fact his plant pests live so closely with him that they are as domesticated as his crops and stock. For example, the English and Scotch emigrated to New Zealand, taking their domesticated apples and their domesticated pests, apple scab and codling moth, with them. Similarly, when they emigrated to New England and Virginia, they took their domesticated smuts and rusts along with the wheat.

Once society sees that pests emigrate, it decides that it needs another law and a set of policemen to stop the emigration—thus a quarantine operation is born. People and their equipages will be stopped at the border or even at the point of origin.

Nearly every country makes some effort to stop agricultural pests and pathogens. In the United States the agency is the Animal and Plant

383

Health Inspection Service with the astonishing acronym of APHIS (a potent pest). In Europe the individual country organizations have joined together in EPPO—the European Plant Protection Organization, as described by Mathys, 1977.

II. OBJECTIVE OF THIS CHAPTER

The material for this chapter originated a few years ago when doubts about the efficiency of its efforts to prevent the entry of exotic agricultural pests and pathogens, coupled with rising volumes of international travel, prompted the officials of the United States Department of Agriculture to look for more efficient alternatives to the present array of quarantine activities. The Department established an Import Inspection Task Force which I was asked to chair.* In 1973 our report was published (see McGregor, 1973). This chapter is an analysis of that report plus some later cogitation. It is a study of the system in the United States, but I hope that it will serve as a sample of the systems of the world.

III. THE ARRIVAL OF IMMIGRANT SPECIES

A. The Potential for Introduction

The continental United States has been and still is particularly prone to pest and pathogen introduction. It is a large land mass settled and developed agriculturally for nearly 500 years. Most of our crops and pests are introductions. The "melting pot" of ethnic groups, it is likewise the melting pot of crops and pests.

In 1919, J. A. Stevenson listed 120 foreign plant diseases known to have been introduced into the country and commented that the list was far from complete because not all those then present were recorded and because those introduced in earliest colonial times were also not recorded. Some rusts and smuts of cereals are in this category.

The appearance of additional foreign pests has continued since that accounting and there is no doubt that introduction and establishment

---

* The other members of the Task Force were as follows: Richard D. Butler, Austin Fox, Donald Johnson, C. H. Kingsolver, Bert Levy, Herbert E. Pritchard, and Reece I. Sailer, all of the United States Department of Agriculture. I would like to extend my thanks to them.

continues. For example, Dutch elm disease was first noted in Ohio in 1931. A recent listing by A. J. Watson (1971) lists 1492 bacterial and fungal diseases foreign to the United States; the listing excludes viruses, nematodes, and all diseases of forest trees.

Since ecosystems evolved long before man became an important force in evolution, their stability was determined by factors of climate, geography, and isolation. During the past 5000 years man has modified the environment at an accelerating rate. Through his agriculture and other impacts on environment he has created new, usually much less complex, ecosystems that became increasingly vulnerable to disruption through disease attack as the area and intensity of agriculture grew. Initially these effects were small and confined to areas where crop plants and livestock were first domesticated. However, the age of discovery that began when Columbus discovered America in 1492 set in motion changes that affected the world biota on every continent and most islands. No continental area has been more affected than the United States.

During the past 480 years the geographic barriers provided by the Atlantic and Pacific Oceans have been breeched by man's commerce, and the ecology of the continent changed by his agriculture and other activities. In developing the agroecosystems that now occupy most of the United States, European immigrants adopted and greatly expanded the culture of such native crop plants as corn, cotton, potatoes, and tobacco. They brought with them wheat and other small grains, forage crops, livestock, vegetables, and fruit trees. Inadvertently they also brought weeds, insect pests, and pathogens. Many thrived because they were unaccompanied by natural enemies that were present in the agroecosystems of Europe. Others failed to overcome the natural ecological barriers to colonization. As commerce to other parts of the world increased, new pests and diseases continued to arrive.

B. The Pathways of Entry

Pathogens foreign to North America have been gaining entry and colonizing favorable habitats within the boundaries of the United States for at least 350 years. The successful immigrants for the most part have been those best adapted to survive in the pathways of entry and fortunate enough to find a favorable environment in which to live and reproduce once they arrived. Initially the successful species were those associated with man and his stored products.

For 1972, the Bureau of Customs reported that about 70 million carriers of persons and merchandise traversed the pathways of entry from other countries to the United States. More than one-half of these were vehicles entering from across the Mexican border. This volume of traffic is steadily increasing.

Since 1940, aircraft have increasingly traveled the paths of entry. The rapidity with which planes move from one part of the world to another allows short-lived, winged adult insects and infective propagules of plant pathogens to survive transit and escape into new geographic areas. The number of planes involved, and the number of distant locations that may be visited on a single flight, increases the magnitude of this threat.

Many of the new exotic additions to our North American insect fauna will arrive by aircraft. Among these insects are those which serve as vectors of plant virus diseases. It is possible that viruliferous insects could survive this pathway to introduce new virus diseases. Short-lived spores of plant pathogens will also enter by this route.

The airplane has accentuated another problem that has long existed. For a variety of reasons biologists frequently wish to use exotic pests and plants as objects of experimentation. Often this is because a foreign scientist has used these species in his research and has developed background knowledge of a kind essential as a starting point for the American scientists' research. In other instances a traveling scientist is tempted to bring breeding stock to his home laboratory because it has characteristics that pique his curiosity. Once such an organism has become the subject of some unusual research contribution, other scientists will often wish to obtain cultures.

We may expect that this avenue of entry will become more important as a result of increased interest in research on genetic control of a variety of pests and diseases. In this case, the danger is often that of increasing the gene pool of an already established pest, but there is also the danger of introducing closely related species having different potential as pests. This could be the result of deliberate introduction for use as sterile hybrids in control experiments or inadvertent introduction of species thought to be the same as an established pest.

Where such research is conducted by competent, responsible scientists under adequate quarantine, no significant hazard will be involved. However, competency, responsibility, and adequacy are all relative terms and any such research should be kept under strict surveillance by competent regulatory personnel.

A more serious danger is the scientist who wishes to bring in an organism but is either unaware of regulations or deliberately chooses to ignore them. He may correctly regard his plant or insect as entirely harmless, but what he may not recognize or have the competence to detect are the associated pathogens. When such efforts to introduce organisms in violation of regulations are detected, they should be investigated. Undetected violations are sometimes discovered later, as the scientist is likely to publish results of research involving the illegally imported organisms. If the violation has resulted in any adverse economic effect, or could have done so, the violator may be held responsible and the matter fully publicized.

C. Colonization and Establishment

It is evident that there is an unknown but very large number of foreign bacteria, fungi, and viruses that are potentially dangerous to the agriculture and environment of the United States. On the basis of past experience we can predict that a certain number of these species may gain entry in a given period of time, but we cannot predict their identity. Thurston (1973) has listed a few.

These potentially dangerous species are ticket holders in a sweepstakes lottery. A relatively small number of the species hold a disproportionate number of the tickets and thus increase their chance of entry. Many of the tickets are lost in the pathways of entry. Since the sweepstakes are illegal in the United States, any tickets found by quarantine inspectors are confiscated and destroyed. Before the final drawing inside the United States the ticket holders are subjected to a series of chance hazards and a final fitness test.

With few exceptions, many of the pathogens on our list will almost surely not become established in the United States in the next few years. Yet, it is almost surely true that some of them will become established here. This paradox points to the futility of gearing countermeasures to a pest-by-pest approach. We cannot have 1000 programs to counter 1000 unlikely pathogens. Some will get through.

Once past the quarantine barrier, the infective propagule must be transported—except for those with airborne spores—to a susceptible host crop and arrival must coincide with environmental conditions required for infection. Probability of success is extremely low, but propagule populations are large and the resistance of some fungal spores and of nematodes to adverse environmental conditions is astounding.

IV. DEFINING THE THREAT

Given the record of establishment of immigrant species and the subsequent importance of many of them as pests, there is good reason to inquire about the additional foreign species that may be able to invade the United States. We need to establish the magnitude of the threat from invasion by additional foreign pests.

A. The Threat of Invasion

There are about 600 plant diseases not present in the United States that may be considered significant. But admittedly our ability to predict the consequences of the introduction of any given foreign pathogen is so poor that any list of allegedly injurious species may provide an inadequate basis for program decisions.

The question boils down to this: How accurate do our predictions concerning the economic significance of particular foreign species introducible into the United States have to be before it makes sense to identify particular species to look for at ports of entry?

A rational program for protection needs some notion of what species a quarantine is trying to keep out. Since it is faced with limited resources and cannot protect the nation against everything, the policeman needs to have some ordering of the potential invaders that provides an opportunity to make choices, however uncertain, in the use of program resources.

As long as commerce exists between the United States and other parts of the world there is a probability of establishment for each potentially dangerous species. The level of probability will be different for each species and may be affected by regulatory activities designed to exclude their entry. Although we are dealing with relatively small probabilities, we should not be misled into thinking that this implies a lack of importance.

This is a situation where a 1% probability may be very high. To illustrate this point, a 1% probability of establishment means that on the average it will be 99.5 years until the first infestation, and a 3% probability means only 32.8 years.

Recorded establishments of plant pathogens have averaged three per year over a 25-year period. From a listing of approximately 2000 foreign plant pathogens 551 were chosen by Dr. C. H. Kingsolver of our Task Force ** as posing significant risk to our agriculture. The selection of this 551 was influenced to a major extent by the economic value of its host or hosts. We thus introduce an additional factor in the quarantine concept. Not only is the probability of entrance to be considered but the possible economic impact as well; e.g., a pathogen of corn is of

---

** Dr. Kingsolver had able assistance from R. W. Beardmore, Magan Golden, Virginia Harrington, K. R. Irish, Bernard Lipscomb, C. G. Schmitt, and Keith Shea. I would like to extend my thanks to them.

greater economic concern than one of geraniums. No consideration of the total universe of fungi, bacteria, nematodes, and viruses was attempted. Some 50,000 parasitic and nonparasitic diseases of plants are listed as present in the "Index of Plant Diseases in the United States" (Anonymous, 1960).

B. A Model for Ranking Importance

1. The Conceptual Design

In our Task Force we considered the desirability of including social and environmental values in the model. However, in this first attempt at ranking it was decided to use only the economic values, since these are quantified and readily available and the use of a single scale of values would simplify the model. The ranking of pests might be quite different if other kinds of values were incorporated.

A three-step procedure was developed for ranking exotic pests. First, estimate the probability of specific exotic pests becoming established in the United States. Second, evaluate the economic impact if those pests were to become established. Third, multiply the first value by the second —that is, the probability of an exotic pest becoming established times the economic impact of the pest if it becomes established. This value constitutes the expected score of economic importance of exotic pests in the United States and was labeled "Expected Economic Impact" (EEI). ("Expected" is used here in the statistical sense, i.e., "average" or "mean.")

In algebraic terms

$$\text{EEI} = P \times E \qquad (1)$$

where EEI = expected economic impact, P, probability of pest becoming established in the United States during the next year, and E = economic impact if pest became established.

The conceptual importance of this procedure is that it yields a quantifiable measure of economic risk. One important limitation of the model is that it is static rather than dynamic. For example, the rate of spread of an immigrant species through its ecological range is not taken into account. This is a significant time-related variable. However, the Task Force excluded it from the model, believing that the estimates required would have a wider confidence interval than the other variables in the model. There is an urgent need to obtain estimates of rates of spread for important pests in their overseas locations, in order to provide a basis for estimates of spread in the United States.

2. Probability of a Pest becoming Established

It was assumed that the probability of a pest becoming established was related to the volume of vector material imported into the United States, the hitchhiking potential of the pest, and the ease with which the pest became established after arrival. It was felt that impressions of the relationship could be estimated empirically with a second degree equation for the general relation.

$$P = G \ (f_1, f_2, f_3) \tag{2}$$

where P = probability of pest becoming established, f_1 = volume of vector material imported into the United States, f_2 = hitchhiking potential of the pest, and f_3 = ease with which a pest becomes established after arrival.

3. Economic Impact If a Pest Becomes Established

The economic impact of an agricultural pest if it becomes established is used here as a marginal measure. It is the expenditures required to maintain production of the host crop. It is the summation of the added cost of pest control on old units, plus the added cost of pest control on new units needed to maintain production, plus the added cost of raising the new units. The formula used is:

$$E = VRT + V \left[\frac{URT}{(100 - U)} \right] + W \left[\frac{URT}{(100 - U)} \right] \tag{3}$$

where E = the economic impact of the established pest, R = the amount of host grown, T = the ecological range of the pest as a percentage of the range of the host, U = percentage loss in yield when normal controls are used, V = added control cost per unit per season for the pest, W = variable cost per unit for host (amount of money required to increase growing area by one acre, herd by one head, etc.).

V. EXOTIC PESTS AND DISEASES

Task Force biologist Dr. C. H. Kingsolver provided essential information on 551 plant diseases and nematodes which are a significant threat to the United States.

The 49 top-ranking exotic pathogens are listed in Table I, in the order of their Expected Economic Impact (EEI). The inclusion of the first 49 species in the table, rather than some other number, is arbitrary. All of the 551 species of plant diseases have been ranked on a computer

TABLE I

The 49 Most Dangerous Exotic Disease Pests
(Millions of Dollars)

Rank [a]	Species	Proba-bility of estab-lishment [b]	EEI Midpoint	EEI range (plus or minus) [c] confidence interval 50%	90%
1	*Rosellinia radiciperda*	H	3126	282	698
2	*Helicobasidium mompa*	M	2703	118	291
3	*Cronartium himalayense*	M	1992	96	238
4	*Poria rhizomorpha*	M	1915	92	228
5	*Cronartium quercuum*	M	1406	68	168
6	*Xanthomonas acernae*	M	1118	54	134
7	*Phytophytora cambivora*	M	1093	81	201
8	*Melampsora pinitorqua*	M	662	32	79
9	*Rosellinia quercina*	M	579	28	69
10	*Phakopsora pachyrhizi*	M	551	35	87
11	*Rhizoctonia lamellifera*	M	496	37	91
12	*Scleroderris abietina*	M	440	33	81
13	*Hypodermella sulcigena*	M	434	21	52
14	*Cenangium kozactstanicum*	M	325	16	38
15	*Heterodera zeae*	L	320	22	54
16	*Heterodera avenae*	M	312	14	34
17	*Heterodera latipons*	H	278	25	62
18	*Brunchorstia pini*	M	258	34	85
19	*Aecidium glycines*	M	222	10	26
20	*Pseudomonas syringae f. populia*	M	221	11	26
21	*Cercospora pinidensiflorae*	M	216	10	26
22	*Sclerospora sacchari*	M	213	10	25
23	*Acanthostigma parasiticum*	M	174	8	20
24	*Sclerospora philippinensis*	M	161	12	30
25	*Sclerospora spontanea*	M	152	7	18
26	Maize streak virus	M	132	8	18
27	*Mycosphaerella sojae*	M	116	8	18
28	Rice dwarf virus	L	115	8	20
29	*Septoria maydis*	H	112	10	25
30	*Synchytrium dolichi*	H	112	10	25
31	*Xanthomonas vaculorum*	M	107	6	12
32	*Synchytrium umbilicatum*	H	104	10	23
33	Datura 437 virus	M	97	6	14
34	*Corynebacterium tritici*	H	92	8	20
35	*Macrophoma mame*	M	91	6	14
36	*Colletotrichum zeae*	M	80	8	22
37	Maize stripe virus	M	71	4	10
38	Soybean yellows mosaic	M	65	4	8

TABLE I (continued)

Rank [a]	Species	Probability of establishment [b]	EEI Midpoint	EEI range (plus or minus) [c] confidence interval 50%	90%
39	*Pythium volutum*	H	61	6	14
40	*Pseudomonas radiciperda*	M	61	3	7
41	*Diplodia zeicola*	L	58	4	10
42	*Cucurbitaria piceae*	L	56	4	10
43	*Sclerophtora raysiae*	M	53	2	6
44	*Chrysomyxa deformans*	M	52	2	6
45	*Chrysomyxa himalensis*	M	52	2	6
46	*Thecopsora areolata*	M	44	2	6
47	*Physopella zeae*	L	43	7	17
48	*Heterodera rostochiensis*	M	40	2	4
49	*Pucciniastrum padi*	M	39	2	5

[a] Rank is based on the midpoint of Expected Economic Impact (EEI).

[b] Probability of establishment is rated as high (25–99%), medium (16–24%), and low (1–15%).

[c] Range is the distance from the midpoint to the maximum or minimum value.

printout, but the inclusion of that large a quantity of information would not serve our purpose. However, the inclusion of this many, rather than a smaller number, provides a display of the variety of organisms involved. In addition, it illustrates how quickly the EEI declines in going down the list; from over \$3 billion for the first-ranked species, to less than \$40 million for the species ranked number 49, a decline of 100-fold. The dollar values are not intended to be reliable estimates of the EEI for a particular species, but rather to serve as a relative scoring device believed to be within reason.

The EEI midpoint is halfway between the maximum and minimum estimates and is followed by the EEI range at 50% and 90% confidence intervals. For example, the EEI midpoint for *Rosellinia radiciperda* (the first-ranked pest) is \$3126 million, and the range is plus or minus \$282 million at the 50% confidence interval, and plus or minus \$698 million at the 90% confidence interval. For about 50% of those listed, the true EEI will lie within the 50% confidence interval (\$2844 million to \$3408 million for *Rosellinia radiciperda*) and for about 90% of the pests the true EEI will be in the 90% confidence interval (\$2428 million to \$3824 million).

Ranking by the EEI midpoint, rather than by the EEI maximum at a selected confidence interval, or by some other method, is an arbitrary choice. If another method were chosen, this same body of information would yield a different ranking. It is important, therefore, that the precise ranking order not be interpreted too literally.

In assembling information on exotic diseases, the Task Force biologist often found that the biological knowledge of key attributes was limited or missing altogether. This lack of information produced uncertainty about the possibilities for international movement and colonization and is responsible for the wide ranges of the EEI that are observable.

A number of the exotic pests with a very high EEI exhibit considerable uncertainty that is associated with the value. This means that while we believe these pests are very important, we are not very certain about precisely how important, and in cases of great uncertainty they may not be important at all. More biological knowledge is needed to provide an improved assessment of the potential danger.

VI. QUARANTINE PROGRAMS

A. Variation among Nations

Almost every nation restricts or prohibits the movement of persons or materials likely to carry pests and diseases affecting domestic plants and animals. Only 11 of the 171 countries reviewed have no form of regulation on either entering travelers or cargo.

While most countries regulate the import of agricultural cargoes, a much smaller number have regulations concerning incoming travelers who may be carrying agricultural materials. As shown in Table II, 82% of the countries regulate cargo, whereas only 22% regulate travelers.

Of the 157 nations that regulate cargo, only 34 of them also regulate travelers. Therefore, as shown in Table III, only 20% of the countries in the world regulate both travelers and cargo; 123 (72%) countries regulate cargo only, and three countries (2%) regulate travelers only.

TABLE II

Regulation of Travelers and Cargo by Countries of the World

	Travelers		Cargo	
	Number	Percent	Number	Percent
Regulations	37	22	157	82
No regulations	134	78	14	18
Total	171	100	171	100

TABLE III

Type of Regulation

Type	Number	Percent
Both travelers and cargo	34	20
Cargo only	123	72
Travelers only	3	2
Total regulated	160	94
Total not regulated	11	6
Total	171	100

No systematic information is available on the extent to which the prescribed regulations are enforced by the nations of the world. However, it is a common observation among United States travelers that baggage inspection of incoming passengers at United States ports of entry is one of the most intensive in the world. The regulations on the importation of agricultural cargoes for 117 countries specifically mentioned that such cargoes are subject to inspection on arrival. However, this provides no clue as to the extent of actual inspection practices. Therefore we do not know the extent of enforcement for either passenger or cargo regulations.

B. Efficacy of Programs

The persons engaged in quarantine programs usually hold that they are efficacious and valuable on the basis of presumptive evidence available for, say, pathogen X: (1) the country seems to provide food and a suitable environment for pathogen X; (2) pathogen X has been intercepted at ports of entry and destroyed; and (3) pathogen X has not been reported in the country. *Quod erat demonstrandum.* But was it?

In correspondence with United States agricultural attaches and quarantine officials of foreign nations, a number of references were made to the efficacy of quarantine efforts. Most of these are simply assertions of its value that are unsupported by any objective evidence.

Worldwide, quarantine programs appear to be based on authority without scientific support or verification. Quarantine actions are a matter of public policy and the usefulness of these activities has not been verified. Most responsible officials assert their validity, an expected attitude for one charged with administering laws and regulations. But skepticism is sometimes expressed, even by regulatory officials.

There are two general philosophical approaches to assessment. First,

the indirect approach—compare events with and without the quarantine
and inspection program. Second, the direct approach—examine the actions of the program itself to assess what is being accomplished.

Since it is not practical to turn regulatory efforts on and off to see
what happens, it was necessary for us to compare the period prior to
the establishment in 1912 of the program in the United States with the
period since 1912. This method introduces many anachronisms, but it is
the best we have.

Has the continuing flow of pathogens and insects into the United
States been altered in any significant way by the establishment of a
quarantine program? The null hypothesis states that the existence of
the quarantine program has not altered the rate of introduction of pests.
For insect pests of crops no evidence could be found to disprove the
null hypothesis. This does not mean that we can say positively that the
program has made no difference, but only that if it has made a difference, no data have been found to demonstrate that difference.

However, we did find data for plant pathogens. The quarantine law
was enacted in 1912 in the United States. How many pathogens were
introduced prior to 1912? How many since? Stevenson (1919) listed 47
that were introduced and established during the 25 years prior to 1912,
and Hunt (1938) listed 75 pathogens introduced and established during
the 25 years after 1912. Those are the numbers—many more pathogens
were introduced per year after the quarantine than before.

The iconoclast would say that, at worst, the quarantine increased
the importations of pathogens or, at best, that 75 pathogens got through
the iron curtain. The defender might say (1) that the data are faulty
even though Stevenson and Hunt both emphasize that their lists were
based on the "best information available" and that they did not consider
them to be entirely accurate; (2) that traffic along the pathways of entry
had increased greatly since 1912; and (3) that until 1930 specific quarantine application was not widespread. In addition, not all recorded entries
were subject to quarantine and the assumption cannot be made that
inspection for all those pathogens was carried out with consistent effect
during the period. In short, there is no valid basis of comparison. The
data do indicate that from 0 to 6 pathogens entered per year during the
1912–1939 period with an average rate of slightly less than three per
year.

The number of interceptions is worthless as a measure of risk. There
are a number of reasons why this is so:

1. The relationship between the number of contraband articles removed from traffic and the number of pest or disease organisms intercepted by that action is not known. This is because only a small propor-

tion of the contraband is examined for pests and diseases, and that is done on an irregular (nonsystematic) basis.

2. A large number of the pest organisms identified are already present in the United States, and their removal from traffic has no effect on the threat from exotic pests and diseases.

3. Among the exotic pests and diseases intercepted and identified there is a wide variation in their ability to colonize and establish themselves in the United States. Without information on the intended destination of the material, it is not possible to assess whether the risk of colonization is high or low.

4. Among the exotic pests and diseases intercepted which may have the ability to establish themselves at the intended destination of the material, there is a wide variation in the damage that would occur. Most exotic pests and diseases, even those that are well known to scientists and pest control experts, have a relatively low capacity for damage.

There is a wide variation in the risk of a given volume of traffic that depends primarily on its point of origin. Other variables include season of the year, degree of infestation at point of origin, and quarantine and inspection procedures at point of origin.

All of this discussion is speculative, with little scientific or statistical basis. It is a series of rationalizations after the fact designed to explain the observed phenomena. Unfortunately, this is all that can be done in the way of evaluation and should be a lesson for the future. Goals, action objectives, and tasks must be specified, for only then does systematic evaluation become possible.

References

Anonymous. (1960). Index of plant diseases in the United States. *U.S., Dep. Agric., Agric. Handb.* **165**.

Hunt, N. R. (1938). Unpublished memorandum. Div. For. Plant Quar., Bur. Entomol. and Plant Quar., U.S. Dept. Agric., Washington, D.C.

McGregor, R. C. (1973). "The Emigrant Pests" (mimeo report to the Adminstrator, Animal and Plant Health Service). U.S. Dep. Agric., Washington, D.C.

Mathys, G. (1977). Society supported disease management activities. *In* "Plant Disease: An Advanced Treatise" (J. G. Horsfall and E. B. Cowling, eds.), Vol. 1, pp. 363–380. Academic Press, New York.

Stevenson, J. A. (1919). Unpublished memorandum to the Federal Horticultural Board, 1919, listing plant pathogens already introduced into the United States. Div. For. Plant Quar., Bur. Entomol. and Plant Quar., U.S Dep. Agric., Washington, D.C.

Thurston, H. D. (1973). Threatening plant diseases. *Annu. Rev. Phytopathol.* **11**, 27–52.

Watson, A. J. (1971). Foreign bacterial and fungus diseases of food, forage, and fiber crops. An annotated list. *U.S., Dep. Agric., Agric. Handb.* **418**.

Author Index

Numbers in italics refer to the pages in which the complete references are listed.

397

Subject Index

A

Abiotic diseases, 6, 339
Abrasion, of leaves, 164
Acanthostigma parasiticum, 391
Adaptation
  concept of, 292
  for dispersal above ground, 193–196
  for sporulation, 14, 194
  for transport by animals and plants, 194, 195
  by wind and water, 195, 196
  for survival, 182
Adult plant resistance, 251, 252
Aecidium glycines, 391
Aerobiological models, 159–180
Aerodynamics
  flow field, of leaves, 168
  of spore dispersal, challenges in, 177–179
  of spore liberation, 160–162, 166–179
Aerosol, dispersal of, 195, 196
After-planting forecasts, 233
Age
  of host
    changes in susceptibility with time, 239–260
    effect on susceptibility to viruses, 255
    physiological, 307
    effect on susceptibility, 255, 256
    use of staggered plant ages in experiments, 248
Age–size structure, of forests, 292

Aggressiveness, of pathogens, 54
Agricultural cargoes, quarantine of, 393
Agricultural crops
  management of disease in, 8, 9, 306–312, 361–378
  mapping of, 343
Agricultural ecosystems, 8, 9, 87
Agricultural intensity, maps of, 356, 357
Agricultural maps, 356, 357
Agricultural practices
  effect on epidemics, 14, 28–31, 330, 361–381
  on soil properties, 328
  goals of, 362
Agricultural profitability, prediction of, 357
Agricultural societies, 6
Agriculture
  management practices in, that favor epidemics, 361–378
  modern, genetic uniformity in, 267, 268, 362, 363, 366–369
  practitioners in, 236
  use of genetic resistance in, 282, 283
Agrobacterium tumefaciens
  distribution of, 197
  map of, 347
Agroecosystems, 8, 9, 87
Airborne diseases, habitat of, 324–328
Air movement, measurement of, 109, 160–177
Air pollutants, 6
  changes in host susceptibility to, 259
Air temperature, in forest stands, 293

Hop
  downy mildew of, 227
  Pseudoperonospora humili on, 44
Horticultural practices that favor epi-
    demics, 370, 371
Host
  alternate, role in epidemics, 364
  alternative, role in epidemics, 364
  differential cultivars, 273
  effect on epidemics, 2, 3, 5, 11, 217,
    362, 363
  on inoculum potential, 147, 148
  nutrition, effects on epidemics, 365
  population, characterization of, 291–
    293
  as subsystems in epidemics, 41
  uncongenial, 272
Host factors, as rate determinants, 52
Host-free period, 364
Host resistance, 38, *see also* Resistance
  effect on inoculum potential, 147
Host susceptibility, *see also* Susceptibility
  changes in, with time, 256, 257
  role in epidemics, 362, 363
  seasonal changes, 239
Humans, importance in dispersal, 197,
    198, 383, 386, 387
Human eye, use as a photocell, 128–130
Humidity, measurement of, 105
  absolute, 106
  relative, 106
    within crop canopies, 327
    effect on epidemics, 85, 325, 333,
      364
    measurement of, 105
      absolute moisture measurement,
        105
      dew point type hygrometers, 108
      hair hygrometers, 106
      psychrometric method, 105
      relative resistance hygrometers,
        107, 108
Hunger, effect of plant disease on, 10,
    19–21, 119
Hurricanes, effect on epidemics, 322
Hydrology, importance in dispersal, 198
Hygrometers, dewpoint type, 108
  condensation type, 108
  saturated lithium chloride type, 108,
    109

Hymenomycetes, role in forest pathology,
    295
Hyperparasites, 307
Hypersensitivity
  to late blight, 251, 252
  to *Phytophthora infestans*, 247
Hypoderma, 296
Hypodermella sulcigena, 391
Hyproxylon pruinatum, 297
Hyproxylon punctulatum, on oak, 303

I

Ignorance, role in plant pathology, 122
Imbalances, as stress factors, 8
Immunity, 307
Incidence, measurement of, 124, 125
Incubation period, 39
Incubation potential, decline in soil, 141
Indigenous disease, definition of, 289
Indigenous species, definition of, 289
Induction, role in epidemiology, 70–73
Inductive method, 342
Infection, 78
  efficiency of, 270
  esodemic, 42
  exodemic, 42
  inoculum and, 142
  latent, 251
  numerical threshold of, Gäumann, 143
Infection chain, 78
  heterogeneous, 42
  homogeneous, 42
Infection court, 138, 140
  fixed, 148
  movement of, 148
  moving, 145
Infection cycle, 78, 82, 87
Infection efficiency, 242
  as measure of susceptibility, 242
Infection gradient, 39
Infection processes, 6
Infection rate, 56, 75
  apparent, 51, 53, *see also* r
  average, 289
  basic, 51, *see also* R
  yearly, 53
Infection studies, 241
Infectious period, 303, 308

A
B 8
C 9
D 0
E 1
F 2
G 3
H 4
I 5
J 6